LOW TEMPERATURE
SOLID STATE PHYSICS

LOW TEMPERATURE SOLID STATE PHYSICS

SOME SELECTED TOPICS

BY

H. M. ROSENBERG

Fellow of Linacre House, Oxford
University Lecturer in Physics
The Clarendon Laboratory
University of Oxford

OXFORD

AT THE CLARENDON PRESS

Oxford University Press, Amen House, London E.C.4

GLASGOW NEW YORK TORONTO MELBOURNE WELLINGTON
BOMBAY CALCUTTA MADRAS KARACHI LAHORE DACCA
CAPE TOWN SALISBURY NAIROBI IBADAN ACCRA
KUALA LUMPUR HONG KONG

FIRST PUBLISHED 1963
REPRINTED LITHOGRAPHICALLY AT THE UNIVERSITY PRESS, OXFORD
FROM CORRECTED SHEETS OF THE FIRST EDITION
1965
PRINTED IN GREAT BRITAIN

PREFACE

THIS is a book of simple explanations. I have written it because the understanding of many of the topics with which solid-state physics is concerned is bedevilled by the fact that the physicist cannot see the physics for the mathematical hedge which besets him on all sides. Because of this, certain subjects are only open to the select few who have mastered some specialized skills. But however important a full mathematical treatment is for the detailed description of a phenomenon, there is often a simple physical model or argument which can be used to gain insight into a problem and it is this approach which has been used wherever possible. In this way I hope that the book will prove useful to the experimentalist who wishes to understand more about the processes which he proposes to investigate and also to the student who has learnt a little and wishes to explore farther.

There may be some criticism over the title and subjects of the book. Why *low-temperature* solid-state physics? Can solid-state physics be divided into hot and cold departments? Possibly not. Nevertheless I think that there are sufficient topics which can be treated quite satisfactorily from the low-temperature viewpoint to warrant the separation. The uncharitable might also suggest that the choice of title for the book was the easiest way for the author to avoid having to include a chapter on the thorny subject of liquid helium!

Inevitably the treatment of the topics which have been chosen for the various chapters reflects the important research programmes of the present day: studies of the Fermi surface, the interactions of paramagnetic ions, the resurgence of work on superconductivity following the publication of the BCS theory, and the development of superconducting devices. Some may mourn the omission of a favourite subject, but for this I make no apologies. The book is long enough as it is.

The general plan of each chapter has been to give as good a physical insight of the selected topic as is possible without using advanced mathematics. There then follows a discussion of the most important types of experiment which have been made, and whilst this treatment is on the lines of a review article no attempt has been made to give a comprehensive cover of all work in the field. The results which are presented are chosen to illustrate the important points and they are not necessarily the most accurate or the most up to date. Wherever possible, reference is made to review articles where modern developments are treated in more detail.

Whilst each of the chapters is more or less self-contained, it has been impossible to avoid some overlap. In addition a choice has had to be made when dealing with a particular type of material, e.g. superconductors, as to whether all the properties should be dealt with in one chapter or not. Inevitably some arbitrary decisions have had to be made. Thus the specific heat of superconductors is dealt with in the superconductivity chapter, whereas the specific heat of semiconductors is described in the specific heat chapter. Cross reference is made in the text to associated topics which are described in other chapters.

Whilst many parts of the book, particularly the introductory sections of the chapters, are suitable for final-year undergraduates in British universities, some prior knowledge of thermodynamics and solid-state physics is essential. In particular, an understanding of the third law of thermodynamics and of the free electron and the band theory of metals is advisable. For Chapter 11 an aquaintance with the elementary concepts of dislocation theory will be useful.

Very few details of experimental technique are given except where this has a direct bearing on the topics discussed. General design data and descriptions of the apparatus which is required for the various types of experiment are given in White's *Experimental Techniques in Low Temperature Physics* (Clarendon Press, Oxford, 1959) and in *Experimental Cryophysics* edited by Hoare, Jackson, and Kurti (Butterworths, 1961).

I have leaned heavily on several of the standard texts for various treatments—in particular, Dekker's *Solid State Physics*, Kittel's *Introduction to Solid State Physics*, and Ziman's *Electrons and Phonons*. In addition I have greatly profited from discussions on certain points with some of my colleagues: Dr. J. M. Baker, Professor B. Bleaney, Dr. R. J. Elliott, Dr. K. Mendelssohn, and Dr. W. P. Wolf. I am particularly indebted to those who have been kind enough to read and criticize certain sections of the book: Dr. A. H. Cooke, Dr. R. W. Hill, Professor W. D. Knight, and Dr. B. V. Rollin. In thanking them, I still retain the author's privilege of being responsible for all errors! I am also grateful to the many authors who have kindly sent me diagrams or advance information of results, and to whom acknowledgement is made in the text.

H. M. R.

Clarendon Laboratory, Oxford, 1962

ACKNOWLEDGEMENTS

GRATEFUL acknowledgement is made for permission to use tables and diagrams which originally appeared in books and periodicals published by the following institutions and companies:

Academic Press, American Association of Physics Teachers, American Institute of Physics, American Society for Metals, J. A. Barth Verlag, Cambridge University Press, Elsevier Publishing Co., Heywood and Co. Ltd., Institute of Physics and the Physical Society, International Business Machines Corporation, Kamerlingh Onnes Laboratory, McGraw-Hill Book Co., Masson et Cie, National Research Council, North Holland Publishing Co., Oxford University Press, Pergamon Press, Physica Foundation, Royal Society, Springer Verlag, Taylor and Francis Ltd., University of Toronto Press, Wiley and Sons, Inc.

CONTENTS

1. SPECIFIC HEATS

2. THERMAL EXPANSION

3. THERMAL CONDUCTIVITY OF NON-METALS

4. ELECTRICAL CONDUCTIVITY OF METALS AND ALLOYS

5. THERMAL CONDUCTIVITY OF METALS, ALLOYS, AND SEMICONDUCTORS

6. SUPERCONDUCTIVITY

8. THERMO-ELECTRICITY

9. MAGNETIC PROPERTIES

10. THE SUSCEPTIBILITY OF METALS—THE DE HAAS–VAN ALPHEN EFFECT

11. MECHANICAL PROPERTIES

The following figures appear as plates:

Plates 1 (Fig. 6.16) and 2 (Figs. 10.5 and 10.6) appear facing pages 178 and 356 respectively.

1

SPECIFIC HEATS

1.1. Introduction

WE begin this book with a description of specific heats at low temperatures because this is a useful manner in which to introduce several of the ideas and systems which will be described at greater length in later chapters. Thermal properties are, of course, determined largely by the spectrum of lattice vibrations and its variation with temperature is very important in many of the topics which we shall discuss. This spectrum may be studied by specific-heat determinations. We deal in some detail with the Debye theory, although conscious of its limitations, because, in the end, even the most high-minded theory which involves lattice interactions is usually forced to descend to the Debye level in order to put its sophisticated ideas on to a practical plane—the Debye θ is ubiquitous in many of the formulae used in later sections.

The spectrum of electron energies may also be studied to some extent by measuring the specific heat and we shall describe the properties of a free electron metal, although we are not unmindful that this now appears to be unrealistic; the properties of real metals are still most simply discussed in terms of the deviations from this simple model.

Other important systems with which we shall deal are those which contain magnetic ions. In particular, the low-lying energy levels which characterize the ions of paramagnetic substances may also be investigated by specific-heat measurements and these may be used in conjunction with magnetic measurements to obtain detailed information about the energies of those states.

If the entropy S of an assembly is a function of temperature then we define the specific heat c_x under a constraint x as

$$c_x = T \frac{\partial S}{\partial T}\bigg|_x . \tag{1.1}$$

Throughout this chapter our constraint will be that of constant volume, with a corresponding specific heat c_v.†

---

† We shall use the symbol c_v for the specific heat at constant volume and c_v for the specific heat per unit volume.

There are two main types of assembly to be considered. In the first S varies over the entire temperature range and so there will be a specific heat at all temperatures. The main systems which exhibit this type of behaviour are the crystal lattice and, in a metal, the conduction electrons. In the second type of system, S only changes appreciably over a restricted temperature region. Under such circumstances the specific-heat contribution will also only be observable in that temperature region. It will occur when there is only a finite number of energy levels for the system. Once the temperature has been raised sufficiently for them all to be equally populated, no further change in S will occur on raising the temperature. If the spacing of the energy levels is kT', where k is Boltzmann's constant, then a specific heat will be observable at temperatures of the order of T'. It is usually detected as a fairly steep hump superimposed on the ordinary specific heat. It is thus the type of observation which is called an anomaly and as such it is keenly sought by research workers. Many anomalies were found in early specific-heat measurements, which have caused considerable discussion, but most of them have more recently been shown to be due to impure specimens. Nevertheless, genuine ones are known to exist. The main types of anomaly with which we shall deal are the Schottky specific heat, which is very important in the study of paramagnetic crystals at low temperatures, and the specific heat associated with co-operative transitions.

Why do we measure specific heats? Historically the first measurements at low temperatures were made by Nernst's school in order to derive the 'chemical constants' which were used to determine the equilibrium conditions for chemical reactions. Whilst such work was important in the development and acceptance of the Third Law of Thermodynamics and is still used by physical chemists, it will not be described in this treatment. For details the reader is referred to Zemansky (1957), chapters 17, 18, 19. Nowadays measurements are usually made in order to check our ideas on the energy states of a system. Sometimes our model is a purely theoretical one as in the Debye theory for the lattice vibrational spectrum. In other cases the model is itself derived from other experimental observations; for example, the electron energy levels in a paramagnetic salt can be deduced from paramagnetic resonance measurements and this scheme can then be checked by measuring the Schottky specific heat as will be described in section 1.19. In metals, the electronic specific heat gives us information about the distribution of the electrons at the Fermi surface.

When we measure the specific heat between two temperatures T_1 and

T_2, then the change in the entropy, S, between those two temperatures can be calculated since

$$S(T_2) - S(T_1) = \int_{T_1}^{T_2} \frac{c_v}{T} \, dT. \tag{1.2}$$

Thus if we wish to know the total entropy change between the absolute zero and any given temperature, the specific heat curves must be extrapolated to $0°\,K$. The lower the temperature to which we can measure the specific heat, the more confidence we can have in the entropy values.

The normal methods for the experimental determination of the specific heat give a measure of c_p, the specific heat at constant pressure, whereas the thermodynamic calculations require the value of c_v. It is therefore necessary to correct for this by using the relation which is derived in any standard text (e.g. Zemansky, 1957, chapter 13). Thus

$$c_p - c_v = \beta^2 \eta V T, \tag{1.3}$$

where β is the volume expansion coefficient, η is the bulk modulus (1/compressibility), and V is the volume of the sample. Whilst this correction is quite small at helium temperatures it gets progressively more important and must be applied as the temperature is increased.

1.2. The lattice specific heat

Before the advent of the quantum theory the values of the specific heats of solids were calculated using the classical theorem of the equipartition of energy. It was assumed that each atom or molecule in a solid was able to vibrate about a fixed point and since this vibration can extend in three dimensions the equipartition of energy theorem ascribes an energy of $3kT$ to each atom or molecule.† Thus the energy of a mole of a substance (containing N atoms or molecules, where N is Avogadro's number) is $3NkT$ or $3RT$, where R is the gas constant per mole; hence from (1.1) the molar specific heat has the constant value

$$3R = 5\cdot96 \text{ cal mole}^{-1} \text{deg}^{-1} = 24\cdot94 \text{ joule mole}^{-1} \text{deg}^{-1}.$$

This is the law of Dulong and Petit, which for most substances is very successful both at and above room temperature.

1.3. *The Einstein theory*

Nevertheless, experiments showed that for some substances (e.g. graphite) the specific heat was smaller than $3R$ at room temperature

---

† The theorem of the equipartition of energy states that for every degree of freedom, a particle has an energy of $\frac{1}{2}kT$ (see, for example, Joos, 1934, p. 560). A particle vibrating in one dimension has *two* degrees of freedom, since its energy depends on both its velocity and its position. Thus the energy of a particle which can vibrate in three dimensions is $3kT$.

and when it became possible to go to lower temperatures with the aid of liquid air, the specific heat of all substances was found to decrease by considerable amounts. For example, the specific heat of copper at room temperature is 24 joule mole^{-1} deg^{-1}, whereas at 100° K it is about 16 joule mole^{-1} deg^{-1}. The first satisfactory explanation for the decrease was put forward by Einstein (1907) who suggested, on the basis of the new quantum theory, that the atoms oscillating at a frequency ν could not vibrate with any arbitrary amplitude but that they could only have discrete values of energy separated from one another by a quantity h, where h is Planck's constant. Wave mechanics shows that the actual energy levels are $(n+\frac{1}{2})h\nu$, where n is zero or any integer.

Let us see what this implies. The frequency of atomic vibration is about 10^{13} sec^{-1} and so the lowest energy which an oscillator can have is $\frac{1}{2}h\nu$ and this is equal to $3\cdot3\times10^{-14}$ erg. If we equate this value to the classical energy of a one-dimensional oscillator, kT, then T comes out to be about 240° K. This means that if at such a temperature (which is, of course, very approximate because we are not sure of the correct value for ν) an oscillator has its classical value kT, then when we cool it below this temperature it cannot reduce its energy any further, because it is already in the lowest possible (or ground) state. Thus it cannot contribute to the specific heat at lower temperatures. On such reasoning the specific heat of the whole substance should be zero below such a temperature. This, in fact, does not occur because at any temperature above the absolute zero there is always a statistical probability for an atom to exist in a higher energy state, and in the Einstein theory this was taken into account. Nevertheless, one can understand why the specific heat decreases as the temperature is reduced and it is clear that, as the absolute zero is approached, the specific heat will tend to zero because the probability that the oscillators will be in the ground state will become much greater. The mean energy $\overline{U}$ of each oscillator at a temperature T is

$$\overline{U} = \tfrac{1}{2}h\nu + \frac{h\nu}{\exp(h\nu/kT)-1} \tag{1.4}$$

and hence the specific heat at constant volume, c_{v}, is

$$c_{\mathrm{v}} = T\frac{\partial S}{\partial T}\bigg|_{\mathrm{v}} = \frac{d\overline{U}}{dT}\bigg|_{\mathrm{v}} = 3R\left(\frac{h\nu}{kT}\right)^2 \frac{\exp(h\nu/kT)}{\{\exp(h\nu/kT)-1\}^2}. \tag{1.5}$$

Whilst the Einstein theory showed why the specific heat decreased at lower temperatures, it was not entirely satisfactory because it assumed that all the atoms vibrated with the same frequency and that they were independent of one another. This is obviously only a very rough approximation. In fact there must be strong forces between the atoms and when one of them moves it will influence the motion of its neighbours. Thus all the atoms of the substance are coupled together and they should be considered as one complete system. Treatments on these lines were developed by Debye (1912), who considered the substance as an elastic continuum, and by Born and von Kármán (1912) who took into account the discrete nature of the crystal lattice. Debye's theory, whilst it is based on a very simple model, has had remarkable success since it enables the specific heat of a substance to be characterized by a single parameter. For this reason it has tended to be used much more than the Born–von Kármán treatment.

1.4. *The Debye theory*

We give here a short account of the Debye theory. A full treatment will be found in any standard text (e.g. Kittel, 1956, chapter 6; Roberts and Miller, 1960, p. 535). In this theory the atomic system is assumed to be an elastic continuum in which only certain frequencies can be excited and maintained. These will be those which are able to set up standing waves in the medium; any others will die out rapidly. Let us assume that the substance is a rectangular prism with sides l_1, l_2, l_3, and that a vibration with a wavelength λ is travelling in any arbitrary direction with respect to the sides of the prism. This will be a standing wave if l_1/λ_1, l_2/λ_2, and l_3/λ_3 are integral or half integral, where λ_1, λ_2, and λ_3 are the components of λ in the directions of l_1, l_2, and l_3 respectively. Besides this particular direction of propagation of the vibration which has been considered, in which the standing wave conditions are satisfied, there will be other directions in which a standing wave with the same value of λ could be maintained. Each of these vibrations associated with a different direction is called a mode. For each direction there will be three modes, corresponding to the two transverse and one longitudinal vibrations which can be excited. It is fairly obvious that the smaller the value of λ the greater will be the number of possible directions in the crystal for which the conditions for a standing wave are fulfilled, i.e. the greater will be the number of modes.

The number of modes in a frequency range between ν and $\nu+d\nu$ is written as $f(\nu)\,d\nu$ and each of these modes is considered as being due to a separate oscillator which will have a mean energy given by (1.4). Thus the internal energy of the system will be

$$U = \int_0^{\nu\text{max}} \overline{U} f(\nu)\,d\nu, \qquad (1.6)$$

where ν_{max} is the maximum frequency which is excited. Such a frequency must clearly exist because otherwise the energy of the system would become infinite. The specific heat is then found by differentiating (1.6) with respect to T.

In the Debye theory the actual distribution function is given by

$$f(\nu)\,d\nu = 4\pi\left(\frac{1}{v_l^3}+\frac{2}{v_t^3}\right)V\nu^2\,d\nu, \qquad (1.7)$$

where v_l and v_t are the velocities of longitudinal and transverse waves in the medium. It is important to note that this theory predicts that f should be proportional to ν^2. The modes which are able to be excited are assumed to be those of the lowest possible frequencies up to the maximum number of $3N$ modes. The maximum frequency of oscillation ν_{max} is determined by ensuring that when (1.7) is integrated between the limits 0 and ν_{max} it will give $3N$ modes,

i.e.
$$3N = \tfrac{4}{3}\pi\left(\frac{1}{v_l^3}+\frac{2}{v_t^3}\right)V\nu_{max}^3. \qquad (1.8)$$

Using (1.8) a substitution can be made for the bracketed term in (1.7) and then with (1.4) the temperature-dependent energy is equal to

$$U = \frac{9N}{V\nu_{max}^3}\int_0^{\nu_{max}} \frac{h\nu^3}{\exp(h\nu/kT)-1}\,V\,d\nu. \qquad (1.9)$$

This expression can be simplified by introducing the quantity θ, known as the Debye characteristic temperature, where

$$\theta = h\nu_{max}/k, \qquad (1.10)$$

and substituting $x = h\nu/kT$ and $x_{max} = \theta/T$. We then obtain

$$U = 9NkT^4/\theta^3\int_0^{x_{max}} \frac{x^3}{e^x-1}\,dx. \qquad (1.11)$$

On differentiation with respect to the temperature we obtain for the specific heat per mole

$$c_v = 9R(T/\theta)^3\int_0^{x_{max}} \frac{e^x x^4}{(e^x-1)^2}\,dx. \qquad (1.12)$$

This is the Debye specific-heat function. It is plotted in Fig. 1.1. Tabulated values are given in Roberts and Miller (1960), p. 537. At high temperatures when x is small, the integral in (1.11) reduces to $\int x^2\,dx$ and hence the energy becomes $3RT$ and this yields the Dulong and Petit value for the specific heat of $3R$. At low temperatures, where x is large, we can approximate by allowing the upper limit in (1.11) to go to infinity. The integral then has the value $\pi^4/15$. On differentiating we obtain the well-known T^3 specific-heat expression

$$c_v = \begin{cases} 464\cdot4(T/\theta)^3 \text{ cal mole}^{-1}\text{deg}^{-1} \\ 1941(T/\theta)^3 \text{ joule mole}^{-1}\text{deg}^{-1}. \end{cases} \tag{1.13}$$

This cubic relationship should hold up to temperatures of about $\theta/10$. At this temperature there is an error of about 2 per cent if (1.13) is used instead of (1.12).

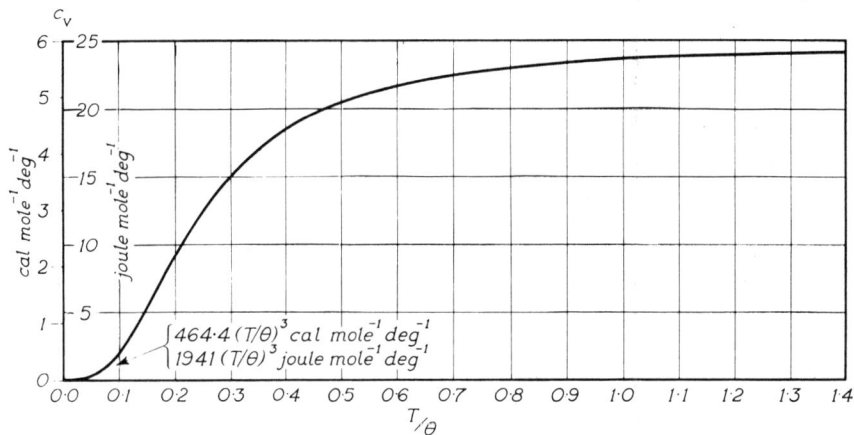

Fig. 1.1. The specific heat per mole according to the Debye theory calculated from (1.12). Below $T/\theta = 0\cdot1$ the curve follows the cubic relation indicated on the figure.

The success with which (1.12) and (1.13) represent the specific heats of most solids is truly remarkable when one considers the simple terms to which the complicated vibrations of the crystal lattice have been reduced. For rough calculations of the lattice specific heat Fig. 1.1 or (1.12) and (1.13) can be used with complete confidence. A table of values for θ for some of the elements is given in Table 1.1.

1.5. Some practical points

It is important to realize how very small the lattice specific heat is at low temperatures. For most ordinary metals used in the construction of an apparatus, θ lies between 300 and 400° K. Thus at 4·2° K, T/θ is about 10^{-2} and hence c_v is about 2×10^{-3} joule mole^{-1} deg^{-1}, i.e. a factor of about 10^4 less than the room-temperature value. This very low specific heat becomes apparent in many ways. Some of the most important practical points are:

(a) Thermal equilibrium is established very quickly at low temperatures. The thermal conductivity, κ, is usually no lower than the room-

TABLE 1.1

Representative values of the Debye θ and γ, the coefficient of the electronic specific heat

Element	$\theta°$ K	$\gamma \times 10^4$ joule mole^{-1} deg^{-1}	Element	$\theta°$ K	$\gamma \times 10^4$ joule mole^{-1} deg^{-1}
A	93		Mn	450	180
Ag	225	6·09	Mo	470	21·1
Al	426	13·6	N	68	
As	275		Na	158	13·7
Au	164	7·0	Nb	250	88·2
Ba	110	27·0	Ne	63	
Be	1160	2·22	Ni	440	72·8
Bi	100		O	91	
C (diamond)	2065 / 2219		Os	500	23·5
			Pb	108	33·6
Ca	229	27·3	Pd	299	99
Cd	188	6·27	Pr	74	
Cl	115		Pt	221	66·3
Co	443	47·5	Rb	55	25·8
Cr	585	15·5	Re	450	24·5
Cs	40	36·3	Rh	478	48·9
Cu	348	7·44	Ru	600	33·5
Dy	182		Sb	201 / 140	
Fe	464	50·2			
Ga	317	6·01	Se	89	
Gd	152		Si	636	
Ge	363		Sn grey	212	
H para	116		Sn white	195	17·5
D ortho	106		Sr	147	36·5
He	28–36		Ta	245	58·5
	(depends on pressure)		Te	153	
Hf	261	26·4	Th	170	46·8
Hg	75		Ti	430	35·5
In	109	18·4	Tl	90	15·2
Ir	420	31·4	U	200	109
K	89	19·7	V	380	92
Kr	63		W	405	12·1
La	130		Zn	310	6·27
Li	369	17·5	Zr	310	30·3
Mg	342	13·65			

temperature value (except in alloys, see section 5.11) and so the thermal diffusivity, $\kappa/(c_v d)$ (where d is the density), which controls the equilibrium time, is very large. Thus systems which would take hours to come to equilibrium at room temperature reach equilibrium in minutes at 4·2° K.

(*b*) The latent heat and the specific heat of liquid helium are about 3 and 0·25 joule cm^{-3} respectively at temperatures above the lambda point (2·19° K). This means that once a large apparatus has been precooled to liquid-air temperature, a relatively small quantity of liquid helium is required to cool it to $4 \cdot 2°$ K, or to further reduce its temperature below $4 \cdot 2°$ K by boiling the helium under reduced pressure.

(c) Parts of the apparatus which are thermally isolated from the main heat sink may have their temperature raised considerably by very small amounts of heat. It is particularly important, for example, when low temperatures are achieved by adiabatic demagnetization, that the stray heat influx should be cut down as much as possible.

1.6. *The Debye θ*

At this stage it is useful to discuss the significance of the parameter θ, because it is constantly referred to in the literature, not merely in connexion with specific heats, but with many other thermal properties as well. Equation (1.10) shows that θ is proportional to the maximum lattice frequency. By analogy with a simple vibrating system, a high value of the frequency, and hence of θ, implies that we are dealing with a lattice which has very strong inter-atomic forces and light atoms. Thus diamond has a value for θ of about 2,000° K. Conversely lead has a low value of θ—about 100° K. $\theta°$ K can be considered as the temperature at which all the oscillators with a frequency ν_{max} are excited. Since (1.7) shows that the number of modes in a given frequency range is proportional to ν^2, this means that at $\theta°$ K we can consider that most of the oscillators will have a frequency near to ν_{max}. (The average frequency is actually $\frac{3}{4}\nu_{max}$.) It is interesting to calculate the wavelength of the vibrations for ν_{max}. For copper θ is about 350° K and the velocity of sound is about 4×10^5 cm sec^{-1}. Thus from (1.10) ν_{max} is $350k/h$ and the wavelength is $4 \times 10^5 \, h/350k = 5 \cdot 4 \times 10^{-8}$ cm, which is about twice the atomic spacing. This is an example of the general rule that at $\theta°$ K the wavelength of the lattice vibrations is of the same order as the lattice constant. This result is important because since the real crystal consists of a discrete number of particles one could not ascribe a unique interpretation to a vibration which had a wavelength which was less than the spacing of the particles. The motion of the particles under these conditions could then be described just as well by a vibration with a longer wavelength. This is illustrated in Fig. 1.2 (b). Thus although the Debye theory uses a continuum model, the boundary conditions at high temperatures are correct for a system of discrete atoms.

1.7. *Wavelength of the lattice vibrations at low temperatures*

It is also important to appreciate the magnitude of the wavelength of the lattice vibrations which are dominant at low temperatures. There is no great virtue in calculating this accurately for the Debye model and only a rough method will be given. For simplicity we may assume that at any low temperature, T, all oscillators with frequencies up to a certain value ν_T will be excited whilst those with higher frequencies are in their ground state. We can also assume that the energy of each mode that is

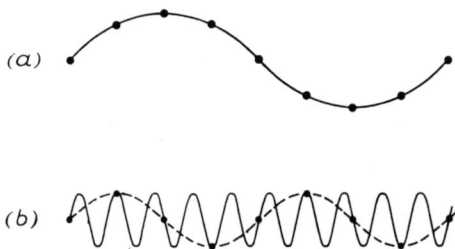

Fig. 1.2. (a) The atomic displacement for lattice vibrations of long wavelength; (b) the atomic displacement when the wavelength is short compared with the atomic spacing. In this case the motion could be described just as well by the dotted line which has a wavelength longer than the atomic spacing.

excited is $h\nu$ (neglecting the zero-point energy). Thus the total energy will be

$$U = \int_0^{\nu_T} h\nu f(\nu)\, d\nu. \qquad (1.14)$$

Using (1.7) and (1.8) this becomes

$$U = \frac{9NVh}{\nu_{max}^3} \int_0^{\nu_T} \nu^3\, d\nu, \qquad (1.15)$$

i.e. U is proportional to ν_T^4/ν_{max}^3. But at low temperatures (1.13) shows that U is proportional to T^4/θ^3, i.e. T^4/ν_{max}^3. Hence ν_T is proportional to T and so ν_T/ν_{max} is proportional to T/θ. A simple calculation shows that the coefficient of proportionality is approximately unity. If λ_{min} is the wavelength corresponding to ν_{max}, then $\lambda_T/\lambda_{min} \doteq \theta/T$. Since we have already seen that λ_{min} is equal to the lattice spacing we finally obtain

$$\lambda_T \approx \theta/T \times \text{lattice spacing.} \qquad (1.16)$$

This means that at low temperatures the dominant lattice wavelength is changing continuously with the temperature and it can be quite large. For copper at 3° K it will be about a hundred atomic spacings long, for diamond at the same temperature it will be about 600 atomic spacings long. The importance of this continuous change in the dominant lattice wavelength will be seen when we consider the thermal conductivity of the lattice.

1.8. *Deviations from the Debye theory*

The success of the Debye theory should not blind us to the fact that over an appreciable temperature range it is not accurate and does not agree with the experimental results. This is usually demonstrated most easily by calculating the value of θ which is necessary to fit the experimental points at each temperature. If the Debye theory is really applicable then, of course, a constant value of θ should be obtained, but in practice this does not occur. As the temperature is reduced the effective value of θ remains approximately constant down to about $\theta/5$, it then passes through a minimum and rises to a constant value at a temperature of about $\theta/100$. Below this temperature the specific heat is strictly proportional to T^3 as the Debye theory predicts, but it will be recalled that such T^3 behaviour should really start at temperatures below $\theta/10$. Thus the theory works well above $\theta/5$ and below $\theta/100$, with, of course, a different value of θ for each range. Naturally deviations from the theory will differ from one substance to another and in some metals such as zinc and cadmium, large changes in the effective value of θ are found. These are shown in Fig. 1.3 (*a*). Other materials (e.g. Fig. 1.3 (*b*)) do not show so wide a variation in θ, but the general shape of the curves for most substances is similar to those shown.

The breakdown of the Debye theory arises from the fact that, except at very low temperatures, we must take into account the discrete nature of the crystal lattice. The theory has been developed by Blackman (1956) for many crystal lattices. The main outcome of his calculations from the low-temperature point of view is that, except where ν is small, the number of modes between ν and $\nu+d\nu$ is proportional, not to ν^2 as in the Debye theory (1.7), but to a higher power of ν. Thus whilst there is agreement with the Debye theory for small ν (and hence at very low temperatures), at higher temperatures when higher frequencies are being excited, more modes (or oscillators) will come into operation than the Debye theory predicts. Thus more energy will be necessary to raise the temperature and so the specific heat increases more rapidly than it otherwise would. Such a rise in specific heat corresponds to a decrease in the value of θ and this is what is observed. (At any temperature the specific heat of a substance with a high θ is less than that of one with a lower θ.) When the temperature has been raised to one where all the modes are excited, further heating does not change the number of modes and so the theory yields a constant value of θ again, in agreement with the Debye model. For further details the review articles by Blackman (1956) and Parkinson (1958) should be consulted.

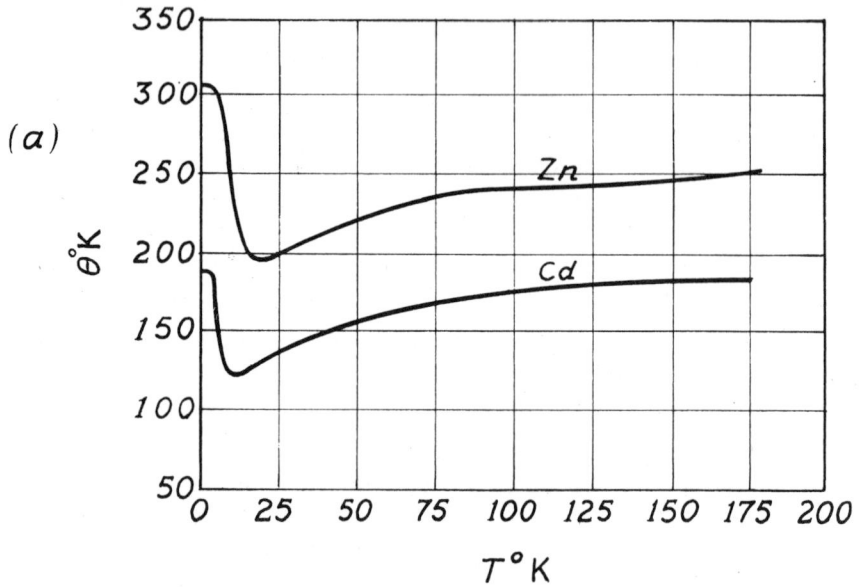

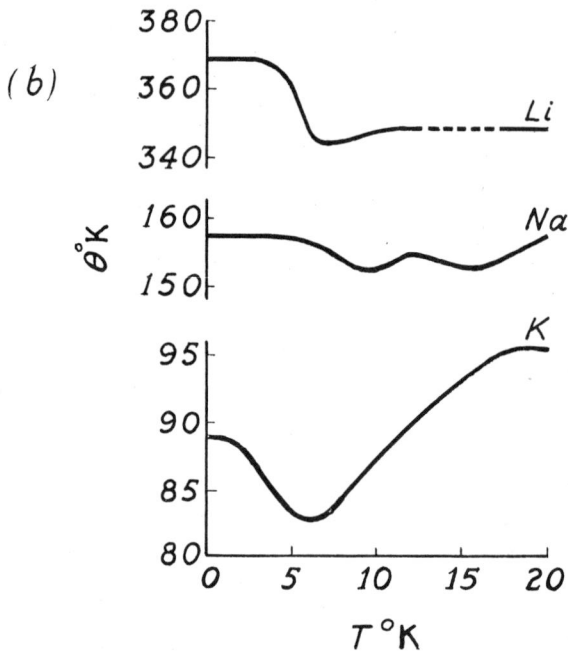

FIG. 1.3. The variation of the effective Debye θ with temperature for (a) zinc and cadmium (Smith and Wolcott, 1956); (b) lithium, sodium, and potassium (Roberts, 1957). Note that a constant value of θ is not achieved until very low temperatures, and hence, until this region, the specific heat will not be proportional to T^3. A similar variation of θ is found in most other substances, although it is not always so marked.

The value of θ corresponding to the specific heat at very low temperatures can be calculated from (1.8) and (1.10) if we know the velocity of sound in the substance or its elastic constants. In the past such calculations did not agree with the measured value of θ because the specific heat measurements had not been taken to a low enough temperature—they had very often been taken only to the minimum in the θ curves of Fig. 1.3 instead of to the constant value at lower temperatures. If this last value is taken, quite good agreement is obtained as is shown in Table 1.2.

TABLE 1.2

Comparison of values of θ obtained from specific-heat measurements with those calculated for the elastic constants

	θ (sp. ht.)	θ (elast.)	Reference		θ (sp. ht.)	θ (elast.)	Reference
Sn	195	202	Chandrasekhar and Hulm (1960)	NaCl	320	322	Barron, Berg, and Morrison (1957)
Cu	348	340	Overton (1960)	NaI	164	163	,,
KCl	235	238	Barron, Berg, and Morrison (1957)	MgO	946	969	,,
KBr	174	173	,,	C (diamond)	2,219	2,240	,,
KI	132	140	,,				

1.9. The specific heat of the electrons

When the properties of metals are studied, we must consider not only the lattice specific heat with which we have dealt in the preceding sections, but also the specific heat of the conduction electrons. If the temperature of a metal is changed there will be an alteration in the distribution of the electrons amongst their possible energy states which will give rise to an extra specific heat. If the electrons are considered as small particles moving freely through the lattice then, according to classical theory, each should have an energy of $\frac{3}{2}kT$ and hence a specific heat of $\frac{3}{2}k$. This would lead to an electronic specific heat of about the same magnitude as that of the lattice and it was one of the great problems how, at room temperatures at least, the specific heat of a metal could be entirely accounted for by considering it to be lattice specific heat. It was not until the Fermi–Dirac statistics were applied to the free electron model that the reason for the small electronic specific heat became evident. A full calculation of the specific heat is long and tedious and would be out of place in this book (see Seitz, 1940, pp. 144 ff.). We give here a simplified derivation which illustrates the main features.

Wave mechanics shows that an electron confined in a metal can only have certain well-defined energies, which are analogous to the special lattice vibrational modes which we discussed in the previous section.

For each possible value of the energy there will be several modes, or states, and the Fermi–Dirac statistics requires that not more than one electron can be present at any time in one state. If $\mathscr{E}$ is one of the permitted energy levels, then the number of states per unit volume with energies between $\mathscr{E}$ and $\mathscr{E}+d\mathscr{E}$, $F(\mathscr{E})\,d\mathscr{E}$ is

$$F(\mathscr{E})\,d\mathscr{E} = 4\pi(2m_e/h^2)^{\frac{3}{2}}\mathscr{E}^{\frac{1}{2}}\,d\mathscr{E}, \tag{1.17}$$

where m_e is the mass of the electron. In many cases a more complicated band scheme may be taken into account by using an effective mass m_e^* instead of the actual free electron mass. At the absolute zero the lowest possible states will be occupied up to a certain maximum energy $\mathscr{E}_0$. This energy may be found by integrating (1.17) between zero and $\mathscr{E}_0$ and equating the result to n, the number of conduction electrons per unit volume, i.e.

$$n = (\tfrac{8}{3}\pi)(2m_e/h^2)^{\frac{3}{2}}\mathscr{E}_0^{\frac{3}{2}}. \tag{1.18}$$

If we insert the value of n for copper ($\sim 10^{23}$) into this equation then $\mathscr{E}_0$ comes out to be about 10^{-11} erg or about 6 electron volts. This is equivalent to the thermal energy at a temperature of about $80{,}000°$ K. Thus when we raise the temperature of the specimen from the absolute zero, only electrons which have energies close to $\mathscr{E}_0$ will be able to go to higher energies. Most of the electrons, however, will need an activation energy of a few electron volts. Since this is so very much greater than kT they will remain in their original states. This is illustrated in Fig. 1.4.

It therefore follows that when we warm a metal only a very small fraction of the electrons are able to change their energy and these will be the only ones which will contribute to the specific heat. Hence this model explains why the specific heat of the electrons is so very much smaller than that which would be predicted by the classical model.

It is quite simple to derive an expression for an approximate value of the electronic specific heat c_e. At any temperature T, electrons whose energies lie within a range kT from $\mathscr{E}_0$ will be excited to higher energy states. If $F(\mathscr{E}_0)$ is the number of possible states per unit energy interval at the energy $\mathscr{E}_0$ (i.e. $F(\mathscr{E}_0)$ is the density of states at $\mathscr{E}_0$) then the number of electrons which are excited will be $kTF(\mathscr{E}_0)$. Each of these electrons will have acquired an extra energy of $\tfrac{3}{2}kT$ and so the increase in energy of the system is $\tfrac{3}{2}k^2T^2F(\mathscr{E}_0)$. On differentiating with respect to T we obtain the specific heat as

$$c_e = 3k^2TF(\mathscr{E}_0). \tag{1.19}$$

A detailed calculation, taking account of the Fermi–Dirac distribution function gives

$$c_e = \tfrac{1}{3}\pi^2k^2TF(\mathscr{E}_0). \tag{1.20}$$

If we assume that $F(\mathscr{E})$ is given by the free electron formula (1.17), we can obtain (using (1.18) for $\mathscr{E}_0$) a simple expression for $F(\mathscr{E}_0)$:

$$F(\mathscr{E}_0) = 3n/2\mathscr{E}_0; \qquad (1.21)$$

thus the specific heat becomes

$$c_e = \pi^2 k^2 n T/2\mathscr{E}_0 \text{ per unit vol}, \qquad (1.22)$$

or
$$c_e = \gamma T, \qquad (1.23)$$

where
$$\gamma = \tfrac{1}{3}\pi^2 k^2 F(\mathscr{E}_0) \text{ or } \pi^2 k^2 n/2\mathscr{E}_0, \qquad (1.24)$$

depending on whether (1.20) or (1.22) respectively is used.

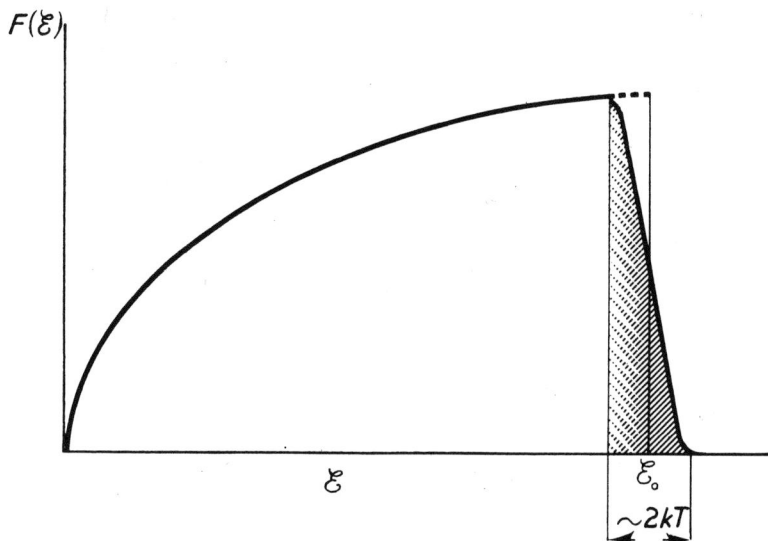

FIG. 1.4. The electron density $F(\mathscr{E})$ as a function of the energy, $\mathscr{E}$, at a temperature T. Only those electrons which have an energy in the shaded region, lying within kT of the Fermi energy $\mathscr{E}_0$, will be able to enter vacant states of higher energy as the temperature is increased.

We have already seen that $\mathscr{E}_0$ is about 10^{-11} erg for copper and hence γ should have a value of $5\cdot 7 \times 10^3$ erg mole^{-1} deg^{-2} or about $1\cdot 35 \times 10^{-4}$ calorie mole^{-1} deg^{-2}. Thus at room temperature c_e should be about $0\cdot 17$ joule mole^{-1} deg^{-1}. This is less than 1 per cent of the lattice specific heat at that temperature. Thus the theory explains why the electronic specific heat is not usually detected from room-temperature measurements. This explanation was indeed one of the great triumphs of the application of wave mechanics and Fermi–Dirac statistics to the theory of metals.

1.10. *The specific heat of a metal at low temperatures*

At low temperatures, however, the situation is different. The lattice specific heat c_g is then proportional to T^3, whereas c_e is proportional to T. At some temperature, therefore, c_e will become equal to c_g and this commonly occurs in the liquid helium region. This is shown in Fig. 1.5. Using our data for copper once again, we see that at $4 \cdot 2°$ K, c_e is about $2 \cdot 4 \times 10^{-3}$ joule mole^{-1} deg^{-1}, which is about the same value as we previously calculated in section 1.5 for the lattice specific heat at that temperature. This means that if we measure the specific heat of metals at low temperatures it should be possible to detect the electronic specific heat very easily. In practice it is the total specific heat, c, which is measured, where

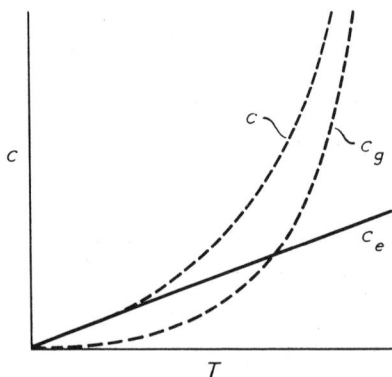

FIG. 1.5. The specific heat of a metal at low temperatures. Since the lattice specific heat, c_g, is proportional to T^3 and the electronic specific heat, c_e, is proportional to T, c_e becomes dominant at very low temperatures.

$$c = c_g + c_e \qquad (1.25)$$

but it is quite straightforward to separate the two terms, since they each have a very simple temperature dependence. From (1.13) and (1.23) we obtain

$$c = \beta T^3 + \gamma' T. \qquad (1.26)$$

Thus a graph of c/T against T^2 should be a straight line with a slope of β and an intercept on the c/T axis of γ. In this way both θ and γ may be calculated from the data. We must, of course, make certain that the measurements are made in the true T^3 region for c_g, i.e. at $\theta/100$ or below. A typical plot of c/T against T^2 is shown in Fig. 1.6 and it will be noted that a good straight line is obtained. Experimental values of γ are given in Table 1.1.

1.11. Experimental results for the electronic specific heat

The prediction that the electronic specific heat is proportional to the absolute temperature has now been verified down to temperatures as low as $0 \cdot 1°$ K for Cu, Ag, Pt, Pd, W, and Mo by Rayne (1954) and there is no doubt that the general features of the theory must be correct. Nevertheless, the numerical values which are obtained warrant some discussion, since in no case does the free electron theory give satisfactory quantitative agreement with the experimental results.

1.12. *The alkali metals*

These metals, which have one 'free' s electron outside a closed shell, are those which should approximate most closely to the free electron model. Nevertheless, the values of γ which have been obtained are not in good agreement with those which can be calculated from (1.24). Even sodium, which is generally considered to be the nearest equivalent to the free electron metal, has a value of γ which is about 60 per cent higher than that predicted by the theory, and lithium and potassium also have

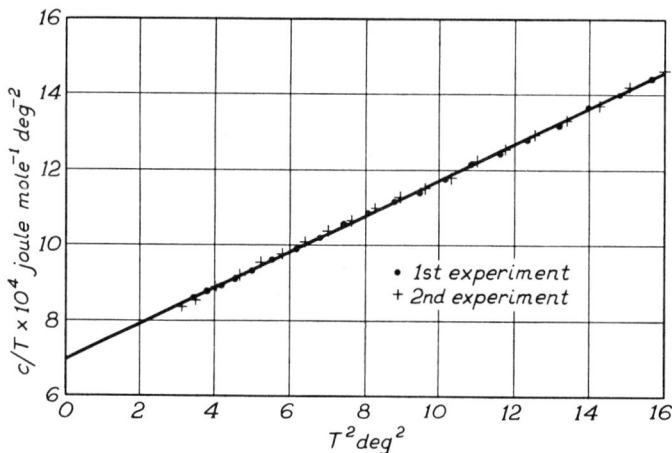

FIG. 1.6. A plot of c/T against T^2 for copper, showing the very good linear relationship which is obtained (Bailey, 1959).

values of γ which are too high (Roberts, 1957). Thus, even with these metals, no reliability should be placed on the results of calculations based on the assumption of the free electron model. Such calculations are the easiest which can be made, and will lead to expressions for γ which are correct to within a factor of, perhaps, two, but no better agreement should be expected.

1.13. *The divalent metals*

The conductivity and general behaviour of these metals is generally ascribed to the fact that the permitted electron energy levels which are occupied lie in two bands of energy which overlap slightly. The metals of this group have a hexagonal crystal structure and the amount of overlap depends on the axial ratio† of the lattice. (If there was no over-

† The axial ratio is the ratio of the spacing between similar hexagonal planes to the spacing of the atoms within those planes. For a close-packed lattice this ratio is 1·633.

lap, only the lower energy band would be filled and the substance would be an insulator.) It will be recalled that c_e depends on $F(\mathscr{E}_0)$, the density of states at the maximum energy, $\mathscr{E}_0$ (1.20), and it is clear that the value of $F(\mathscr{E}_0)$ must be dependent on the amount of overlap which occurs.

$F(\mathscr{E}_0)$ is very sensitive to the value of the axial ratio when this ratio is small, but for large values of the ratio, $F(\mathscr{E}_0)$ varies much more slowly.† These theoretical predictions have been verified in work on Be and Mg (which have small axial ratios) and on Zn and Cd (which have large axial ratios). The experimental results are shown in Table 1.3, from which it will be noted that the values of γ for Be and Mg are very different from one another. Those for Zn and Cd, however, are the same, although their axial ratios still differ.

TABLE 1.3

The electronic specific heats of some divalent metals

	γ joule mole^{-1} deg^{-2}	axial ratio	Reference
Be	$2 \cdot 22 \times 10^{-4}$	1·57	Hill and Smith (1953)
Mg	$13 \cdot 65 \times 10^{-4}$	1·62	Smith (1955)
Zn	$6 \cdot 27 \times 10^{-4}$	1·86	Smith (1955)
Cd	$6 \cdot 27 \times 10^{-4}$	1·89	Smith and Wolcott (1956)

1.14. *The transition metals of the first group* (Ti, V, Cr, Mn, Fe, Co, Ni, Cu)

The electronic specific heat of these elements can also be explained in terms of their band structure. When we consider the isolated atoms of these elements, they are characterized by the fact that the energy of the $4s$ state is below that of the $3d$ state. The $4s$ sub-shell contains its full quota of 2 electrons and as we pass from Ti to Ni the $3d$ state fills up from 2 to 8 electrons (for Cr there is a slight variation, there being 5 electrons in the $3d$ state and only one in the $4s$ state). In the metal the situation is rather different. As the atoms are brought close together, the wave functions of the states overlap and this will produce the characteristic broadening of what were unique energy levels, into energy bands. The wave function for $4s$ is much more extended than that for $3d$ and hence it overlaps its partner in a neighbouring atom more than does the $3d$ function. Hence the $4s$ band (as it now becomes) is broader and covers a much wider energy range than does the $3d$ band. This occurs to such an extent that some of the states in the $4s$ band now have higher energies than those of the $3d$ band, although in the free atom the $4s$ energy was

---

† For a detailed discussion on overlapping bands and the effect of the axial ratio on $F(\mathscr{E}_0)$ the reader is referred to Hume-Rothery (1946), chapter 28.

lower than that for the $3d$ level. This effect is shown diagrammatically in Fig. 1.7 (a).

How do these overlapping bands affect the specific heat? We should note that the $3d$ level contains 10 degenerate levels and hence when it becomes a band it will have $10N$ states if there are N atoms in the metal. It is a narrow energy band and so the density of states will be high. The $4s$ level, on the other hand, contains only two states in the free atom and hence $2N$ states in the metal. It has a much larger energy spread

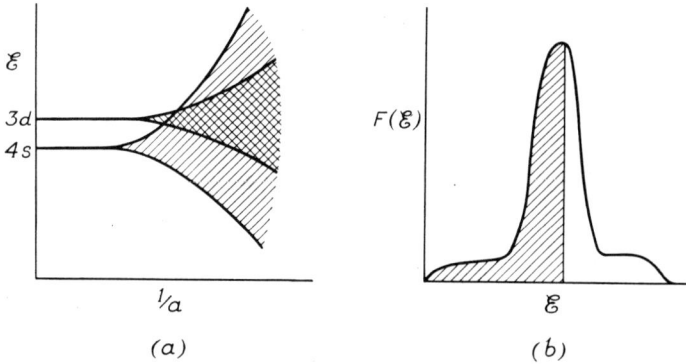

FIG. 1.7. (a) The effect of the atomic spacing a on the $3d$ and $4s$ energy levels showing how the $4s$ band becomes much broader than the $3d$ band and completely overlaps it. (b) The combined density of states function for the $3d$–$4s$ bands. For most of the transition group the Fermi energy lies in the peaked part of the distribution and this leads to a large specific heat.

and hence the density of states is low—much smaller than that for the $3d$ band. Diagrams of the density of states against energy for the two bands are shown in Fig. 1.7 (b). Thus the composite curve for $F(\mathscr{E})$ will have a very high maximum in it, as is shown in the figure. The electrons, as usual, will fill up the lowest possible energy states and this will mean that for nearly all the transition metals, the maximum occupied energy level, $\mathscr{E}_0$, will fall within the high portion of the curve, which was determined by the $3d$ band. Thus $F(\mathscr{E}_0)$ will be much larger than usual and hence c_e will be large. This occurs for all the metals of the group except copper. Copper has 11 electrons per atom and the band is filled beyond the $3d$ portion. Table 1.4 shows that the general features of this model are confirmed satisfactorily by the experimental results, although the change in γ in going from one metal of the series to the next is by no means as smooth as that which one would deduce from the rather idealized curve of Fig. 1.7 (b). It is not to be expected, however, that the band structure for each metal will be exactly alike. As the table shows, they do not all have the same crystal structure and thus the overlapping of the wave functions and the consequent spreading of the energy levels will not be the same in each case. An additional complication arises from the fact that some of these metals are ferromagnetic and hence there are strong interactions between neighbouring ions (see section 9.19). This will tend to broaden the energy bands, and hence the density of states is reduced.

TABLE 1.4

(compiled from Wolcott, 1955)

	Ti	V	Cr	Mn	Fe	Co	Ni	Cu
$\gamma \times 10^4$ joule mole^{-1} deg^{-2}	33·5	92	15·5	180	50·2	50·2	72·8	7·44
Total electrons in $3d + 4s$ states	4	5	6	7	8	9	10	11
Usual crystal structure	hcp	bcc	bcc	complex	bcc	hcp	fcc	fcc

hcp = hexagonal close packed; bcc = body centred cubic; fcc = face centred cubic.

Experiments show that the metals of the second and third transition groups also have high values of c_e and it is probable that these may also be explained in terms of high, narrow bands of energy states (Wolcott, 1955). In these cases they will be bands which are derived from the $4d$ and $5d$ levels.

1.15. *The electronic specific heat of alloys*

The preceding section suggests that it might be possible to investigate the shape of an energy band by measuring c_e for a series of samples which have as nearly as possible the same band structure, but that each has a different number of conduction electrons. By adding electrons the band can be filled to a higher energy, i.e. we change $\mathscr{E}_0$, and so $F(\mathscr{E}_0)$ can be found from c_e. For a simple interpretation of the results it is necessary that the energy band should remain the same for all samples. The most satisfactory specimens for such experiments would seem to be a series of alloys of two elements each of which has a different number of electrons in the outer shell and which form a continuous range of solid solutions with one another over a wide range of composition. It would also be an advantage for the crystal structure of the alloys to be the same as that of one of the elements. Several such series of experiments have been undertaken including the following alloy series: palladium-silver (Hoare and Yates, 1957), nickel-copper and nickel-iron (Keesom and Kurelmeyer, 1940), copper-zinc (Bailey, 1959; Rayne, 1957), and silver-cadmium (Montgomery and Pells, 1959).

Fig. 1.8 shows the values of $F(\mathscr{E}_0)$ for silver-cadmium alloys plotted against the concentration of cadmium. It is assumed that the addition of a cadmium atom contributes an extra conduction electron to the system and hence the 5s band should be filled up as more cadmium is added. The results confirm this simple hypothesis and, as the figure shows, are in fairly good agreement with the theoretical curve for a free electron model.

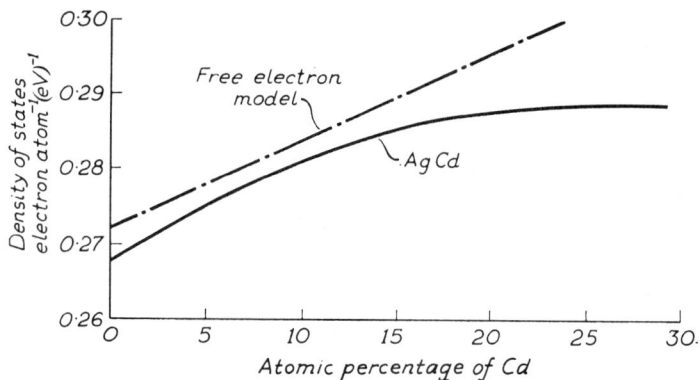

FIG. 1.8. The density of states for alloys of AgCd (Montgomery and Pells, 1959) as calculated from specific heat data. This work compares well with the curve calculated for the free-electron model.

1.16. *The lattice specific heat of alloys*

Since when we alloy a metal we, in general, alter the elastic constants, it follows that the lattice vibrational spectrum and hence the specific heat, c_g, will change. This is a difficult problem to treat theoretically and a satisfactory solution has not yet been achieved. Experiments show that for substitutional solid solutions c_g does not change very much with alloy concentration. When there is a phase change, however, there will be a corresponding change in c_g. Fig. 1.9 shows the variation of the Debye θ with zinc content for copper-zinc alloys. There is a drop in θ as the β (bcc) phase is approached and then a rapid rise in the γ phase (complex). It is clear that no general rules can be given for the behaviour of c_g nor, at present, can useful information be deduced from it.

1.17. The specific heat of semiconductors

Semiconductors differ from metals in that they only have a small density of free current carriers. These may be either electrons or holes which are excited from bound states by thermal activation (see section 7.1. for a fuller description of the level scheme). Hence as the temperature is reduced, their density decreases rapidly and their contribution to the specific heat cannot be detected. The semiconductors which have been studied most intensively have been the elements germanium and silicon and the compound InSb. These all have the diamond lattice crystal structure and hence one would expect that the lattice vibrational spectrum would be of a similar form for each. This is confirmed by the fact that θ has the same type of temperature dependence for each material. The lattice specific heat behaves very similarly to that already

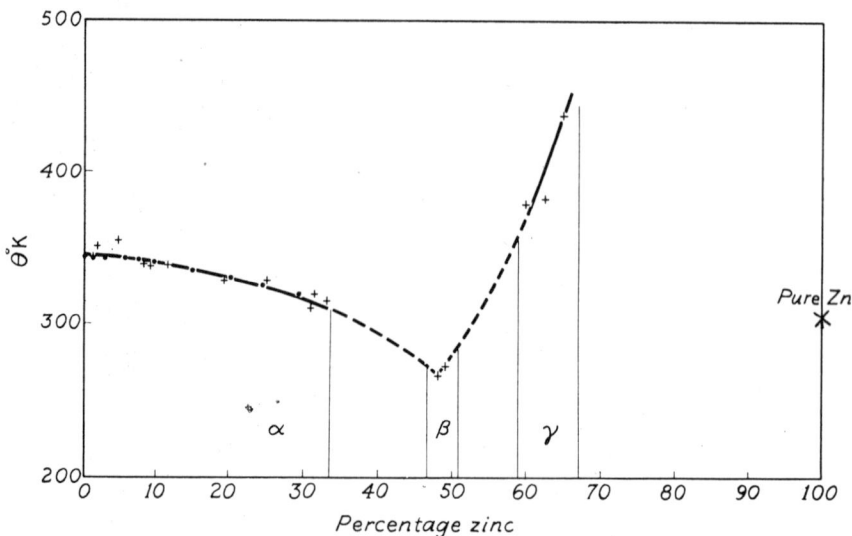

FIG. 1.9. The variation of the Debye θ for CuZn alloys, showing the changes which occur for compositions of different structure (Bailey, 1959).

described in section 1.8. A true T^3 dependence is observed only at temperatures below $\theta_0/100$, where θ_0 is the limiting value of θ at very low temperatures. Above this, θ drops sharply and passes through a minimum at about $\theta_0/20$, with a value of $0.7\theta_0$ in a manner similar to that shown in Fig. 1.3. θ_0 has the values $363°$ K for Ge, $636°$ K for Si, and $200°$ K for InSb. Similar behaviour has also been observed for the isomorphous crystals grey tin ($\theta = 212°$ K) and diamond ($\theta = 2220°$ K).

The specific heat of the free carriers has only been observed in heavily doped Ge and Si (both n- and p-type) containing $\sim 10^{19}$ carriers per cm^3 (Keesom and Seidel, 1959). In these specimens the impurity states overlap the conduction or the valence band so that free carriers are present even without thermal activation. The specific heat at low temperatures can then be expressed in the form $c = \beta T^3 + \gamma T$, in the same way as it is in a metal, where the γT term is ascribed to the specific heat of the carriers as in section 1.10. Using expressions very similar to those quoted in that section, the effective masses, m_e^*, of the carriers have been calculated with the aid of the experimentally derived values of γ. For Ge very good agreement is obtained between the values of m_e^* derived from these specific heat experiments and those determined from measurements of cyclotron resonance (section 7.21). For Si the agreement is not quite so good but since the cyclotron resonance experiments are made on relatively pure samples in which there is no overlap of the impurity states, it is not surprising that the detailed structure of the band edges (which as we have seen is one of the main things determining c_e) should be different in the two cases.

Since, due to the rather low value of θ for Ge, the lattice specific heat was very much larger than γT above 1° K, it was necessary to take specific heat measurements down to 0·5° K (with the aid of a helium 3 cryostat) in order to determine the electronic term accurately. For Si with its larger value of θ it was only necessary to take measurements to about 1·3° K.

Further details concerning the specific heat of semiconductors are given by Johnson and Lark-Horovitz (1957).

1.18. Anomalous specific heats

In addition to the specific heats with which we have already dealt, there are in some substances extra contributions to the specific heat which are present over a limited temperature range. These are often evident as sharp peaks in the total specific-heat curve. Such specific heats are of two main types, the Schottky specific heat, which is usually associated with paramagnetic substances, and the specific heat due to co-operative transitions.

1.19. *The Schottky specific heat*

This occurs when the lowest energy state of an ion in a crystal is composed of two or more levels which are separated by a small energy, kT'. At temperatures very much below T' the ions cannot be excited to the upper levels and they will all be in the lowest state. As the temperature T' is approached, some of the ions will be excited to the higher states. This will require energy and hence there will be an extra specific heat in this region (Schottky, 1922). At temperatures much greater than kT', all levels will be equally populated and so there will then be no further excitation. Hence there will be no further contribution to the specific heat. It is this change in population of low-lying levels which is important for the interpretation of paramagnetism (see chapter 9) and hence measurements of the Schottky specific heat are always made on paramagnetic crystals. It is usually assumed that in such crystals the paramagnetic ions are sufficiently far away from one another that there is no magnetic interaction between them. They may then be considered as an assembly of independent systems. In order that this condition may be better achieved in practice, the specimens are sometimes diluted with an isomorphous diamagnetic salt.

The general expression for the specific heat is quite simple to calculate. Let there be n energy levels above the ground state and let each level have a degeneracy g and an energy ϵ, above the ground state energy. Then using the Maxwell–Boltzmann distribution function, the fraction of atoms in the jth level with energy ϵ_j will be

$$\frac{g_j \exp(-\epsilon_j/kT)}{\sum\limits_{i=0}^{i=n} g_i \exp(-\epsilon_i/kT)}. \tag{1.27}$$

If N is the total number of atoms, then the extra energy due to the jth state being excited will be

$$\frac{N\epsilon_j g_j \exp(-\epsilon_j/kT)}{\sum\limits_{i=0}^{i=n} g_i \exp(-\epsilon_i/kT)} \tag{1.28}$$

and hence the total energy of excitation U_{Schottky} is

$$U_{\text{Schottky}} = \frac{\sum\limits_{i=0}^{i=n} N\epsilon_i g_i \exp(-\epsilon_i/kT)}{\sum\limits_{i=0}^{i=n} g_i \exp(-\epsilon_i/kT)}. \tag{1.29}$$

The specific heat is obtained by differentiating (1.29) with respect to the temperature.

In practice the number of levels which are involved is few and it is instructive to calculate the extra specific heat for a system consisting of two levels, the ground state and one higher level, which are separated by an energy ϵ. We then obtain

$$U_{\text{Schottky}} = \frac{N\epsilon}{1 + (g_0/g_1)\exp(\epsilon/kT)}, \tag{1.30}$$

where the subscripts 0 and 1 refer to the ground and upper states respectively. On differentiation, the specific heat is

$$c_{\text{Schottky}} = \frac{N\epsilon^2}{kT^2} \frac{g_0}{g_1} \frac{\exp(\epsilon/kT)}{[1 + (g_0/g_1)\exp(\epsilon/kT)]^2}. \tag{1.31}$$

At low temperatures, such that $\epsilon \gg kT$ this reduces to

$$c_{\text{Schottky}} = \frac{N\epsilon^2}{kT^2}\frac{g_1}{g_0}\exp(-\epsilon/kT),\tag{1.32}$$

and at high temperatures where $\epsilon \ll kT$ we get

$$c_{\text{Schottky}} = \frac{N\epsilon^2}{kT^2}\frac{g_0}{g_1(1+g_0/g_1)^2}.\tag{1.33}$$

Thus at very low temperatures the Schottky specific heat rises exponentially in accordance with (1.32), it passes through a maximum and at

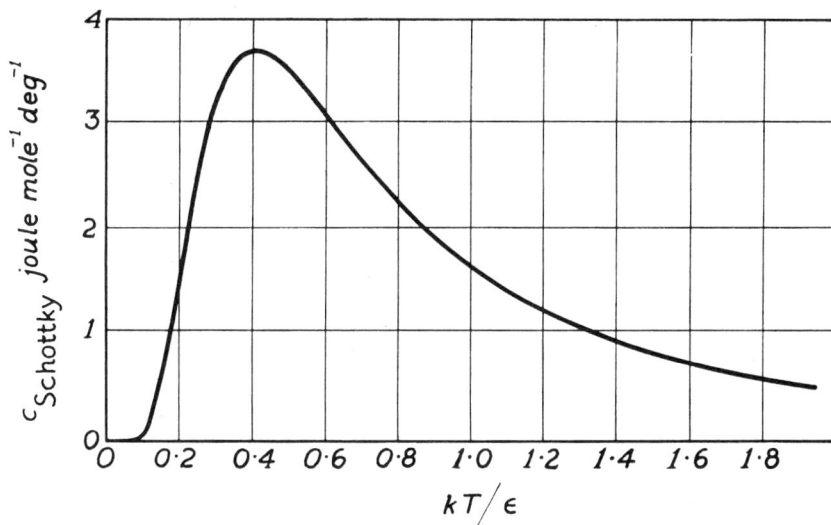

FIG. 1.10. The theoretical curve for the Schottky specific heat, when there are two levels of equal degeneracy.

high temperatures it decreases with a T^{-2} dependence (1.33). The theoretical curve for the case when $g_0 = g_1$ is shown in Fig. 1.10. By differentiating (1.31) with respect to T and equating the result to zero we obtain the condition for the specific heat to be a maximum. This is

$$\exp(-\epsilon/kT) = \frac{g_0(\epsilon/kT-2)}{g_1(\epsilon/kT+2)}.\tag{1.34}$$

When $g_1/g_0 = 1$ the maximum occurs for $\epsilon/kT = 2\cdot4$ and this corresponds to a maximum in c_{Schottky} of $3\cdot64$ joule mole^{-1} deg^{-1}. The values of ϵ/kT and c_{Schottky} for some of the cases when $g_1 \neq g_0$ are shown in Table 1.5. When any of these values of c_{Schottky} are compared with the magnitudes of the electronic or the lattice specific heats at liquid-helium temperatures (of the order of 10^{-4} joule mole^{-1} deg^{-1}) it will be appreciated that a Schottky anomaly which occurs at low temperatures will, over a certain temperature range, be the major contribution to the specific heat.

TABLE 1.5

Schottky specific-heat data for two levels separated by an energy ϵ

(Hill, 1952)

g_1/g_0	Maximum value of c_{Schottky} joule mole^{-1} deg^{-1}	ϵ/kT at the maximum of c_{Schottky}	Total entropy of the anomaly per mole (1.40)
0·5	2·00	2·23	$R \log_e(3/2)$
1·0	3·64	2·40	$R \log_e 2$
1·5	5·06	2·54	$R \log_e(5/2)$
2·0	6·31	2·65	$R \log_e 3$
5·0	12·0	3·13	$R \log_e 6$

The use of (1.31), (1.32), or (1.33) in the appropriate temperature ranges enables us to obtain the value of ϵ. In particular, from (1.32) for very low temperatures, we have

$$\log_e[c_{\text{Schottky}} kT^2/N] = \log_e[g_1\epsilon^2/g_0] - \epsilon/kT. \qquad (1.35)$$

Hence a graph of the L.H.S. against $1/T$ will have a slope of $-\epsilon/k$ and so ϵ can be determined without a knowledge of g_0/g_1. This ratio can then be found from the intercept of the curve on the L.H.S. axis. In many cases, however, the Schottky anomaly occurs at such a low temperature that it is difficult to take measurements in the region described by (1.32) and it is then necessary to use the other expressions ((1.31) or (1.33)).

The preceding analysis has been made for a system of two energy levels, but similar results are obtained for more complicated systems. It is necessary to start with some scheme of levels (derived either theoretically, or from susceptibility and paramagnetic resonance data— or probably from a mixture of all three) and the specific-heat data are then compared with that to be expected from the proposed model. If the agreement is poor then the model is incorrect, or incomplete. Further discussion on the utilization of the various types of experimental data is given in section 9.37.

In order to obtain an accurate value for c_{Schottky} it is necessary to know the lattice specific heat c_g throughout the region where the anomaly occurs. This often causes some difficulty. One way of overcoming this is to extend the measurements up to higher temperatures where c_{Schottky} is proportional to $1/T^2$. If we assume that in this region c_g is proportional to T^3 we then have for the total specific heat, c, an expression of the form

$$c = A/T^2 + \beta T^3 \qquad (1.36)$$

and hence a graph of cT^2 against T^5 will have a slope of β and so c_g may be determined. Unfortunately, the $1/T^2$ region of the Schottky anomaly is very often at too high a temperature for c_g to be assumed to be of the form βT^3 and so this method is not always satisfactory. Under such circumstances it is common practice to use for c_g the specific heat of a salt of similar composition and the same crystal structure, but which does not have an anomaly. In general the anomaly is so large that any errors in the value of c_g, when it is derived in this manner, are quite small. In some cases it is possible by paramagnetic relaxation measurements to determine the Schottky specific heat separately, without the lattice contribution. This technique is described in section 9.28.

1.20. *Entropy associated with a Schottky anomaly*

According to statistical mechanics, the expression for the entropy S of a system is
$$S = k \log_e W, \tag{1.37}$$
where W is the 'statistical probability' or the number of possible states of the system. The additional entropy associated with the low-lying energy levels which we have been considering may therefore be calculated very simply. Let us suppose that we have Z separate energy levels for each ion. Then the probability for each ion is Z and if there are N ions the total probability for the whole system is Z^N. The extra entropy associated with these Z levels will therefore be
$$S_{\text{Schottky}} = Nk \log_e Z, \tag{1.38}$$
or
$$S_{\text{Schottky}} = R \log_e Z \tag{1.39}$$
for one mole of the substance, where R is the gas constant. From the curve of the Schottky specific heat one can calculate the extra entropy and hence from (1.39) this can be used to determine the number of levels which are being occupied.

In some cases one or more of the different energy levels might have associated with it more than one state (i.e. it might be degenerate). In this case the probability of such a level must be weighted, since an ion will now stand a greater chance of being in that level. If there are $g_0, g_1, g_2, ..., g_n$ states for the energy levels 0, 1, 2,..., n, then the probability for each atom is
$$W = \sum_{i=0}^{i=n} (g_i/g_0), \tag{1.40}$$
where g_0 is the degeneracy of the ground state. Thus if we had a system with two non-degenerate states the extra entropy would be $R \log_e 2$ per mole. If the lower state was triply degenerate the entropy would be $R \log_e 1{\cdot}33$ per mole. If, however, it was the upper state which was

triply degenerate, then the entropy would be $R \log_e 4$ per mole. For other examples see the last column of Table 1.5.

1.21. *Experiments on the Schottky specific heat*

Nearly all low-temperature investigations of the Schottky specific heat have been made on paramagnetic salts and these have been used together with the results of susceptibility and resonance experiments to obtain information about the number and the spacing of energy levels (see section 9.37). In early experiments it was not possible to take measurements to a temperature which was low enough to cover the entire range of the anomaly (in many cases of interest the maximum in the specific heat is below $1°$ K). For this reason the coefficient of the $1/T^2$ term was used from the data at higher temperatures (1.33), but this limited the amount of information which could be obtained. The entropy, for example, could not be calculated. Nowadays, with improvements in calorimetric techniques, it is possible to take measurements to much lower temperatures and in many cases the full Schottky curves may be determined. Nevertheless, even if the entire specific-heat data are available it will be appreciated (as we have mentioned earlier) that it can be interpreted with confidence only if we have some prior knowledge of the probable number and spacing of the levels. This information can be obtained from theoretical considerations of the ground state of the ion and of the crystal structure of the salt (which will determine how the lower energy levels will be split) and from paramagnetic resonance measurements, which give a direct measure of the number of levels and of their spacing. However, it is distressing to admit that even if a complete picture of the energy levels and their spacings is available it is rare that the measured specific heat is in full agreement with the calculated value. The main reason for this is that in our calculation of c_{Schottky} we have assumed that the ions were independent of one another, i.e. if one of them was in a certain energy state, ϵ', it would not affect the probability that another one would be in a state ϵ''. Since, however, the ions we are considering are magnetic dipoles whose orientations will be different in different energy states, it is clear that the orientation of one ion will influence the orientation of its neighbours and hence, except to a first approximation, the assembly cannot be considered as a collection of statistically independent particles. A correction can be made for this (Van Vleck, 1937) but it is not entirely satisfactory.

As an example of good agreement between theory and experiment we show in Fig. 1.11 (*a*) the specific-heat curve for ferric methylammonium sulphate. Paramagnetic resonance experiments have shown that the ground state of the Fe^{3+} ion consists of three levels (each a doublet). In this salt the middle level is separated from those on either side by energies of 1·05 and 0·58° K, but it was not known which of these levels was above the middle one. The two possible energy schemes are shown

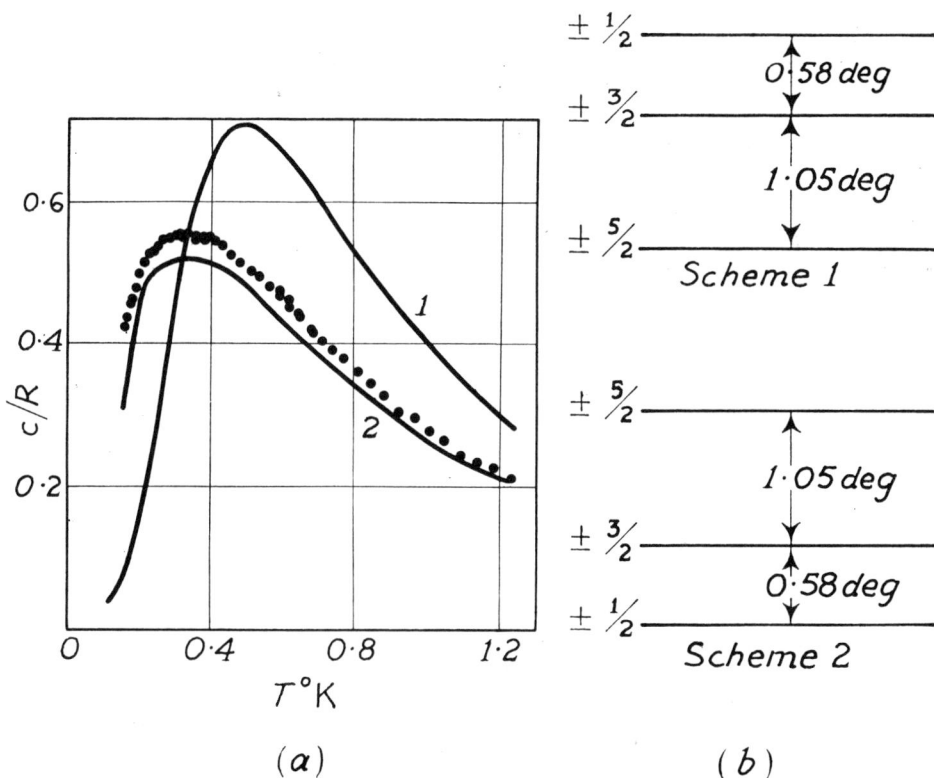

Fig. 1.11. (a) The specific heat of ferric methylammonium sulphate. The points are the experimental results and the two curves are from theoretical calculations using the two energy schemes shown in (b). It is clear that scheme 2 is the correct one (Cooke, Meyer, and Wolf, 1956).

in Fig. 1.11 (b). The curves marked 1 and 2 in Fig. 1.11 (a) show the calculated specific heats assuming schemes 1 and 2 respectively. It is clear that there is very close agreement between the experimental points and the calculations for scheme 2 (0·58° K level below the 1·05° K) and thus the true level scheme may be established.

In Fig. 1.12 the anomalous specific heat of cerium ethylsulphate is shown. This substance has two low-lying energy levels (each of them doublets) which have an energy separation of 6·7° K. The value for

$c_{Schottky}$ should be the same as that which we have already calculated for two levels (1.31); and which is shown in Fig. 1.10. This theoretical curve is redrawn in Fig. 1.12 where it will be seen that there is a considerable discrepancy with the experimental results for which, at the moment, there is no satisfactory explanation. Such discrepancies are more the rule than the exception in this type of work.

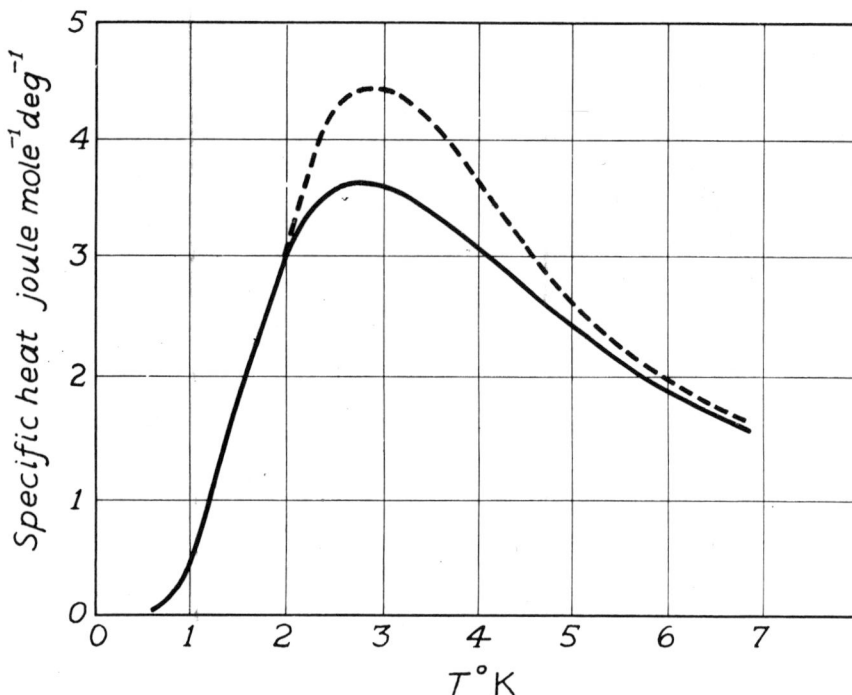

FIG. 1.12. The specific heat of cerium ethylsulphate. The experimental results are given by the dashed curve whilst the continuous curve is the theoretical Schottky anomaly calculated for an energy separation of 6·7° K (Meyer and Smith, 1959).

1.22. The nuclear specific heat

In the previous section we have not discussed how the level splittings arise in the various salts we have mentioned. This is described in detail in Chapter 9. Suffice to say that in most of the materials which are investigated the splittings are caused by the interaction of the crystalline electric field (from the surrounding ions) on the electron magnetic dipole of the paramagnetic ion. This will cause a preferential alignment of the dipole along certain directions with a consequent lowering of its energy.

If, however, the nucleus of an ion has a spin, and hence a magnetic moment, then due to the mutual interaction of the nuclear and the

1.22 THE NUCLEAR SPECIFIC HEAT 31

electron dipoles, there will be an extra splitting of the ionic levels, the hyperfine splitting (HFS). The depopulation of the higher HFS levels will give rise to another specific heat anomaly.

Thus at a temperature below the ordinary Schottky anomaly of a paramagnetic there will often be a further specific-heat anomaly associated with the region where the alignment of the nuclei takes place.

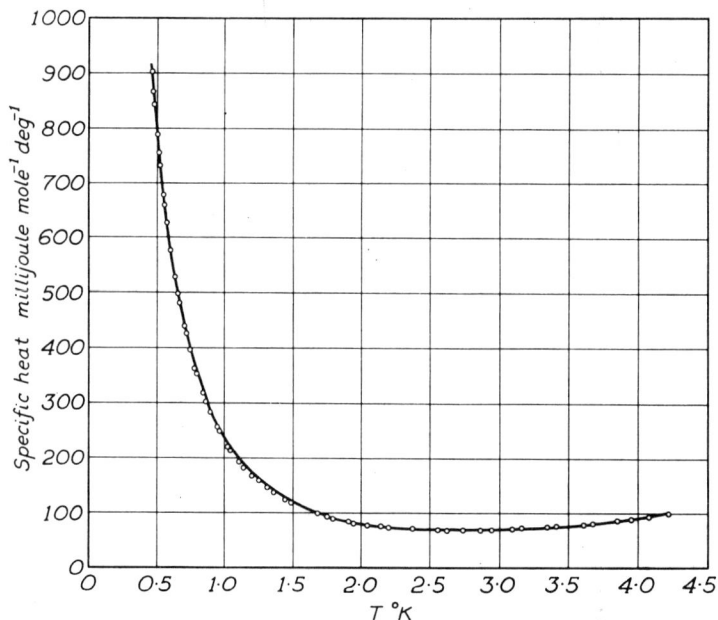

FIG. 1.13. The specific heat of terbium metal at very low temperatures, which shows the very large rise below 1° K due to the nuclear hyperfine interaction (Lounasmaa and Roach, 1962).

This usually happens well below 1° K. This HFS anomaly is most easily detected in ferromagnetic metals in which the alignment of the electron spins (and the consequent specific heat anomaly) has already occurred at much higher temperatures. The effect was first observed in cobalt (Arp, Kurti, and Petersen, 1957), and later experiments have shown it to exist in several of the rare earth metals, Tb, Dy, Sm, and Er. It has also been observed in an antiferromagnetic MnF_2 (Cooke and Edmonds, 1958). The specific heat anomaly in Tb metal is shown in Fig. 1.13. So far it has not been possible to take measurements on the low-temperature side of the maximum.

It is interesting to note that the magnitude of the HFS interaction, which may be calculated from the T^{-2} tail of the specific heat anomaly

(1.33), is very similar in the ferromagnetic metal to that which is found in its paramagnetic salts from paramagnetic resonance data (section 9.34). This indicates that the immediate interaction of a nucleus in a ferromagnetic is with its own electron dipole system rather than with the whole assembly of aligned dipoles (Erickson and Heer, 1957).

Even in a diamagnetic material there will be a nuclear specific heat if the nucleus has a spin. The orientation may occur either by the inter-action of the nuclear electric quadrupole with the gradient of the internal electric field of the crystal (section 9.18), or by the direct magnetic interaction of neighbouring nuclei. The quadrupole interaction should give a specific heat anomaly at 10^{-1} to 10^{-3} °K, whereas the nuclear magnetic dipole interaction should be very much lower—at about 10^{-6} °K.

1.23. *The specific heat associated with co-operative phenomena*

In section 1.21 we mentioned the fact that the ions were not strictly independent of one another, and hence the form of the Schottky anomaly ought to be modified. In co-operative phenomena we deal with effects where there is very strong interaction between the atoms or ions. This causes the temperature range over which transitions occur to be very much reduced.

In a substance which exhibits a co-operative transition it is no longer meaningful to speak of the energy states of individual atoms or ions since the coupling between them is so strong that we must consider only the energy of the assembly as a whole. For this reason we do not talk about the energy state or level of a particular atom, but rather of its configura-tion (e.g. orientation of its axis of angular momentum or magnetic moment) with respect to that of its neighbours. Thus instead of saying that a certain atom goes to a higher or lower energy state, we should say that the change in configuration raises or lowers the energy of the entire assembly.

There are several examples of co-operative effects—the onset of ferro- and antiferromagnetism (sections 9.19 and 9.20), order-disorder pheno-mena, the change from 'ordinary' liquid helium to liquid helium II (which exhibits superfluidity) and in certain aspects, the onset of super-conductivity. The specific heat of superconductors is described in section 6.16. All of these transitions are accompanied by a very sharp rise in the specific heat over a narrow temperature range, the shape of the curve approximating to that of the Greek letter lambda (hence the term 'lambda point' applied to liquid helium at 2·19° K). On the low-temperature side of the transition the specific heat rises first gradually and then rapidly; when it has reached its maximum value (which is usually very difficult to determine) it drops very rapidly (in some cases almost discontinuously) to a much lower value. The general form of these curves is explained by assuming that once the transition to configurations of higher energy begins, then the number of atoms or ions which change their configuration is dependent on the number which have already done so. This is the reason for the rapid rise which is observed after the relatively slow increase at the beginning of the transition.

It should be noted that co-operative phenomena differ from ordinary phase transitions in that the change takes place not at a unique temperature but over a small but finite temperature range. A true phase transition is accompanied by a latent heat. In a co-operative transition the latent heat can be considered as being spread out over the transition range to give the co-operative specific-heat anomaly.

It is clear that there can be no simple general theory for these transitions because the form of the specific-heat curve must depend intimately on the type of interactions which exist between the atoms, molecules, or ions. The only general rule which may be formulated is that for the entropy associated with the transition. This will be the same as that already quoted for the Schottky anomaly (1.39) since the entropy takes no account of the reasons or manner in which the different configurations are populated but depends only on the number of types of configuration which are involved.

The main experiments on co-operative phenomena which have been made at low temperatures have been concerned either with an ordering of the rotational axes of a molecule or with magnetic ordering, and in particular, with the onset of anti-ferromagnetism. As examples of rotational ordering we can cite, first, the anomaly in solid ortho-hydrogen (Mendelssohn, Ruhemann, and Simon, 1931; Hill and Ricketson, 1954) at $\sim 2°$ K which is due to an alignment of the axes of rotation of the H_2 molecules (Fig. 1.14 (a)). Secondly, there is an anomaly in the specific heat of methane at $\sim 20°$ K (Clusius and Perlick, 1934) which is ascribed to the onset of hindered rotation, i.e. the molecule is only able to oscillate backwards and forwards about its axis of rotation instead of rotating completely through 360°

In recent years, however, the emphasis has been on measurements connected with the transition to the anti-ferromagnetic state. This occurs when the magnetic dipoles of a paramagnetic salt spontaneously orient themselves relative to one another. In some cases the lowest energy is achieved if the orientation of one dipole is the same as that of its neighbours (ferromagnetism) and in other cases the energy is lower when neighbouring dipoles point in opposite directions (anti-ferromagnetism) (see sections 9.19 and 9.20). At low temperatures antiferromagnetism seems to be the general rule. Fig. 1.14 (b) shows curves

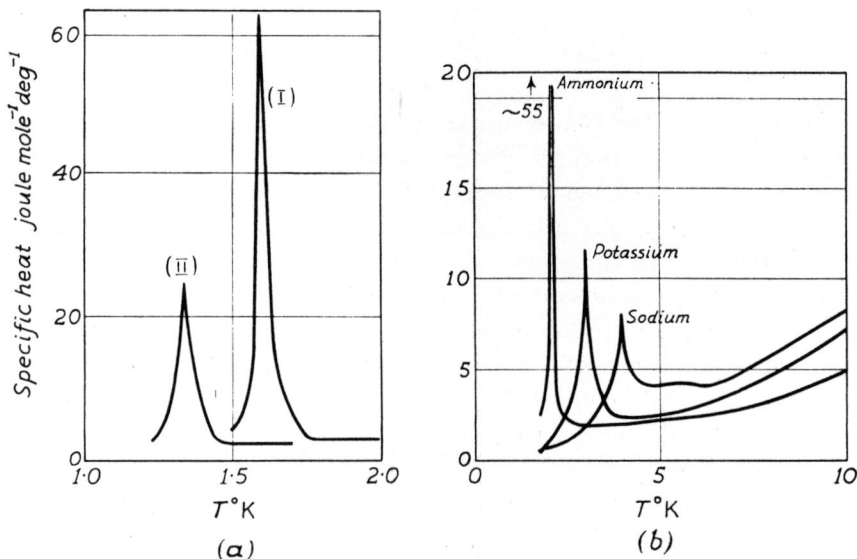

FIG. 1.14. Co-operative specific-heat anomalies. (a) Due to rotational ordering in solid hydrogen (I) 74 per cent, (II) 66 per cent orthohydrogen (Hill and Ricketson, 1954). (b) Due to the onset of anti-ferromagnetism in ammonium, potassium, and sodium chloroiridates (Bailey and Smith, 1959).

for the specific heats of the chloroiridates of ammonium, potassium, and sodium all of which become anti-ferromagnetic at low temperatures. It will be noted that the curves are much steeper than those for the Schottky specific heat shown in Fig. 1.11 and 1.12. In each of the three substances of Fig. 1.14 (b) there are two possible configurations for a pair of adjacent ions. The magnetic moments may be parallel (higher energy) or antiparallel (lower energy). This is confirmed by the fact that the total entropy associated with the anomaly is $R \log_e 2$ per mole, in accordance with (1.39). The transition involves an increase in the antiparallel configuration as the temperature is reduced.

It is clear that the transition temperature T_{tr} must be connected with the energy ϵ_c associated with the ordering of the assembly. A simplified theory by Bragg and Williams (1934, 1935) shows that

$$\epsilon_c = RT_{\text{tr}}/2 \text{ per mole} \tag{1.41}$$

and hence, if the co-operative anomaly is due to a simple two configurational scheme, the energy difference between the two types of configuration for the whole system can be estimated. Except for this information it is not possible from the specific-heat results to obtain any further data on the energy levels or on the interaction between the ions. The main importance of specific-heat measurements of co-operative effects is to show that such a transition exists over a certain temperature range and that it is of a co-operative nature. The detailed nature of the transition must be investigated by other means.

2

THERMAL EXPANSION

2.1. Introduction

IN the next chapter we need to use the concept of anharmonic lattice vibrations in order to discuss the manner in which they influence the heat conductivity of dielectrics. Anharmonicity manifests itself most clearly in the phenomenon of thermal expansion and it is for this reason that we are devoting a short chapter to the subject.

2.2. *A simple model of the expansion mechanism*

Let us consider two atoms in a lattice whose equilibrium spacing at $0°$ K is a and whose separation at any arbitrary time and temperature is y. Then the potential energy, $\mathscr{V}$, of the atoms may be written as a function of their displacement from their original equilibrium spacing, i.e. it will be $f(y-a)$. This can be written as a power series

$$\mathscr{V} = A(y-a)^2 + B(y-a)^3 + \dots. \tag{2.1}$$

There is obviously no term in $(y-a)$, because otherwise the system would be unstable. The coefficient A is positive and this, by itself, would lead to a simple harmonic oscillation of the atoms. The coefficient B takes account of the mutual repulsion of the atoms when they approach one another. It will be negative and it is the first anharmonic term.

For very small oscillations (i.e. at very low temperatures), the B term may be neglected. The atomic motion is harmonic and hence the mean separation of the atoms remains equal to a; i.e. the thermal expansion is zero. At higher temperatures where the amplitude of vibration is such that the cubic term in $\mathscr{V}$ cannot be neglected, the oscillations will no longer be symmetrical about the original equilibrium positions of the atoms, because $\mathscr{V}$ is then no longer symmetrical about those positions. The mean separation of the atoms will therefore increase; i.e. there will be a thermal expansion. This is shown in Fig. 2.1.

2.3. *Thermodynamic derivation*

Whilst this argument illustrates the manner in which thermal expansion arises it is not a very satisfactory one to carry over when we consider an assembly of atoms in a lattice. It then becomes much more convenient to use a thermodynamic derivation.

We first recall the properties of the Helmholtz free energy F,

$$F = U - TS, \tag{2.2}$$

where the symbols have their usual meaning. The pressure p is given by

$$p = -\partial F / \partial V |_T \tag{2.3}$$

and

$$U = -T^2 \frac{\partial (F/T)}{\partial T}\bigg|_V. \tag{2.4}$$

Let us take a crystal at $0°$ K and warm it up to a temperature T, keeping its volume constant. Let the free energy be $F_0(T)$. Since in

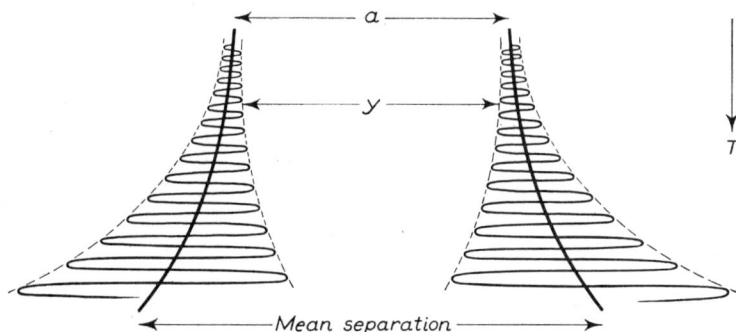

FIG. 2.1. The amplitude of vibration of a pair of atoms as a function of temperature. At higher temperatures the vibrations are not symmetrical about the equilibrium position for $T = 0°$ K, but instead, their mean separation increases.

order to do this the crystal must be constrained by an external pressure, when we relax this pressure the body will expand by an amount ΔV until its free energy $F(T)$ is a minimum. We may write $F(T)$ as an expansion in powers of the change in volume ΔV,

$$F(T) = F_0(T) + \Delta V F_0'(T) + \tfrac{1}{2}(\Delta V)^2 F_0''(T) + ..., \tag{2.5}$$

where F' and F'' are written for $\partial F / \partial V |_T$ and $\partial^2 F / \partial V^2 |_T$ respectively. We should note the relation between $F_0''(T)$ and the bulk modulus, η, defined as $-V \, dp/dV$. Using (2.3), η is $V F_0''(T)$. The equilibrium volume of the crystal will be determined by the conditions necessary for the free energy to be a minimum, i.e. for $\partial F / \partial V |_T$ to be zero. Note that by (2.3) this is the same condition as that for which the pressure is zero. Thus

$$\frac{\partial F}{\partial V}\bigg|_T = 0 = F_0'(T) + \Delta V \, F_0''(T) + ... \tag{2.6}$$

Hence the increase in volume at $T°$ K is given by

$$\frac{\Delta V}{V} = -\frac{F_0'(T)}{V F_0''(T)}. \tag{2.7}$$

In order to estimate $F_0'(T)$ we assume that the energy U_0 of the solid, when its volume is constrained to that at $0°$ K, may be calculated by the Debye theory (section 1.4), but that the value of θ is dependent on the volume. From (1.11) we see that U_0 is then of the form $T \times$ function (T/θ) and TS will also be of this character. Hence from (2.2) we may write $F_0(T)$ as

$$F_0(T) = Tg(T/\theta) \tag{2.8}$$

and

$$F_0'(T) = \left.\frac{\partial F_0}{\partial V}\right|_T = T \left.\frac{\partial g}{\partial (T/\theta)}\right|_V \left.\frac{\partial (T/\theta)}{\partial V}\right|_T \tag{2.9}$$

i.e.

$$F_0'(T) = -\frac{T^2}{\theta^2} \left.\frac{\partial g}{\partial (T/\theta)}\right|_V \left.\frac{\partial \theta}{\partial V}\right|_T. \tag{2.10}$$

But from (2.4)

$$U_0 = -T^2 \left.\frac{\partial g}{\partial T}\right|_V. \tag{2.11}$$

Hence from (2.10) and (2.11)

$$F_0'(T) = \frac{U_0}{\theta} \left.\frac{\partial \theta}{\partial V}\right|_T. \tag{2.12}$$

Now α, the coefficient of linear expansion, is $(1/3V)(dV/dT)$ and so from (2.7) and (2.12),

$$\alpha = \frac{-c_v}{3V F_0''(T)\theta} \left.\frac{\partial \theta}{\partial V}\right|_T, \tag{2.13}$$

where c_v is the specific heat at constant volume. Since, as we have seen, $V F_0''(T)$ is the bulk modulus, η, we then have

$$\alpha = \frac{-c_v}{3\eta V} \left.\frac{\partial (\log \theta)}{\partial (\log V)}\right|_T. \tag{2.14}$$

We define the Grüneisen constant, γ, by†

$$\gamma = -\left.\frac{\partial (\log \theta)}{\partial (\log V)}\right|_T \tag{2.15}$$

and hence

$$\alpha = \frac{\gamma c_v}{3V\eta}. \tag{2.16}$$

Thus the coefficient of thermal expansion is proportional to the specific heat (since η does not change very much with temperature, see section 11.2) and hence at low temperatures α should vary as T^3. This trend towards zero as $T \to 0°$ K may also be deduced from the third law of

---

† Care should be taken not to confuse the Grüneisen constant with the coefficient, γ, for the electronic specific heat.

thermodynamics since $\partial V/\partial T|_p = -\partial S/\partial p|_T$ and this must become zero as T approaches $0°$ K.

Our definition of γ (2.15) might also have been written, using (1.10), in the form $\gamma = -\partial(\log \nu_{max})/\partial(\log V)$. The original formulation by Grüneisen was less general and γ was defined as $-\partial(\log \nu)/\partial(\log V)$, i.e. γ was assumed to be the same for all lattice vibrational modes. With his definition it can be shown after rather a lengthy calculation which we shall not reproduce here (see Roberts and Miller, 1960, p. 561) that γ is zero if the potential is harmonic, i.e. if B in (2.1) is zero. Thus γ may be considered as a measure of the anharmonicity of the vibrations.

2.4. Experimental results

Measurements of the thermal expansion are in quite good agreement with the general form of (2.16). γ is independent of the temperature

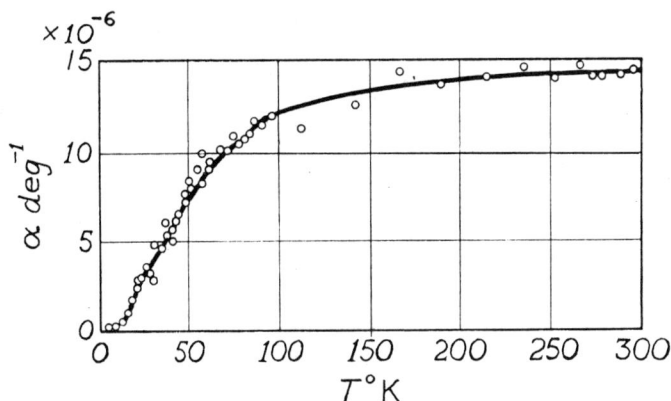

Fig. 2.2. The coefficient of linear expansion of gold as a function of temperature (Fraser and Hollis Hallett, 1960).

and has a value of about 2. At high temperatures where the specific heat is constant, α is temperature-independent and at low temperatures it decreases in accord with the theory (Fig. 2.2). Unfortunately, owing to the very small displacements which have to be measured at low temperatures, it has been very difficult to make measurements of very high accuracy in the liquid-helium region, although recent techniques have improved the situation. White (1961) has developed a very sensitive apparatus which is capable of detecting changes of length of 10^{-9} cm by means of a capacity measurement. He has shown that for copper the Grüneisen γ is indeed constant down to $10°$ K (see Fig. 2.3).

A more careful analysis of the low-temperature data, however, shows that the expansion coefficient is not exactly proportional to T^3, which is what we should expect from (2.16) in the T^3 specific-heat region. Instead it is of the form

$$\alpha = AT^3 + BT. \tag{2.17}$$

It is suggested that the term proportional to T is due to an expansion effect which is caused by the variation of the free energy of the *electrons* with volume. If the density of states $F(\mathscr{E})$ varies with volume, then by

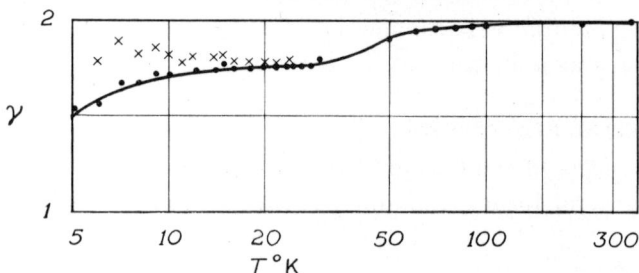

FIG. 2.3. The variation of the Grüneisen γ with temperature for copper. The circles show the results as calculated in the normal way, using (2.16). The crosses are calculated from $3(\alpha - \alpha_e)V\eta/c_v$. The drop in the circled points is ascribed to an electronic contribution to the expansion (White, 1960).

(1.20), so will the electronic specific heat. We can define an electronic Grüneisen constant, γ_e, by analogy with (2.15) as

$$\gamma_e = \frac{\partial \log F(\mathscr{E}_0)}{\partial \log V} \tag{2.18}$$

and an electronic expansion coefficient, α_e, from (2.16) as

$$\alpha_e = \frac{\gamma_e c_e}{3V\eta}, \tag{2.19}$$

where c_e is the electronic specific heat. For a free electron metal γ_e is equal to $\frac{2}{3}$. The theory has been developed by Mikura (1941), Visvana-than (1951), and Varley (1956). The drop in the value of γ shown in Fig. 2.3 at the lowest temperatures is ascribed to the electronic effect.

Table 2.1 gives values of the ordinary lattice-expansion coefficients and Grüneisen constants, which are now denoted by α_g and γ_g respectively, together with α_e and γ_e for several metals. The largest electronic effects are given by the transition elements. Chromium seems to have a negative value of α_e (as also does invar); this might be due to some spin interaction, but the effect has not yet been fully resolved.

Measurements of expansion coefficients have also been made on some of the alkali halides. These show a marked drop in γ at temperatures

TABLE 2.1

Linear expansion data ($1 \cdot 5°$ K to $\theta/20$)

(White, private communication)

Element	$10^{10}\alpha_e/T$	γ_e	$10^{11}\alpha_g/T^3$	γ_g	γ (room temp.)
Ag	..	..	13 ± 1	$2\cdot2\pm0\cdot1$	$2\cdot4$
Al	$9\cdot1\pm0\cdot3$	$1\cdot8\pm0\cdot1$	$2\cdot6\pm0\cdot2$	$2\cdot6\pm0\cdot3$	$2\cdot3$–$2\cdot6$
Co	25 ± 1	$1\cdot9\pm0\cdot1$	$0\cdot9\pm0\cdot1$	$1\cdot6\pm0\cdot2$	$1\cdot9$
Cu	2 ± 1	$0\cdot6$ to $1\cdot2$	$2\cdot7\pm0\cdot2$	$1\cdot8\pm0\cdot1$	$2\cdot0$
Cr	-35 ± 5	$-8\cdot5\pm1$	..	..	..
Fe	29 ± 2	$2\cdot1\pm0\cdot2$	$1\cdot1\pm0\cdot2$	$2\cdot0\pm0\cdot3$	$1\cdot7$
Nb	22 ± 2	$1\cdot6\pm0\cdot2$	$2\cdot3\pm0\cdot3$	$1\cdot1\pm0\cdot2$	$1\cdot6$
Ni	38 ± 2	$2\cdot0\pm0\cdot1$	$0\cdot95\pm0\cdot1$	$1\cdot7\pm0\cdot2$	$1\cdot9$
Pb (normal)	20 ± 5	$1\cdot7\pm0\cdot4$	165	$2\cdot6\pm0\cdot2$	$2\cdot7$
Pd	37 ± 2	$1\cdot8\pm0\cdot1$	$4\cdot3\pm0\cdot5$	$2\cdot3\pm0\cdot3$	$2\cdot3$
Ta	12 ± 1	$1\cdot35\pm0\cdot1$	$3\cdot4\pm0\cdot3$	$1\cdot7\pm0\cdot2$	$1\cdot7$
V (normal)	38 ± 2	$1\cdot6\pm0\cdot1$	$0\cdot7\pm0\cdot1$	$0\cdot9\pm0\cdot1$	$1\cdot2$

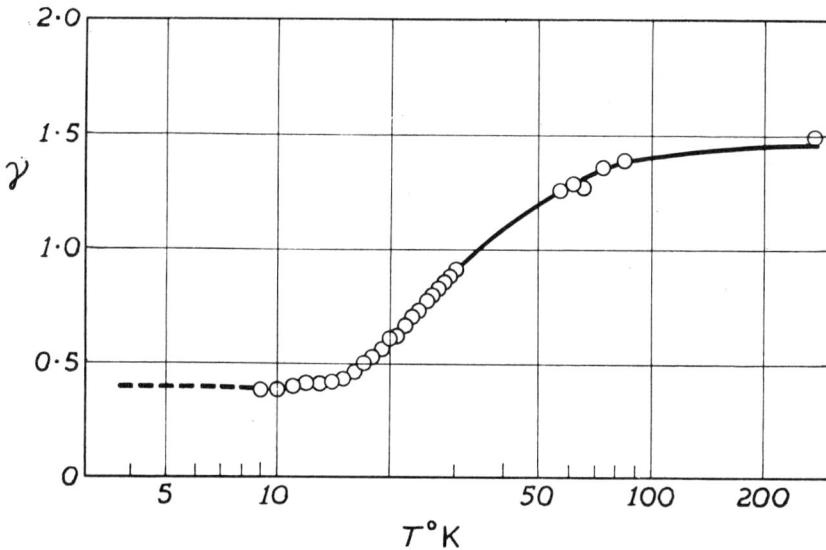

FIG. 2.4. The temperature dependence of the Grüneisen γ for KCl (White, 1962).

of the order of $\theta/10$ which is not observed in metals. An example is shown in Fig. 2.4.

2.5. Considerations of thermal expansion in the design of apparatus

The change in dimensions in the low-temperature region itself is so small that for nearly all practical purposes it may be ignored. On going from $0°$ to $50°$ K the fractional change in length is usually less than one part in 10^4. From $50°$ K up to room temperature the change is $\sim 0\cdot2$ per cent. In the design of cryostats which are suspended from tubes at room temperature the effects of expansion should be considered, particularly if the tubes are of different materials, because the differential expansion can be quite troublesome. Useful design data are given by White (1959), p. 285.

3

THE THERMAL CONDUCTIVITY
OF NON-METALS

3.1. Introduction

THERE are two processes by which heat may be transported through a solid. Due to the strong coupling between the atoms in a crystal lattice, conduction can take place by means of the lattice vibrations (or phonons). These can be considered as waves travelling through the material. This is the mechanism of heat transport in dielectrics. The heat may also be carried by any 'free' electrons in the solid and hence in metals this constitutes a second conduction process. In fairly pure metals nearly all the heat is, in fact, carried by the electrons and very little by the lattice vibrations. In semiconductors the heat transport will be shared between the two mechanisms, depending on how many conduction electrons are present. In alloys too (although for somewhat different reasons), both processes of thermal conduction must usually be considered. Experiments show that the heat transport is proportional to the temperature gradient along the specimen and from this observation the heat conductivity, κ, is defined as

$$\kappa = \frac{\dot{Q}L}{A\,\Delta T} \tag{3.1}$$

where $\dot{Q}$ is the rate of flow of heat along a specimen of cross-sectional area A, which produces a temperature difference ΔT over the length L.

The conduction processes are limited by various scattering mechanisms without which infinite conductivity would result, and the theoretical problem at the outset is to decide what types of scattering are possible, which ones are the most effective (i.e. which ones give an appreciable thermal resistance), and finally to calculate that thermal resistance. The problem is usually simplified by assuming that one scattering process is not influenced by the others which are occurring at the same time. This is not rigorously true, but, in general, the errors involved are small. This simplification (which in the case of electrical conductivity is called Matthiessen's rule) means that the thermal resistivity due to each scattering process may be calculated separately and the sum of these resistivities will give the total thermal resistivity for that conduction mechanism. If we have two such mechanisms, that by the electrons and that by the phonons, then they will have thermal resistivities W_e and W_g respectively. These thermal resistivities will each be the sum of a number of resistivities, e.g. $W_e = W_A + W_B + \ldots$ due to the scattering mechanisms A, B,.... The thermal conductivity in a dielectric will be determined by the value of W_g, but in substances where both processes are effective, the total heat transport will be the sum of the heat transported by the electrons and by the phonons and hence the total thermal conductivity, κ, is the sum of the conductivities due to the electrons and the phonons, i.e.

$$\kappa = \kappa_e + \kappa_g, \tag{3.2}$$

where $\kappa_e = 1/W_e$ and $\kappa_g = 1/W_g$.

In this chapter we shall describe the behaviour of the phonon conductivity with particular reference to heat transport in non-metals. The most important resistive mechanisms which have been observed are:

(a) the interaction of the phonons with one another (umklapp-processes);

(b) scattering by point defects (impurity atoms, isotopes, etc.);

(c) scattering by the boundaries of the specimen or the crystallites;

(d) scattering by dislocations.

These interactions will be considered in turn in the succeeding sections.

3.2. The Debye theory of lattice conductivity

The early work of Euken (1911) showed that when non-metal single crystals were cooled their thermal conductivity *increased* and appeared to be roughly proportional to $1/T$ down to liquid-hydrogen temperature. For amorphous solids (such as glass), on the other hand, κ was much less than that for single crystals and it *decreased* as the temperature was reduced, being approximately proportional to the specific heat. In order to explain these results Debye (1914) developed a theory in which he suggested that the heat was transported through the substance by travelling elastic waves. Whilst the details of this theory have been outmoded it is still useful to discuss some of the concepts which he introduced.

Let us assume the solid to be a continuous elastic medium. If one end is heated then this may be considered as the introduction of an oscillatory disturbance at that end. The disturbance will be propagated through the medium in a manner analogous to that in which ripples are emitted from a disturbed region of a pool of water. If the wave front is plane then the disturbance which started out from the hot end of the solid will arrive at the other end, under ideal conditions, with undiminished amplitude. The temperature of any region of the specimen before and after the wave has passed through it will be the same. Thus if the far end of the specimen was in good contact with a thermal reservoir so that all the heat reaching it was absorbed, then the passage of heat down the specimen would not affect the temperature of the specimen, i.e. heat energy would have been transported along the specimen under zero temperature gradient. From (3.1) a zero temperature gradient would imply an infinite conductivity. Since this is not observed in practice, we must assume that the amplitude of the disturbance is diminished as it passes down the specimen. Debye suggested that this attenuation is produced by the interaction of the particular wave which we are considering with all the other waves which will be present in the medium and whose frequencies and amplitudes will be dependent on the specimen temperature. No interaction will occur, however, if the wave motion is perfectly harmonic.† This is analogous to the way in which two sets of water ripples intersect each other and then carry on without any change in their patterns. Any interaction between the waves is due to the anharmonicity in their motion.

3.3. *Anharmonicity*

Debye treated the effect of the thermal vibrations of the solid by assuming that they created regions of varying density. The velocity, v, of a transverse wave in a medium of density d and modulus of rigidity η may be determined from the well-known relation

$$v = (\eta/d)^{\frac{1}{2}}. \tag{3.3}$$

If the atoms of which the medium is composed move in a strictly harmonic manner, then it is very simple to show that if, by compression, d is increased, then η will be increased by exactly the same factor. Thus v remains constant. If, however, there are anharmonic terms present in the expression for the forces on the atoms, i.e. if the potential energy is of the form $Ax^2 + Bx^3 + \ldots$ for a displacement, x, from the equilibrium position, where $B \neq 0$, then the change in d on compression is *not* exactly compensated for by the change in η and so the velocity of the waves in the medium will change. Thus due to the anharmonicity, the waves will be scattered as they pass through regions of different density. The mean

---

† By a harmonic wave motion we mean one in which the potential energy is proportional to the square of the displacement.

free path, l, of such a wave was defined as the distance it had to travel for its energy to be reduced by $(\exp)^{-1}$. The theory showed that l is inversely proportional to T. Debye also derived the formula which can also be obtained from kinetic theory and which is most useful in all discussions on thermal conductivity

$$\kappa = \text{constant} \times c_v vl, \tag{3.4}$$

where c_v is the specific heat of the heat carriers per unit volume (the lattice vibrations in this case), and v is their velocity. In Debye's theory the constant has the value $\frac{1}{4}$, but if we are dealing with particles (as in the kinetic theory) or with phonons, the constant is $\frac{1}{3}$. These deductions can be used to explain the results of Euken's experiments in the following manner. Since the wave velocity is approximately constant and l varies as T^{-1}, (3.4) shows that in the region where c_v is not changing very much, κ is proportional to $1/T$, in agreement with the experiments on crystals. For amorphous substances there is no regular crystal lattice and the waves will be rapidly attenuated. l will be a constant with a value of the order of the interatomic distance and thus κ will be proportional to c_v in accord with the observations.

The actual calculation of the magnitude of the anharmonicity of the lattice vibrations and the manner in which it should be incorporated into explicit theoretical expressions is rather uncertain. The most straightforward evidence for lattice anharmonicity is that crystals expand when they are heated (see section 2.2), and it is from measurements of the coefficient of linear expansion, α, that a parameter which gives a measure of the anharmonicity is derived. This is γ, the Grüneisen constant (2.15) and (2.16). It is often used (for the want of a better parameter) in calculations. For this reason it will appear in the coefficients of several of the expressions in the succeeding sections.

3.4. The thermal conductivity of crystals

Whilst Debye's theory is useful in giving us a mechanism of heat transport which can be easily visualized, it is not entirely satisfactory. It assumes that the crystal is a continuum and is at rest, except for the wave train we are considering, whereas the scattering points (i.e. the regions of different density) are actually moving with approximately the same velocity as that wave train. A satisfactory theory must take account of the discrete nature of the crystal lattice, the frequency spectrum of the waves and, particularly at low temperatures, the quantum effects of a vibrating system.

Such a theory was first developed by Peierls (1929). A detailed exposition of this work is outside the scope of this book and the interested reader is referred to Ziman (1960). An outline of the main points of the theory will be presented.

3.5. *The vibrational spectrum of a discrete lattice*

When we discussed the Debye theory of the lattice specific heat (section 1.4) it was assumed that the crystal could be considered as an

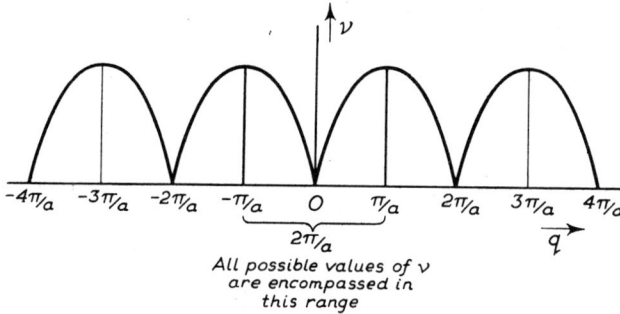

FIG. 3.1. The relationship between the frequency, ν, and the wave number, q, for a linear chain of identical atoms. Note that the whole spectrum of possible waves, travelling in either direction, is covered within a range of q equal to $2\pi/a$. The choice of the range indicated is quite arbitrary.

elastic continuum in which the velocity of the waves was independent of their frequency, i.e. ν was proportional to λ^{-1}. When calculations are made for a *discrete* lattice of identical atoms, this assumption is no longer justified. Fig. 3.1 shows the relationship between ν and q, the wave number (which is equal to $2\pi/\lambda$), for a one-dimensional discrete lattice of identical particles. Instead of the simple linear relationship, the graph is a succession of the positive parts of a sine curve. The positive and negative values of q represent waves which are travelling in opposite directions. From this diagram the following points should be noted.

(a) For values of q which have a magnitude greater than π/a, where a is the lattice spacing, the values of ν repeat; thus the complete spectrum of vibrations for the two possible directions of propagation in the one-dimensional lattice, is encompassed within the range $-\pi/a \leqslant q < \pi/a$.

(b) The maximum frequency of the lattice waves is a natural result of the shape of the curves. There is no need for the arbitrary cut-off at ν_{max} which was necessary in the Debye theory (section 1.4).

(c) The group velocity, v, which is the velocity at which energy is transported by the wave train, is given by

$$v = 2\pi \, dv/dq. \qquad (3.5)$$

It is therefore proportional to the slope of the curves. Thus at low frequencies the group velocity is approximately independent of v, whereas at frequencies approaching v_{max} it is very much smaller, becoming zero at $q = \pm\pi/a$. Beyond $\pm\pi/a$ it changes its direction.

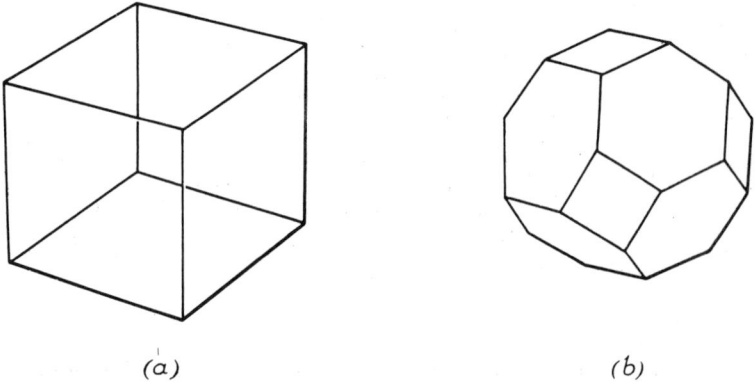

(a)   (b)

FIG. 3.2. Diagrams of the first Brillouin zone for (a) a simple cubic and (b) a face-centred cubic lattice.

Because of (a) it is not necessary to consider values of q which lie outside the range $-\pi/a \leqslant q < \pi/a$. The choice of a region centred on $q = 0$ is quite arbitrary, but it is convenient because the sign of q then indicates the direction of propagation of the wave. We could, however, use any other part of the curve provided that it spanned a region in which q increased by $2\pi/a$. It will be seen that values of q within a given zone are separated from equivalent values of q outside (i.e. those which have the same value of v and the same direction of propagation) by an integral $\times 2\pi/a$, where $2\pi/a$ is the 'repeat distance' of the curves.

In a three-dimensional lattice an analogous situation arises. $\mathbf{q}$ is now a vector whose direction is determined by the direction of propagation of the wave and the significant values of $\mathbf{q}$ lie within a region, called the first Brillouin zone, whose boundaries depend on the crystal structure. For the simple cubic lattice the zone is bounded by a cube of side $2\pi/a$, whilst for a face-centred cubic structure, the zone is a dodecahedron (Fig. 3.2). Mott and Jones (1936), chapter 5, give further examples of the structure of Brillouin zones. As $\mathbf{q}$ increases and approaches the zone boundary the group velocity tends to zero. Beyond the boundary, the velocity is reversed in direction and the relationship between $\mathbf{q}$ and ν repeats itself as it does in the one-dimensional case. The vectors $\mathbf{q}_0$ which join any point in the first zone to equivalent points in the next zone are called basic vectors of the reciprocal lattice. These are analogous to the 'repeat distance' of $2\pi/a$ in the one-dimensional case. In the simple cubic lattice, the basic vectors, $\mathbf{q}_0$, are $\pm 2\pi/a$ with directions parallel to the cube axes.

3.6. *Umklapp-processes*

At low temperatures we must take account of the quantum nature of vibrational energy. We can consider a disturbance to consist of a

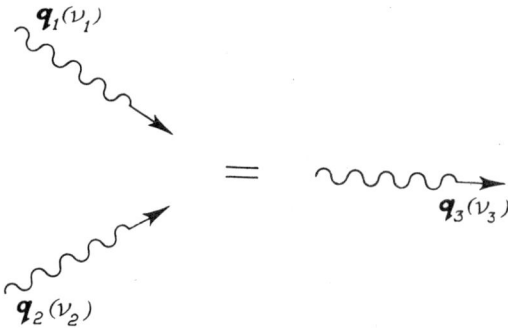

FIG. 3.3. The interaction of two phonons, $\mathbf{q}_1$, $\mathbf{q}_2$, to give a resultant $\mathbf{q}_3$, in a normal-process. The direction of energy flow is unchanged in this interaction.

number of wave packets of energy $h\nu$ and wave vector $\mathbf{q}$. Each of these packets is called a phonon. The passage of heat through the specimen can be thought of as being due to a flow of phonons and in order that there may be a thermal resistance, phonons must interact with one another. The probability for such an interaction, as we have already seen, will depend on the anharmonicity of the waves. The simplest type of interaction which can occur is shown in Fig. 3.3. This is

$$\text{phonon } 1 + \text{phonon } 2 \to \text{phonon } 3, \qquad (3.6)$$

i.e. a phonon with wave vector $\mathbf{q}_1$ and frequency ν_1 combines with a phonon with wave vector $\mathbf{q}_2$ and frequency ν_2 to form a third phonon. This is called a three-phonon process. Peierls showed that in such a reaction the following laws must be obeyed. Firstly the conservation of energy between the initial and final phonons,

$$h\nu_1 + h\nu_2 = h\nu_3, \qquad (3.7)$$

and secondly the conservation of the wave vectors

$$\mathbf{q}_1 + \mathbf{q}_2 = \mathbf{q}_3. \qquad (3.8)$$

The interpretation of (3.8) has led to considerable confusion, particularly when $\mathbf{q}_1$ and $\mathbf{q}_2$ are of sufficient magnitude for $\mathbf{q}_3$ to lie outside the zone boundary. First of all, however, let us consider the case where $\mathbf{q}_1$ and $\mathbf{q}_2$ are both fairly small so that $\mathbf{q}_1+\mathbf{q}_2$ does not fall outside the first zone. Such interactions are called normal-processes or n-processes. After one of these interactions the vector sum of the $\mathbf{q}$'s remains unchanged. Recalling that within the first zone the direction of $\mathbf{q}$ is the direction of the phonon velocity, we see that if there had been a net flow of energy

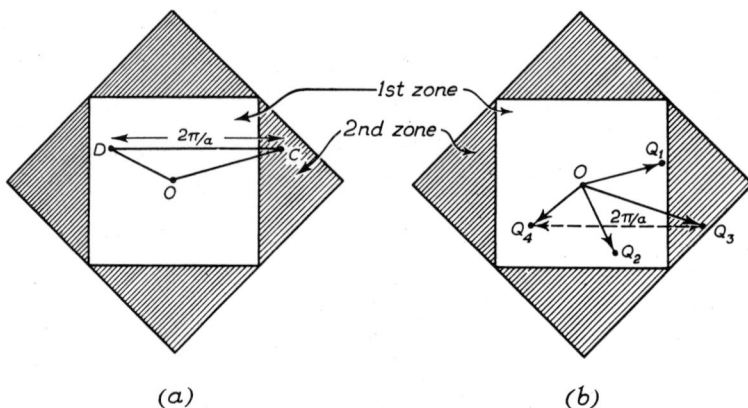

(a)   (b)

FIG. 3.4. An umklapp-process illustrated using the first two Brillouin zones in a simple, square, two-dimensional lattice. (a) The vector OC which extends into the second zone is equivalent to the vector OD in the first zone and differs from it by the basic vector, $2\pi/a$. (b) In an umklapp-process OQ_1 and OQ_2 combine to give OQ_3. This is equivalent to OQ_4 in the first zone and hence the direction of energy flow has been changed.

in a given direction, the flow in this direction would persist after an n-process. Thus n-processes do not give rise to any thermal resistance.

What happens if $\mathbf{q}_1+\mathbf{q}_2$ is sufficiently large so that $\mathbf{q}_3$ falls *outside* the zone boundary? In order to determine the actual direction of the motion associated with $\mathbf{q}_3$ we must first find the vector within the first zone to which it is equivalent. This equivalent position will be obtained by adding to (or subtracting from) the vector $\mathbf{q}_1+\mathbf{q}_2$ the appropriate basic vector of the reciprocal lattice, $\mathbf{q}_0$, which will bring the final vector within the zone boundary. An example of how this may be done is illustrated in Fig. 3.4. Fig. 3.4 (a) shows in two dimensions two neighbouring zones for a simple square lattice. The wave vectors of all the phonons in the crystal can be represented by lines starting from the centre O of the inner square and extending out as far as its boundary. A vector, such as OC which extends into the next zone, must have an equivalent vector within the first zone. This vector will be OD, where

CD is equal to the length of the side of the first zone (a basic vector). We now return to the interaction of the two phonons $\mathbf{q_1}$ and $\mathbf{q_2}$ which are represented in Fig. 3.4 (b) by OQ_1 and OQ_2 respectively. Their vector sum is OQ_3 which falls in the outer zone. The equivalent vector in the first zone is OQ_4, where $Q_3 Q_4 = 2\pi/a$. Thus the two phonons $\mathbf{q_1}$ and $\mathbf{q_2}$ whose motion was directed towards the right-hand side of the diagram, have combined to form $\mathbf{q_3}$ whose motion is approximately in the opposite direction. The energy flux from left to right has been diminished, i.e. the process gives rise to a thermal resistance. Such a mechanism, which is formally not different from the normal processes, has been called an umklapp-process (from the German, 'flipping over') or u-process. The change in direction of $\mathbf{q_3}$ is analogous to the Bragg reflection of X-rays by the crystal lattice. A u-process can therefore be considered as an interference effect in which the phonons are reflected by the crystal lattice as a whole.

In the literature it is customary to write (3.8) in a slightly different form in which the R.H.S. is modified so that the value of the resultant $\mathbf{q_3}$ used is its value in the first Brillouin zone. Thus for a u-process we have

$$\mathbf{q_1} + \mathbf{q_2} = \mathbf{q_3} + \mathbf{q_0}. \tag{3.9}$$

The addition of one of the basic vectors $\mathbf{q_0}$ brings $\mathbf{q_3}$ back into the first zone as was shown in the change from OQ_3 to OQ_4 in Fig. 3.4 (b). For a normal process (3.8) remains unchanged.

Since, except when it has a small value, $\mathbf{q}$ is not proportional to ν, we should note that if the energy relationship (3.7) holds in any interaction, then it will not, in general, be possible to satisfy the wave-vector equation (3.8). We have so far, however, assumed that there is only one relationship between $\mathbf{q}$ and ν; i.e. that all phonons are of the same type. In reality some of them will be associated with longitudinal waves and these have a higher velocity than the transverse phonons, with a consequent different dependence of $\mathbf{q}$ on ν. Thus whilst (3.7) and (3.8) cannot hold simultaneously if $\mathbf{q_1}$, $\mathbf{q_2}$, and $\mathbf{q_3}$ have the same polarization, interaction is possible if one of them has a different polarization. A more detailed consideration of the problem (Herpin, 1952; Klemens, 1958) shows that other restrictions on the interacting phonons also have to be imposed. The only possible interactions are

$$\text{transverse} + \text{transverse} \leftrightharpoons \text{longitudinal} \tag{3.10}$$

and
$$\text{transverse} + \text{longitudinal} \leftrightharpoons \text{longitudinal}. \tag{3.11}$$

It may be shown (see Klemens, 1958, p. 27) that these restrictions also apply to u-processes (3.9) as well as to n-processes. An important consequence of (3.10) and (3.11) is that whilst n-processes do not give rise to a thermal resistance they do play a very important part in establishing thermal equilibrium between the various vibrational modes. It will be seen in a later section that this becomes important when the scattering of phonons by point defects is considered.

A crystal lattice containing more than one atom per unit cell has an additional vibrational spectrum which embraces a region of very high frequencies. This forms what is usually termed the optical branch of the spectrum, but the energies involved are so high that these modes are not excited at low temperatures. Hence they need not be considered here. The low-energy band (the acoustical branch) is similar in general form to that which we have already discussed.

Whilst we have shown that if a u-process occurs it results in a scattering of the incident phonons, we have still to calculate the probability of such a process. We shall only do this in a very approximate manner. Since in a u-process the resultant wave vector must fall outside the zone boundary, at least one of the phonons must have a wave vector which is greater than half the maximum possible value of $\mathbf{q}$ within the first zone. Thus its frequency must be (approximately) more than one-half the maximum vibrational frequency of the lattice, i.e. $\frac{1}{2}\nu_{max}$. If we use the Debye theory of specific heat to get a rough value of ν_{max}, then from (1.10)

$$\tfrac{1}{2}\nu_{max} = \tfrac{1}{2}k\theta/h. \tag{3.12}$$

The probability of exciting a phonon of this frequency at low temperatures is $\exp(-\tfrac{1}{2}h\nu_{max}/kT)$ and this is equal to $\exp(-\tfrac{1}{2}\theta/T)$. We must also take account of the probability that the second phonon will have a value of $\mathbf{q}$ which is large enough to make $\mathbf{q}_1 + \mathbf{q}_2$ fall outside the zone boundary. This depends very much on the detailed shape of the zone. It is usually assumed to be proportional to some power, n, of the temperature (Klemens, 1958, suggests that $n = -3$). Thus the thermal resistance due to u-processes, W_u, can be written

$$W_u = \text{constant} \times T^n \exp(-\theta/gT) \quad \text{for } T \ll \theta, \tag{3.13}$$

where g is a numerical factor whose value is about 2 and the constant contains terms which are a measure of the lattice anharmonicity. Thus as the temperature falls the thermal conductivity should rise approximately exponentially. Whilst this type of behaviour is observed in some cases, the rise is usually so rapid that the data cannot be used to determine n with any accuracy.

We have only considered the simplest possible form of phonon interaction, that involving three phonons in all. Other more complicated processes might occur in which, for example, two phonons interact to form two other phonons. It is not thought that such four-phonon processes are very important, but the theory has not been worked out in any detail.

At high temperatures ($T \approx \theta$ or $T > \theta$) most of the phonons will have wave numbers which will be large enough to enable u-processes to occur. The probability of a u-process will then be proportional to the number of phonons, i.e. to the temperature. One estimate (Leibfried and Schlömann, 1954) of the absolute value of W_u is

$$W_u = \frac{5}{12}\frac{\gamma^2}{4^{\frac{1}{3}}}\left(\frac{h}{k}\right)^3\frac{T}{Ma\theta^3} \quad \text{for } T > \theta, \tag{3.14}$$

where M and a are the mean atomic mass and spacing respectively. Whilst (3.14) does give an order of magnitude agreement with experiment it tends to underestimate the thermal resistance by a factor of two or three.

3.7. Boundary scattering

The decrease in W_u at low temperatures which is predicted by (3.13) is not observed below a certain temperature. It is found that as the temperature is reduced the thermal resistance decreases and then it starts to increase again (e.g. see Fig. 3.7). Peierls suggested that this increase occurred when the mean free path of the phonons became as large as the specimen diameter. At any lower temperature the mean free path had to remain constant. The theory of the effect which was first developed by Casimir (1938) may also be obtained from a consideration of the kinetic theory expression (3.4). This shows that when l is constant the conductivity must be proportional to the specific heat; hence the thermal resistivity will rise as the temperature is lowered. This boundary scattering resistance, W_B, is therefore of the form

$$W_B = 3/(c_v vl), \tag{3.15}$$

where l is now the diameter of the specimen. For a specimen of square cross-section with a side p, l is replaced by $1 \cdot 1p$. Thus we now have the curious circumstance that W_B is not a fixed quantity for a certain material. It depends on the actual size of the sample. The larger the specimen, the smaller is W_B. At very low temperatures where c_v is proportional to T^3, W_B is proportional to T^{-3}. If the specimen is a polycrystal, l will tend to be the size of the crystallites.

The two resistivities, W_u and W_B, are those which would be expected from an ideal crystal. We must now consider what other scattering processes are observed in practice. In non-metals these have been found to be the scattering of phonons by point defects, isotopes and dislocations. In metals and semi-metals, the phonons will also be scattered by the conduction electrons.

3.8. *Point defect and isotope scattering*

A defect of atomic dimensions, such as a vacant lattice site, an interstitial atom, an impurity atom, or an isotope of the specimen whose atomic weight is different from that of the majority of atoms in the specimen, will upset the regularity of the crystal lattice. At low temperatures this irregularity will be very much smaller than the phonon wavelength (section 1.7) and the scattering which it produces is analogous to the effect in optics known as Rayleigh scattering. Rayleigh's treatment (1896) may be used and this gives a scattering probability which is proportional to q^4. The actual magnitude of the scattering will be determined by the differences in density and in elastic constant which are caused by the presence of the scattering centre in the lattice.

A difficulty arises when one calculates the thermal conductivity because it is necessary to integrate the inverse of the scattering probability (i.e. $1/q^4$) over all the phonon modes. For the long wavelength phonons the density of states is proportional to v^2 (1.7), i.e. to q^2, and hence the integral will contain a term of the form $\int q^2/q^4 \, dq$. This integral diverges at the lower limit and gives an infinite conductivity for the very long wavelength phonons as $q \to 0$. In actual fact this catastrophe will be avoided by the onset of boundary scattering (section 3.7) and also by the presence of n-processes, which, whilst they do not give rise to any thermal resistance, do help to maintain equilibrium between the modes, (3.10) and (3.11). They will ensure that the energy flux is shared out with the phonons of higher q and these, of course, are scattered by the point defects. For a full discussion on these lines showing how the problems of divergence may be avoided, see Ziman (1960), chapter 8, and Callaway (1959). The following argument, however, which we shall call the 'dominant phonon' argument, and which we shall have occasion to use again in later sections, whilst not rigorous, will indicate the type of result which one might expect.

Instead of considering all the phonons, we only take account of those with a wave vector q which are dominant at a particular temperature. We have already shown (1.16) that the wave vector of these dominant phonons is proportional to T. Hence, since the scattering probability for this particular group of phonons is proportional to q^4, it will also be proportional to T^4, i.e. l will be proportional to T^{-4}. In the region where the specific heat is proportional to T^3, the kinetic theory expression (3.4) then yields a thermal resistivity due to point defects, W_D, which is proportional to T. A more rigorous treatment (see, for example, Callaway, 1959) gives a slightly modified power of T so that

$$W_D = \text{constant} \times T^{\frac{3}{2}}. \tag{3.16}$$

At high temperatures where the phonon wavelength is short, the Rayleigh formula does not apply and the scattering becomes less frequency-dependent. This, coupled with the fact that the specific heat also becomes constant, leads to a value of W_D which does not depend on the temperature.

The calculation of the constant in (3.16) cannot be made with any precision because it involves a knowledge of the change of the force constants and the atomic spacing due to the presence of the defect. The only case in which calculations may be made with some confidence is when the point defect is an isotope of the same element as the rest of the specimen, but whose mass is different from that of the neighbouring atoms. Neither the elastic constants nor the atomic spacing should be altered by the isotope and the only scattering will be due to the mass difference ΔM between the isotope and its neighbours. Ziman (1960), chapter 8, shows that the mean free path of a phonon of wave vector q is then

$$l_q = \frac{4\pi d^2}{N_{\text{def}}(\Delta M)^2 q^4}, \tag{3.17}$$

where d is the density of the material and N_{def} is the number of scattering centres per unit volume. Unless the specimen is almost isotopically pure this mean free path is comparable with or is smaller than that due to u-processes, particularly at low temperatures. Thus in general, unless a specimen is isotopically pure, the exponential rise with conductivity, due to the reduction in u-processes, will not be observed; isotope scattering will tend to be the dominant resistive mechanism, until boundary scattering overshadows it at the lowest temperatures.

3.9. Dislocation scattering

A dislocation is an imperfection in the crystal lattice which has extension in one dimension. The atomic irregularity with which it is associated may be divided into two regions. Firstly, there is a region of gross mismatch running along the centre, or core, of the dislocation, and secondly, there is a strain field in the crystal which surrounds this core. Whilst it is in the core that the disorganization of the crystal lattice is most serious, its effect is not so great as that of the outer strain field because the effective diameter of the core will only be a few atomic distances. At low temperatures, where the phonon wavelength is long compared with the atomic spacing, the scattering by the dislocation core may be considered as Rayleigh scattering by a long cylinder. This gives a mean free path which is proportional to q^3 and, following a 'dominant phonon' line of reasoning as was used in the previous section, this leads to a temperature-independent thermal resistance. The absolute magnitude of this resistance is much smaller than that due to the scattering of the phonons by the region of elastic strain around the dislocation. The magnitude of this strain is of the order of b/r, where b (which is approximately an atomic spacing) is the magnitude of the Burgers vector of the dislocation. (A definition of b is given in section 11.1.) Thus the strain field extends a long way from the dislocation and it is able to scatter phonons which are at distances r, considerably greater than the phonon wavelength, from the centre of the dislocation. As we have already seen (section 3.3) the lattice anharmonicity will produce a change in the phonon velocity in these strained regions and the phonons will be refracted. If the velocity in the unstrained lattice is v_0, then in the region where the lattice strain is $\pm b/r$ the velocity becomes

$$v = v_0(1 \pm \gamma b/r).\tag{3.18}$$

If the wavelength of the phonon is short compared with its distance from the centre of the dislocation, the lattice can be treated as a continuum. It can then be shown that a phonon whose closest distance of approach to the centre, O, of a dislocation is r_0 (Fig. 3.5) will be scattered through an angle of about $\gamma b/r_0$. If the incident phonon flux is homogeneously distributed and it is travelling in the direction indicated by the arrow with a density F cm^{-2}, then the amount which will pass within the region between r and $r+dr$ from O per unit length of dislocation can be written as $F\,dr$. After being scattered through an angle ϕ, the flux at a plane such as AB which is at right angles to the original direction of F will be

$$F\,dr\cos\phi.\tag{3.19}$$

The change in flux, or the number of phonons scattered, is

$$F\,dr\,(1-\cos\phi),\tag{3.20}$$

and since ϕ is small this is equal to

$$F\,dr\,\tfrac{1}{2}(\gamma b/r_0)^2.\tag{3.21}$$

Hence, on integration, the number of phonons which are scattered is $\frac{1}{2}F\gamma^2 b^2/r_0'$ where r_0' is the shortest distance from O for which this treatment is valid, i.e. $r_0' = \lambda$. Thus the scattering (i.e. l^{-1}) is proportional to $\gamma^2 b^2/\lambda$. Using the dominant phonon argument once again, l^{-1} will be proportional to T and hence in the region where the specific heat varies as T^3, we obtain for W_{dis}, the thermal resistivity due to dislocations,

$$W_{\text{dis}} = DT^{-2}. \tag{3.22}$$

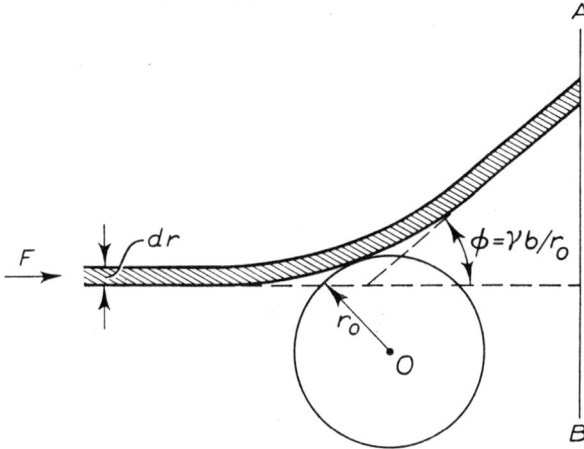

Fig. 3.5. The deflexion of phonons by the strain field around a dislocation whose core is at O. The phonon flux at right angles to AB is decreased by a factor $\cos \phi$.

We have not considered here differences in the strain field depending on whether the dislocation is of edge, screw, or mixed character. Nor have we taken into account dislocations which are not at right angles to the phonon flux. Klemens (1958), p. 60, suggests that the most satisfactory expression for the calculation of D for a random array of dislocations is

$$D = \frac{h^2 v \gamma^2 b^2}{28k^3} N_{\text{dis}}, \tag{3.23}$$

where N_{dis} is the density of dislocations per unit area.

The scattering processes which we have considered appear to be those which are the most important in determining the thermal conductivity of non-metals. Other lattice defects which should in principle give a thermal resistance are stacking faults and small angle grain boundaries. Their resistance is thought to be rather small, however, and there is no conclusive experimental evidence that the effect of either has been observed. It is also possible for phonons to interact with the spin systems of paramagnetic crystals. This also produces an extra thermal resistivity, but a treatment of this subject is outside the scope of this book.

3.10. A summary of the behaviour of the phonon conductivity

Before describing the results of experiments we shall attempt to collate the results of the preceding sections. In an ideal pure crystal the thermal resistance over most of the temperature range will be determined by u-processes. At low temperatures, however, due to the exponential decrease in the probability of these processes, the mean free path of the phonons will increase until it becomes limited by the dimensions of the specimen, whereafter it must remain constant (Fig. 3.6 (a)). The phonons

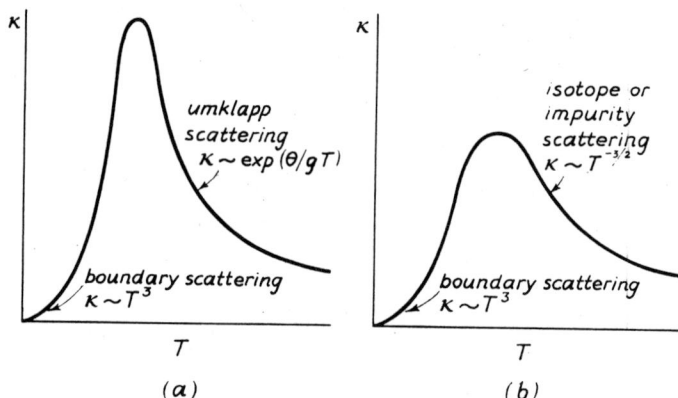

FIG. 3.6. Diagrammatic representation of the low-temperature thermal conductivity of dielectric crystals. (a) An isotopically pure material shows a steep exponential rise in the conductivity due to the reduction in the number of umklapp-processes, followed by a decrease due to the T^3 boundary scattering term. (b) In a material of mixed isotopic content (or in an impure specimen) the conductivity rises less steeply (proportional to $T^{-\frac{3}{2}}$) before the boundary scattering becomes dominant as in (a).

will then be scattered by the boundaries of the specimen and at sufficiently low temperatures this gives a conductivity which is proportional to T^3, and which also depends on the size of the specimen.

Point defects and mixtures of isotopes will upset this behaviour. The exponential increase in the conductivity due to the u-processes will be suppressed in favour of a conductivity which is proportional to $T^{-\frac{3}{2}}$ (Fig. 3.6 (b)) although at sufficiently low temperatures boundary scattering will again become dominant. If the specimen is polycrystalline, the size of the individual crystallites will determine the conductivity.

Large amounts of impurity and dislocations will decrease the conductivity. The dislocations will be effective at low temperatures whilst the impurities tend to scatter only at higher temperatures. Table 3.1 gives the temperature dependence and some approximate expressions for the thermal resistance for phonon scattering.

3.11. Comparison of the theory with experiment

The various phonon scattering mechanisms which have been dealt with in the preceding sections have all been observed in thermal conductivity experiments, and, in general, the temperature dependence and the order of magnitude of the thermal resistivity are in quite good agreement with the theoretical predictions.

TABLE 3.1

Scattering mechanism	Thermal resistance for $T \leqslant \theta/10$
Specimen and crystallite boundaries of dimension l	$3/(c_v vl) = 1 \cdot 55 \times 10^{-3}\theta^3 \, Md^{-1}v^{-1}l^{-1}T^{-3}\,\mathrm{watt}^{-1}\,\mathrm{cm\,deg}$
U-processes	$\dfrac{4\gamma^2}{7 \cdot 05aM}\left(\dfrac{h}{k}\right)^3\left(\dfrac{\theta}{T}\right)^3 \exp(-\theta/gT)$
Isotopes	$\mathrm{constant} \times \dfrac{N_{\mathrm{def}}(\Delta M)^2}{d^2} \, T^{\frac{3}{2}}$
Point defects	$\mathrm{constant} \times T^{\frac{3}{2}}$
Dislocation strain field	$\dfrac{h^2v\gamma^2b^2N_{\mathrm{dis}}}{28k^3} \, T^{-2}$
Conduction electrons	$3 \cdot 2 \times 10^{-3}\alpha\theta^4n_0^{4/3}T^{-2}$

a = atomic spacing; b = Burgers vector; d = density; l = mean free path in boundary scattering (diameter of specimen); M = molecular or atomic mass; N_{def} = density of isotopic defects; N_{dis} = dislocation density; n_0 = number of conduction electrons per atom; v = velocity of sound; α defined by (5.9).

3.12. Umklapp-processes and boundary scattering

Unless one is specifically interested in the effect of crystal imperfections it is necessary that specimens for heat-conductivity experiments should be made from single crystals which are as perfect as possible. One substance which is very good in this respect is synthetic sapphire (corundum, Al_2O_3). It also has another advantage in that it is very nearly isotopically pure. Fig. 3.7 shows some conductivity curves for specimens of different diameters. The rapid increase in conductivity as the temperature is reduced below $100°$ K corresponds to the exponential decrease in W_u (3.13). The very high values of the conductivity (> 50 watt cm^{-1} deg^{-1}) which are attained should be noted. Beyond the region of the maximum in the conductivity curve, boundary scattering becomes evident and at lower temperatures, as the conductivity starts to decrease, it becomes the dominant scattering process. At temperatures approaching $2°$ K, κ is proportional to T^3, in agreement with the expression for W_B (3.15). It will also be noted that in the boundary-scattering region the conductivity for the larger specimens is greater than that for the smaller ones and that it is approximately proportional to the specimen diameter.

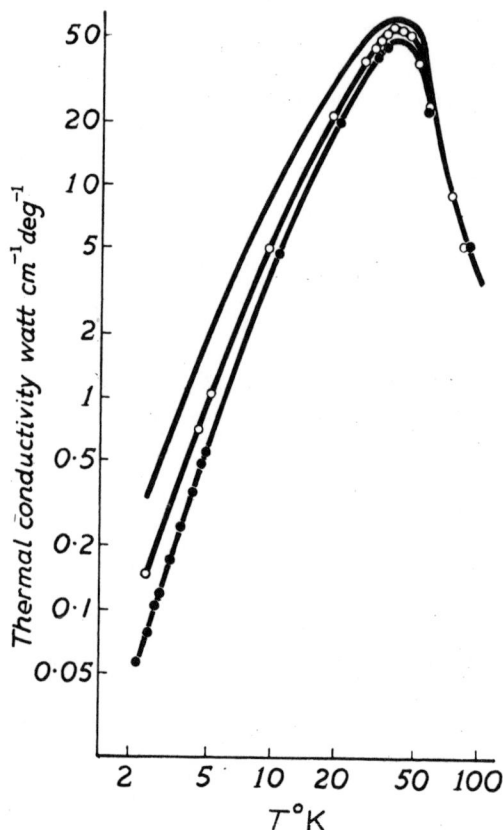

FIG. 3.7. The thermal conductivity of a synthetic sapphire crystal which was ground down to successively smaller diameters. The plain line relates to the original specimen which was 3 mm in diameter. Open circles, 1·55 mm, closed circles 1·02 mm diameter. The conductivity at the lowest temperatures, since it is limited by boundary scattering, is approximately proportional to the specimen diameter. The rapid rise in conductivity towards the high-temperature side of the maximum is due to the reduction in the number of umklapp-processes (Berman, Foster, and Ziman, 1955).

At very low temperatures the kinetic theory equation (3.4) works well. For example, in one specimen of synthetic sapphire (Berman, Foster, and Ziman, 1955) the value of the mean free path, using measured values of c_v, v, and W_B, is found to be 2·23 mm, which is in good agreement with the expected value (from the specimen diameter) of 2·40 mm. In some cases, however, the mean free path, when calculated in this way, is *greater* than the specimen diameter. This usually occurs when the specimen surface is highly polished, so that the surface imperfections are small compared with the phonon wavelength. The phonons will then not be scattered diffusely at the surface, but instead, they will be specularly reflected and hence the energy flux will not be decreased.

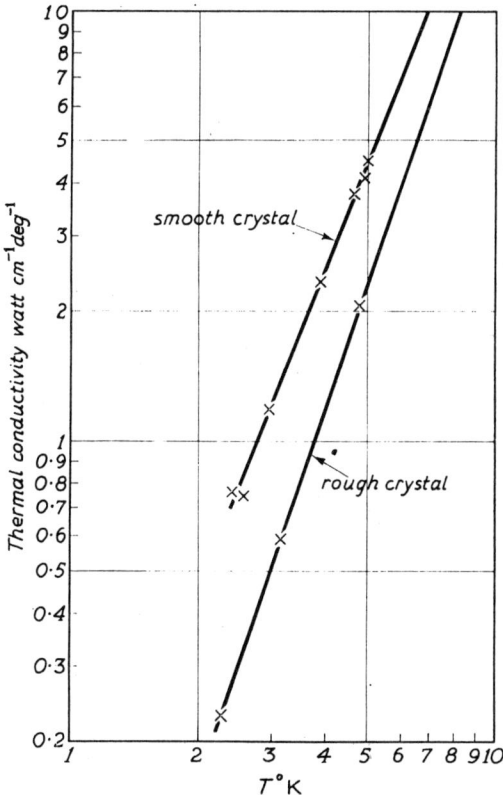

FIG. 3.8. The effect of the surface finish on the heat conduction of a synthetic sapphire crystal. The smooth crystal has a much higher conductivity because the phonons are specularly reflected and thence their mean free path is increased (Berman, Foster, and Ziman, 1955).

Fig. 3.8 shows the thermal conductivity of a synthetic sapphire, first with a polished surface and, secondly, the same crystal with a ground finish. The higher conductivity when the specimen is polished demonstrates very clearly the effect of specular reflection. In some cases mean free paths of two or three times the specimen diameter can be obtained.

Diamond is another crystal which has been used in particular to demonstrate boundary scattering effects (Berman, Simon, and Ziman, 1953). It has the advantage of having a very high value of the Debye θ

($\sim 2{,}000^\circ$ K). It therefore follows that W_u is very small and also that its specific heat is proportional to T^3 at temperatures below about 20° K. Thus, over quite a wide and easily attainable temperature range, pure boundary scattering can be studied. Difficulties in obtaining many good specimens of a suitable size, however, necessarily limit the experiments which can be made, but as in the case of synthetic sapphire, the theoretical predictions for W_B are well substantiated.

It is not so easy to investigate the effect of the relevant parameters on the umklapp resistance, W_u, because some imperfections are always present in the crystal which will tend to introduce an extra thermal resistance. This will prevent κ on the high-temperature side of the conductivity maximum from rising as rapidly as it otherwise would. There is one substance, however, which is particularly suitable for the investigation of u-processes. This is solid helium. It is very compressible and its density can be increased by a factor of approximately two. The change in the atomic spacing and hence in the elastic constants leads to a fourfold increase in the value of the Debye θ (Webb and Wilks, 1953) and hence, from (3.13), one would expect the umklapp resistance to decrease. Fig. 3.9 shows the results of some experiments on the heat conductivity of solid helium for various densities. It will be noted that at a given temperature the conductivity is much greater for the specimens of higher density and θ (i.e. the umklapp scattering is less) and this is in accord with our discussion. From these experiments the value of the parameter g in (3.13) was about 2·3.

It is sometimes useful to know at what temperature the maximum in the thermal conductivity occurs, since this gives a rough indication as to when boundary scattering starts to become important. Examination of the experimental data shows that the maximum is usually at a value of T which lies in the range of approximately $\theta/20$ to $\theta/30$.

3.13. *Isotope scattering*

In practice there are very few crystals which are isotopically pure and which show the exponential umklapp behaviour. Many of them tend to have a conductivity which is proportional to $T^{-\frac{3}{2}}$ at temperatures above the conductivity maximum and Slack (1957) suggested that this was indicative of isotope scattering. Below the maximum, boundary scattering becomes dominant in the same way as has been described in the preceding section.

Several series of experiments have been made (Berman, Nettley, Sheard, Spencer, Stevenson, and Ziman, 1959; Geballe and G. W. Hull,

1958) which confirm that a $T^{-\frac{3}{2}}$ thermal resistance may be observed which is due to an isotope effect. Fig. 3.10 shows the thermal conductivity of two samples of germanium, which in its naturally occurring state has a large number of isotopes. Both specimens had the same dimensions. The lower curve is for natural germanium, whereas the

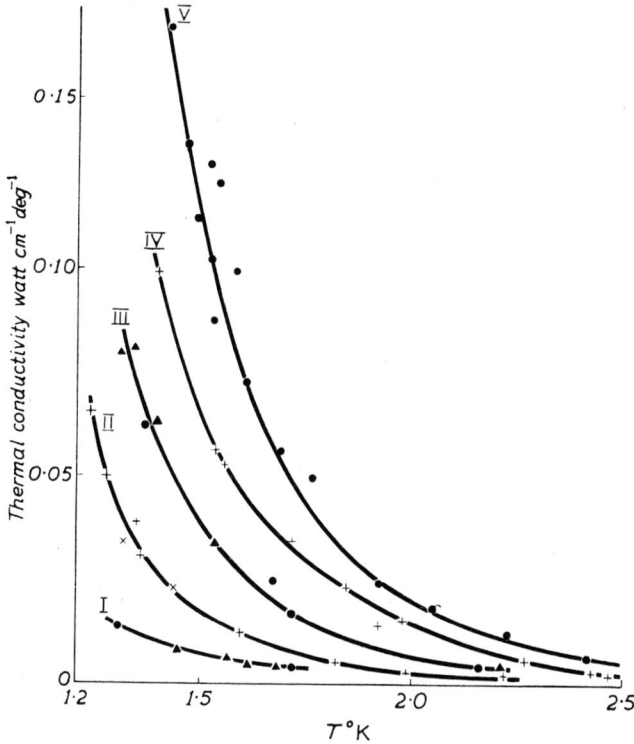

Fig. 3.9. Umklapp-processes. The variations of the heat conductivity of solid helium with temperature and density. I, 0·194; II, 0·203; III, 0·208; IV, 0·214; V, 0·218 g cm⁻³ (Webb, Wilkinson, and Wilks, 1952).

upper curve is for an enriched sample which was 95·8 per cent Ge⁷⁴. The effect of this enrichment is clearly shown, the conductivity in the region of the maximum being about three times greater than that of the natural sample. Nevertheless, a true u-process, exponential curve, was not obtained, because the specimen still contained some other isotopes and also, perhaps, some impurity atoms.

Fig. 3.11 shows the reduced thermal conductivity, normalized at $\theta/10°$ K, plotted against the reduced temperature, T/θ, for various types of crystal which have been reported in the literature. The relative values of the slopes of the curves are a good measure of the isotopic purity of the samples. Specimens such as Al_2O_3 and solid helium which are isotopically pure have high slopes which approach an exponential behaviour. Then come specimens such as LiF which have a little isotope admixture. This tends to modify the umklapp behaviour as is shown by the smaller slope of the curves. Specimens with large mixtures of

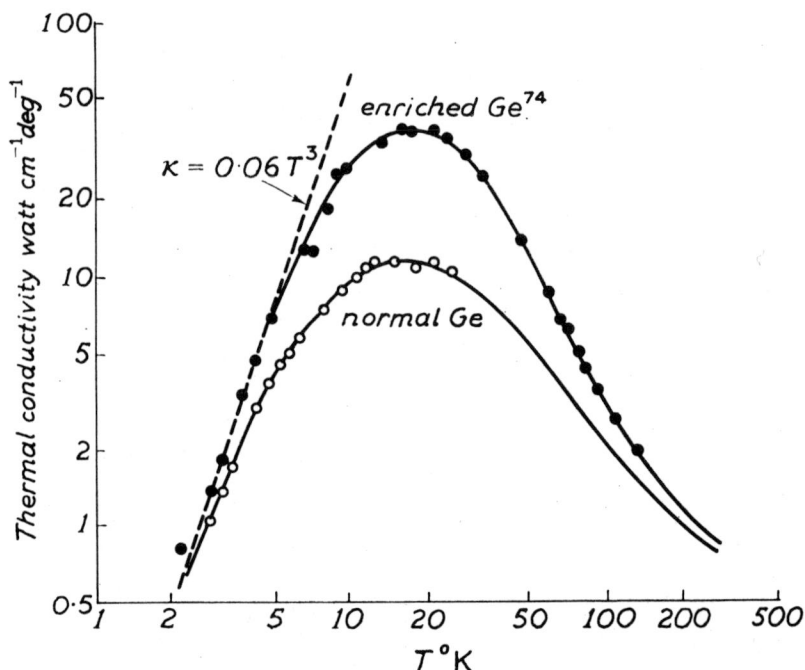

FIG. 3.10. The isotope effect. The enriched sample of Ge^{74} has a much higher conductivity than that with the naturally occurring abundance (Geballe and G. W. Hull, 1958).

isotopes such as Ge, Si, and KCl have the lowest slopes of all. The umklapp scattering is then completely overshadowed and the conductivity is approximately proportional to $T^{-\frac{3}{2}}$.

3.14. Defect scattering; irradiation damage

Apart from the investigations concerning isotopes very little systematic work has been done on the effect of point defect scattering on phonon conduction. There has been, however, a considerable amount of research on specimens which have been irradiated by neutrons and by gamma rays. It is to be expected that high-energy neutron irradiation will produce rather large regions of damage which would cause scattering of phonons of both long and short wavelengths, and heat-conduction measurements confirm this. One series of experiments on neutron irradiated Al_2O_3 is shown in Fig. 3.12 (a). The cumulative doses after each irradiation were 1·5, 8·9, 20·2, and $50·2 \times 10^{17}$ fast neutrons cm^{-2} respectively. It will be noted that both the boundary and the umklapp

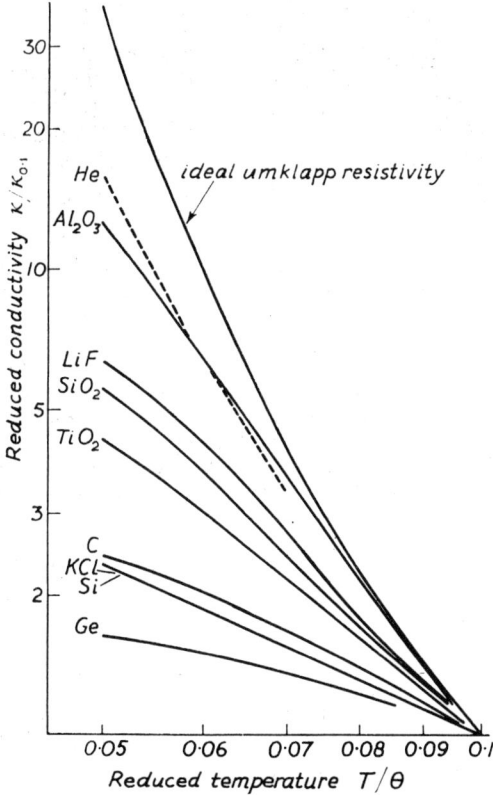

FIG. 3.11. The reduced thermal conductivity as a function of the reduced temperature for some crystalline solids. This illustrates that those materials which have only small amounts of isotopic admixture approach the ideal umklapp curve, whereas those with large numbers of isotopes have much lower conductivities (Berman, Foster, and Ziman, 1956).

scattering are modified; Fig. 3.12 (b) shows the extra thermal resistivity which is introduced by the irradiation. This is approximately constant at higher temperatures and hence it cannot be due to defects of spherical shape (because it should then be proportional to $T^{\frac{3}{2}}$ (3.16)) but rather to scattering by cylindrical regions (see section 3.9). This is indeed the shape of the damaged material which one would expect from a fast neutron collision. Rough calculations from the thermal resistivity increase (assuming Rayleigh scattering) show that the size of the

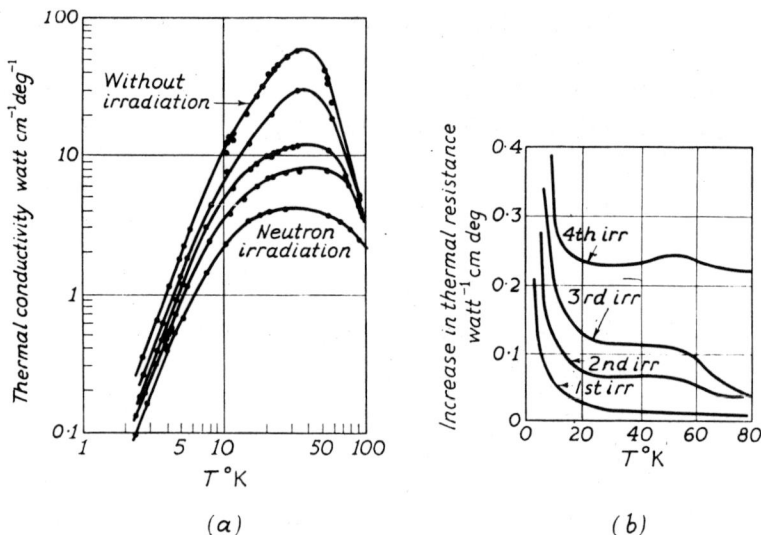

FIG. 3.12. The effect of neutron irradiation on the heat conductivity of synthetic sapphire. (a) The heat conduction curves. Starting from the top curve the cumulative irradiation is 1·5, 8·9, 20·2, and 50·2 × 10^{17} fast neutrons cm^{-2}. (b) The extra thermal resistance introduced by the irradiation. The rather extended region where the resistance is approximately constant should be noted (see text) (Berman, Foster, and Rosenberg, 1955).

damaged regions is equal to the volume of about 1,000 atoms. This is in good agreement with estimates of neutron damage as deduced by Seitz (1949).

The preceding paragraph shows that heat-conductivity measurements can be a useful tool in the investigation of irradiation damage. It should, however, be pointed out that the results of many experiments do not lead to such simple interpretations as in the example which has been selected. Irradiation usually produces a complicated mess in the crystal, with regions of damage of different sizes and shapes. The phonon scattering which these produce will not be of just one type that can be easily analysed from the experimental data. It might be thought that gamma irradiation, which should only affect individual atoms, might yield results which are easier to understand but in practice it has been found that such experiments are also difficult to interpret satisfactorily (e.g. see Berman, Foster, Schneidmesser, and Tirmizi, 1960; Cohen, 1957).

3.15. Dislocations

The mechanical properties of most dielectric crystals are such that at room temperature it is not very easy to deform them and thereby

introduce dislocations. Most of the experiments involving the scattering
of phonons by dislocations have been made on metals. A full discussion
is therefore deferred until section 5.16 ff., but it should be mentioned
here that the theoretical predictions of section 3.9 have been very
satisfactorily verified.

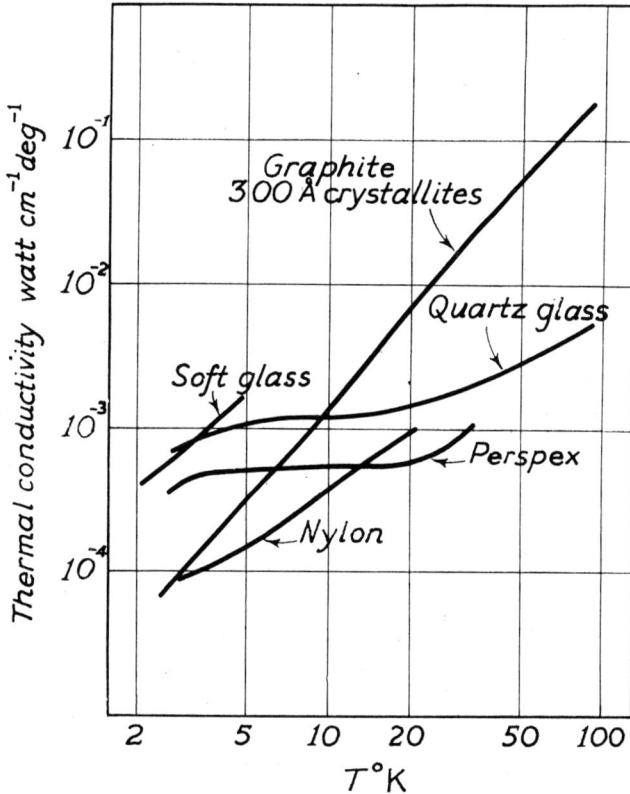

FIG. 3.13. The temperature variation of the thermal conductivity of micro-crystalline
and glassy materials (Berman, 1953; Berman, Foster, and Rosenberg, 1955a).

3.16. Glasses and polycrystalline specimens

As was discussed in section 3.3, the mean free path of the phonons
in a glass will be very small and will tend to be temperature-independent.
Similar considerations apply to specimens which are composed of very
small crystallites. The conductivity will therefore be very small and it
will be proportional to the specific heat, i.e. it will decrease as the
temperature is reduced. Fig. 3.13 shows the conductivities of some
glasses and some microcrystalline substances (graphite, nylon). The
very small values of the conductivity, compared with the data from
earlier figures in this chapter, should be noted.

THE ELECTRICAL CONDUCTIVITY
OF METALS AND ALLOYS

4.1. Introduction

In the previous chapter we considered the various ways by which
phonons could be scattered in a crystal lattice. In this chapter we study
the methods by which electrons may be scattered. Such a scattering
gives rise to the electrical resistivity, ρ, and also, since nearly all the heat
in pure and fairly pure metals is transported by the electrons, to an
electronic thermal resistivity whose behaviour we shall discuss in
Chapter 5.

The detailed theory of the scattering of electrons in a crystal is in a
far less satisfactory state than is that for the scattering of phonons.
Not only is the mathematical treatment much more complicated, but
it yields results which are usually only in rough qualitative agreement
with experiment. One reason for this is that the electronic properties
tend to be very dependent on the details of the electron energy spectrum
and the usual assumption of an energy distribution which is applicable
to a free electron model of a metal is an oversimplification. Thus whereas
in the case of phonons a Debye-type spectrum gives a tolerable model
for most substances (particularly at very low and at high temperatures),
in real metals the electron energy spectrum does not usually approximate
to that which we assume in the simple free electron model. It is only
in a few metals that we have any detailed knowledge about the occupied
electron energy levels and calculations of the transport processes using
these data have not yet been made. Another difficulty arises when one
calculates the electron scattering probability by various types of defect.
Electrons are scattered at regions where the general pattern of potential
variation is upset. We do not know, however, the details of the potential
distribution around most defects and this can also lead to considerable
errors.

We shall assume that the reader is familiar with the basic ideas underlying the free electron model and band theory which can be found in
many texts, e.g. Dekker (1957), chapters 9 and 10, Kittel (1956), chapters
10 and 11. In this section we shall merely give an outline of the main

points of these theories with emphasis on those aspects which are important for the understanding of the transport phenomena.

4.2. *The electron distribution*

It is assumed that in a metal the electrons in the outermost (uncom-
pleted) shells of the atoms are free to move through the crystal lattice and
that they are influenced only by the attractive forces of the ions. These
ions are presumed to be sited so that they form a geometrically perfect
lattice which repeats itself, with no faults, throughout the crystal. Under
the influence of the ions, therefore, the electrons move in a potential

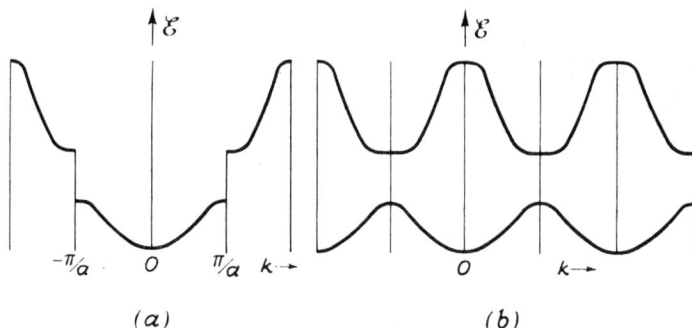

FIG. 4.1. (*a*) The extended zone scheme and (*b*) the repeated zone scheme, for
a one-dimensional electron system.

which varies from point to point (depending on the position relative to the
neighbouring ions), but whose variation is not random; it has a periodicity which is exactly the same as that of the ionic lattice. Under these
ideal conditions the electrons experience no resistance to their motion.
Scattering will only occur when the periodicity of the potential is upset.

In this assembly the relationship between the energy, $\mathscr{E}$, and the
wave vector, $\mathbf{k}$, of the electron may be represented in several ways (see
Ziman, 1960, chapter 2). These different representations arise because,
for a given $\mathscr{E}$, $\mathbf{k}$ is not uniquely defined, but may always be changed by
the addition of an integer $\times$ a basic vector of the reciprocal lattice (see
section 3.5). The general type of behaviour for an electron in a one-
dimensional lattice is often represented by a diagram similar to that
shown in Fig. 4.1 (*a*). The form of the curve is parabolic, but at certain
values of $\mathbf{k}$ ($\pm n\pi/a$) there is a break in the curve which gives rise to an
energy gap for which there are no permitted values of $\mathscr{E}$. This type of
representation is called an extended zone scheme. Since, however, $\mathbf{k}$ is
really multi-valued, for any $\mathscr{E}$ we may equally well repeat the curves
of Fig. 4.1 (*a*) beyond the regions where the energy gaps occur, in the

manner shown in Fig. 4.1 (b). This gives the repeated zone scheme. It is analogous to the $\nu \sim q$ curve for phonons shown in Fig. 3.1.

In a three-dimensional system the values of **k** at which the energy discontinuities occur lie on surfaces which define the Brillouin zones, whose shapes are determined by the crystal structure (Fig. 3.2). The existence of these zones would have little significance, however, unless there was a reasonable probability that some electrons would actually have values of **k** which were large enough to lie near their boundaries.

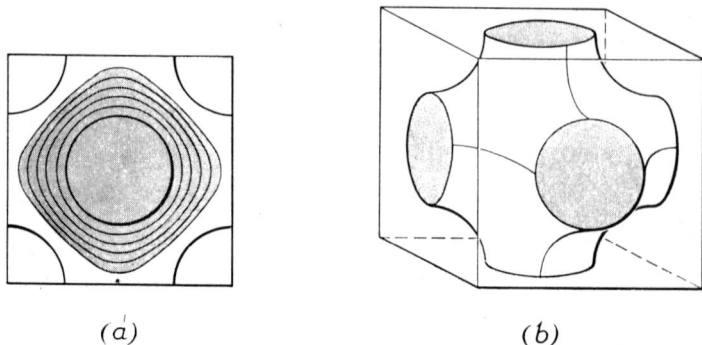

(a)   (b)

FIG. 4.2. (a) A cross-section through a simple cubic zone showing the surfaces of constant energy. The occupied states are shown shaded. (b) A simple cubic zone in which the Fermi surface touches the zone boundary.

For a crystal containing N ions with one ion per unit cell, there are $2N$ discrete energy states within the first zone. Since the electrons obey the Fermi–Dirac statistics each must be in a different energy state. At the absolute zero the states with the lowest possible energy will be occupied and hence if there was one free electron per atom (as in the monovalent metals), the N electrons would half fill the first zone, as is shown by the shading in Fig. 4.2 (a), and all the other states would be empty. The energy of the highest occupied state is called the Fermi energy, $\mathscr{E}_F$, and the surface in k-space which has this energy is called the Fermi surface. $\mathscr{E}_F$ is usually of the order of a few electron volts. Thus even at room temperature, $\mathscr{E}_F$ is very much higher than the energy of thermal vibration, which is only about $1/40$ eV. Since electrons may only be excited into vacant sites this comparatively small thermal energy will only enable those with energies which are close to $\mathscr{E}_F$ to go to higher states (Fig. 1.4). Thus at any easily attainable temperature the electron distribution will differ only very slightly from that which exists at the absolute zero. It is for this reason (as was discussed in section 1.9) that the electronic specific heat is so small.

In the simple cubic zone the surfaces of constant energy are as shown (for a cross-section through the centre of the zone parallel to a cube face) in Fig. 4.2 (a). For small values of the energy they are approximately spherical, but for higher values the Fermi surface tends to have lobes which extend towards the zone boundary. Thus in three dimensions the Fermi surface for a metal might look something like Fig. 4.2 (b). In this diagram the lobes are shown as actually touching the zone boundary. This is not particularly unrealistic since a sphere which has half the volume of a given cube has a diameter $0.985 \times$ the cube edge and so very little distortion has to occur before contact is made with the boundary. If more electrons were added, the corners of the zone would fill up, so that more and more of the Fermi surface was in contact with the boundary. This seems to be characteristic of many metals.

4.3. *The basic concepts of conductivity*

When dealing with conductivity problems, we require an expression for the velocity of the electrons. In a wave-mechanical treatment, the position of an electron is determined by the position of a wave packet, and its velocity, v, by the velocity of that wave packet. This velocity is equal to the group velocity of the waves, i.e. $v = 2\pi \, dv/dk$. Since $\mathscr{E} = h\nu$ we obtain

$$v = (2\pi/h) \, d\mathscr{E}/dk. \qquad (4.1)$$

Thus the velocity of an electron of energy $\mathscr{E}$ will be in the same direction as $\mathbf{k}$[†] and it will be proportional (in the one-dimensional case) to the slope of the curves shown in Fig. 4.1. If the zone is filled up to a given energy level then, by symmetry, for each electron with a given value of $\mathbf{k}$ in one direction there will be another electron with the same value of $\mathbf{k}$ in exactly the opposite direction. Hence for the complete assembly of electrons there will be as many travelling with a given velocity in one direction as there are travelling with the same velocity in the other direction. Thus there will be no net flow of charge.

This is a rough picture of the equilibrium state of the electrons. What happens when we apply a potential to the ends of the metal, so that an electric current is made to flow? We should first recall the de Broglie relation that the momentum of a particle is equal to $h\mathbf{k}/2\pi$. Since the force on a particle is equal to the rate of change of momentum, we may write[‡]

$$e\mathbf{E} = (h/2\pi) \, d\mathbf{k}/dt \qquad (4.2)$$

---

† For free electrons only; more generally $\mathbf{v} = (2\pi/h)\mathrm{grad}_{\mathbf{k}} \, \mathscr{E}$.

‡ Whilst this argument is only valid for a free particle, (4.2) also holds for electrons in a periodic lattice, see Ziman (1960), p. 92.

where $e\mathbf{E}$ is the force on an electron with charge e when it is in an electric field $\mathbf{E}$. Thus when we apply a constant field the values of $\mathbf{k}$ for all the electrons will change at a constant rate given by (4.2). At any given time after the application of the field the distribution of the electrons in k-space will no longer be symmetrical as it was in Fig. 4.2 but it will be as is shown in Fig. 4.3. There will not now be a balance between the electrons flowing in opposite directions; instead there will be a net flow of charge in the direction of force, $e\mathbf{E}$.

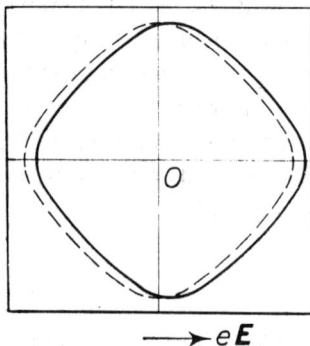

FIG. 4.3. The shift in the electron distribution under the influence of an electric field $\mathbf{E}$. The dotted line shows the Fermi surface in zero field.

This description of the manner in which a current is made to flow is not complete because, if it were, the current would not remain constant for a given $\mathbf{E}$ but it would continually change. This is contrary to experience. In order to see how a unique current is obtained for a certain value of $\mathbf{E}$, we must take account of the fact that our crystal is not as perfect as we have hitherto assumed. It will be distorted by various types of defects and impurity atoms, as well as by the thermal vibra-tions of the lattice. In certain regions the periodic potential will be upset and this will cause transitions between the electron energy states. In particular, there might be sufficient interaction for an electron to lower its energy by going into some of the now empty states on the left-hand side of Fig. 4.3. If it does this the net flow of charge will be decreased. In practice a dynamic equilibrium is set up between the increase in the value of $\mathbf{k}$ due to the applied field, and the decrease due to the effect of the perturbations of potential. For small fields, $\mathbf{k}$ increases slowly and so during the time taken for a given change in $\mathbf{k}$, there is a greater pos-sibility that the states which have been vacated will be refilled by the influence of the perturbation. Thus a small current flows. In high fields where $\mathbf{k}$ increases rapidly, there is less likelihood of this occurring and so the resultant current is larger. The stronger the perturbation of the periodic potential, the greater will be the probability for transitions to occur, so the current produced by a given field will be smaller. This means that the conductivity will be reduced by the addition of im-purities and also by an increase in temperature. In more pictorial terms we can say that the electrons are scattered by the static defects and by the thermal vibrations of the lattice. It is, of course, one of the main objects of transport theory to calculate the amount of electron scattering which occurs, i.e. to determine the probability of transitions between the various states due to the perturbations.

4.4. *Effect of the zone boundary*

Before we describe the effect of these transitions in more detail, however, there are still a few points in our simplified description of band theory with which we should deal. What happens if, when we apply a field, some of the electrons with the highest values of **k** reach the zone boundary ? From Fig. 4.1 we see that at the boundary the curves have zero slope and hence the electron has zero velocity. The electron has two choices. It might continue to move in the same direction that it was moving before it reached the boundary, but for this to happen, since **k** is still increasing as determined by (4.2), it must jump to the higher curve where the slope will still be positive. In general, however, the energy gap between the two curves is about an electron volt or more and hence at ordinary temperatures and with normal fields such a transition is very unlikely to occur. The electron must therefore stay on the lower curve and hence as **k** increases its velocity is now reversed. This is exactly the same kind of mechanism as the umklapp-processes which we have already discussed in connexion with phonon scattering (section 3.6). The electrons are reflected at the zone boundary and this will tend to decrease the current flow.

If we have a metal for which, even in zero field, some of the electrons have values of **k** which lie on the zone boundary (as is shown in Fig. 4.2(b)), then when a field is applied the total possible flow of charge is considerably reduced, because in that band across the zone which is terminated by contact with the boundary it will not be possible to produce an unbalance in the numbers of electrons which are travelling in opposite directions. Thus the conductivity will be determined by the area of the Fermi surface which does *not* contact the boundary. As an extension of this idea we see that if there are sufficient electrons for the zone to be completely filled then there can be no electrical conduction and the substance will be an insulator. Only at sufficiently high temperatures or in very high fields such that excitation can occur across the energy gap into the higher zone would there then be any conduction. We have already noted that the first zone when completely filled can accommodate two electrons per atom and this would lead one to suppose that the divalent elements should therefore be insulators. Since, in fact, these elements are conductors, it is assumed that there is an overlap of the energy bands† for certain directions in k-space, so that in zero field some of the higher states in the first zone will have higher energies than some of the lower states in the second zone. As the electrons will occupy the lowest possible energy states, the first zone will not be completely filled. Conduction can then occur, although one would expect it to be rather small. A small energy gap between the zones, across which thermal excitation can take place, gives rise to certain types of semiconducting behaviour (see Chapter 7).

We now see, even from this brief description, why we stated at the beginning of this chapter that the conductivity is very dependent on the details of the electron energy distribution. In order to obtain an accurate theoretical estimate of the conductivity it is necessary to know the shape of the Fermi surface, particularly when parts of it lie close to, or actually touch, the zone boundary, because the conductivity is really determined by the acceleration and the scattering of the electrons of maximum energy. Detailed information is necessary even in some monovalent metals for which at one time it was thought that the assumption of a spherical Fermi surface was sufficient.

4.5. Resistivity mechanisms

We must now return to our description of the behaviour of the electrical resistivity, and in particular to the way in which it depends on temperature. From simple kinetic theory (see, for example, Ziman, 1960, p. 258) the electrical resistivity may be written in the form

$$\rho = \frac{m_e v_F}{ne^2 l}, \tag{4.3}$$

where m_e is the mass of the electron, v_F is its velocity at the Fermi surface, and l is its mean free path; n is the number of electrons per unit volume. The main problem in transport theory is to calculate l or the relaxation time, τ, which is defined by

$$l = \tau v_F. \tag{4.4}$$

Formally τ may be determined by using time-dependent perturbation theory. This gives the probability that an electron in state $\mathbf{k}$ will find itself in state $\mathbf{k}'$ due to the influence of a perturbation, U, in the potential energy. $1/\tau$ always contains a term of the form

$$1/\tau_{\mathbf{k}} \propto \left| \int \psi_{\mathbf{k}}^* U \psi_{\mathbf{k}'} \, d\mathbf{k}' \right|^2, \tag{4.5}$$

---

† The divalent hexagonal metals have two atoms per unit cell and hence each zone contains only 1 electron per atom. Thus the first two zones are filled and we assume that there is some overlap into the third zone.

where $\psi_{\mathbf{k}}$ and $\psi_{\mathbf{k'}}$ are the wave functions of the electrons with wave vectors $\mathbf{k}$ and $\mathbf{k'}$ respectively. Whilst we shall not have occasion to calculate this expression a knowledge of the form of (4.5) is sometimes very useful in helping us to determine the behaviour of τ.

The resistivity of a metal may be considered as being due to two main contributions: (a) the intrinsic or 'ideal' resistivity, ρ_i, due to the scattering of the electrons by the thermal vibrations of the lattice, which

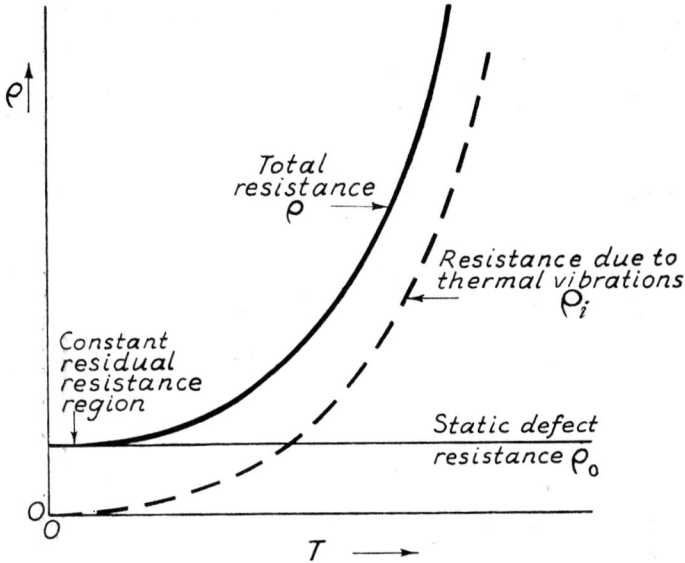

FIG. 4.4. A diagram showing the addition of the impurity and the thermal scattering resistance to give the total resistance of a metal.

we should expect to become smaller as the temperature is reduced, and (b) the resistivity, ρ_0, due to static lattice defects such as impurity atoms, vacancies, and dislocations which one would not expect to be dependent on the temperature. This separation of the resistivity into two independent components is very well justified from the analysis of experimental data and is one example of what is known as Matthiessen's rule. It is not strictly accurate but the deviations are not usually more than 1 or 2 per cent. Fig. 4.4 shows the type of behaviour which one might expect for the variation of the resistivity of a metal with temperature. ρ_0 remains constant whereas ρ_i decreases, becoming zero at $0°$ K. The total resistivity shown by the upper curve will thus decrease as the temperature is reduced, but it will flatten out to a constant value ρ_0 at very low temperatures. For this reason ρ_0 is called the residual resistance and its value will depend on the physical and chemical purity of the particular sample of metal which has been measured. ρ_i, on the other hand, is determined only by the lattice vibrational spectrum of the metal and it should therefore be independent of the detailed condition of the specimen.

4.6. The residual resistivity

In our description of ρ_0 we shall first describe the effects which are due to impurity atoms. These can be studied from two main viewpoints. (1) The effect on ρ_0 of the addition of a certain atomic fraction of impurity atoms of the various different elements, and (2) the behaviour of ρ_0 when the concentration of a particular impurity atom is changed. We shall begin by discussing (1).

There are four main reasons why ρ_0 might be influenced by the presence of an impurity atom in the crystal lattice. (a) If the charge on an impurity atom is different from that of the solvent atoms, then the potential around it will also be different and so it will scatter the electrons. (b) The introduction of an impurity atom might change the number of conduction electrons and hence the position of the Fermi surface in the zone might be altered. This would be particularly important if the surface was already near the zone boundary. (c) The impurity atoms might themselves affect the dimensions and geometry of the zones. (d) Over- or undersized impurity atoms will distort the lattice and will therefore upset the periodicity of the potential. It is only possible to discuss the effect of (a) in any detail.

4.7. *Effect of impurity atoms; the valency rule*

The charge on the nucleus of an impurity atom will be counteracted by the charge on the bound electrons in the closed shells around it. The net charge which is visible to the system outside will be equal (numerically) to the number of free electrons which it contributes to the conduction band, i.e. to its valency. Thus, if the impurity atom A has the same valency as the pure metal, B, the potential in its neighbourhood should not be very different from that around any solvent atom and hence the resistivity increase should not be very large. Nevertheless, in the region very close to an atom of type A, the field will not be exactly the same as that in a corresponding position near an atom of type B and hence some scattering must be expected. In addition to this there will always be some distortion of the crystal lattice around an impurity atom, although the amount of such distortion and the resultant scattering would be difficult to calculate. A much larger effect is produced, however, if there is a valency difference ΔZ between the atoms A and B. In the simplest case this would produce an extra perturbing potential $\Delta Z e^2/r$ at a distance r from the impurity atom. In practice the potential falls off much more rapidly than this because of the screening effect of the electron gas, but it is clear that the perturbation will still be proportional to ΔZ. In the most elementary treatment the situation is exactly the same as that treated by Rutherford for the scattering of alpha particles by nuclei. He showed that the scattering cross-section (i.e. $1/l$) was proportional to $e^4(\Delta Z)^2/\mathscr{E}$, where $\mathscr{E}$ is the energy of the incident particle, and this result also applies to the situation which we are considering.

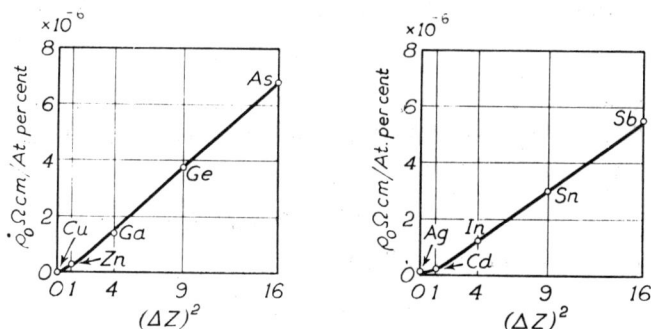

FIG. 4.5. The increase in the resistance of copper due to 1 atomic per cent of various metals in solid solution plotted against $(\Delta Z)^2$ (Linde, 1932).

This dependence on $(\Delta Z)^2$ can also be deduced from (4.5) because since U is proportional to ΔZ and the ψ do not depend on it, the scattering probability will vary as $(\Delta Z)^2$.

Such a dependence is shown very well in copper when elements from groups 1 to 5 are present in solid solution. Results from the classic experiments of Linde (1932) are shown in Fig. 4.5 for the resistivity due to 1 atomic per cent of solute and it will be seen that a linear dependence on $(\Delta Z)^2$ is established very satisfactorily. It should be noted, however, that alloys with the transition metals do not obey the rule (see section 4.9).

The magnitude of the resistivity for 1 atomic per cent of solute is very approximately $0 \cdot 3(\Delta Z)^2 \times 10^{-6}$ ohm cm and whilst this is lower than the value obtained from the original theory of Mott (1936) more recent improvements (see Ziman, 1960, p. 342) give results which are in good agreement with the experimental values.

4.8. Variation of the amount of a given impurity

We have so far considered the effect of a fixed amount of solute in the parent metal. We must now study how the resistivity changes when the concentration, x, of a given solute is changed. This problem was first treated by Nordheim (1931). If we have a very small concentration of an element A in element B, then the electrons may be considered as travelling through the perfect lattice of B but occasionally they are scattered by the abnormal potential around the A atoms. We should then expect the scattering probability to be proportional to the number of A atoms, i.e. to the concentration x. If x is not so small, however, we cannot consider the alloy as consisting of perfect regions with the characteristics of B throughout which small point defects of A are interspersed. It is more realistic to assume that there is an average periodic potential, U_0, for the alloy which is of the form

$$U_0 = xU_A + (1-x)U_B, \tag{4.6}$$

where U_A and U_B are the potentials around an A and a B atom respectively. Then around each A atom there will be a deviation from this average potential equal to

$$U_0 - U_A = (1-x)(U_B - U_A) \tag{4.7}$$

and around a B atom we shall have

$$U_0 - U_B = x(U_A - U_B). \tag{4.8}$$

As in (4.5) the scattering probability for an atom depends on the square of the perturbing potential. Thus the chance of an electron being scattered by an A atom will be proportional to

$$C_A = (1-x)^2 \left| \int \psi_{\mathbf{k}}^* (U_B - U_A) \psi_{\mathbf{k}'} \, d\mathbf{k}' \right|^2 \tag{4.9}$$

and by any B atom it will vary as

$$C_B = x^2 \left| \int \psi_{\mathbf{k}}^* (U_A - U_B) \psi_{\mathbf{k}'} \, d\mathbf{k}' \right|^2. \tag{4.10}$$

Since there are a fraction x of A atoms and $(1-x)$ of B atoms the probability of an electron being scattered by either an A or a B atom will be proportional to

$$xC_A + (1-x)C_B = [x(1-x)^2 + (1-x)x^2] \left| \int \psi_{\mathbf{k}}^* (U_B - U_A) \psi_{\mathbf{k}'} \, d\mathbf{k}' \right|^2 \tag{4.11}$$

$$= x(1-x) \left| \int \psi_{\mathbf{k}}^* (U_B - U_A) \psi_{\mathbf{k}'} \, d\mathbf{k}' \right|^2. \tag{4.12}$$

Thus the residual resistivity should be of the form

$$\rho_0 = \text{constant} \times x(1-x). \tag{4.13}$$

This parabolic expression, which is sometimes known as Nordheim's rule, holds quite well in certain cases. It will, of course, only be valid if the two metals form a continuous range of solid solution and if no ordering or phase changes occur. One would also not expect it to be valid if the electron density or the Fermi surface changed with composition. Never-theless, it does give an idea of the kind of behaviour which one might expect. Fig. 4.6 (a) shows the results on disordered copper-gold alloys from the work of Johansson and Linde (1936) which demonstrates the validity of (4.13).

Equation (4.13) suggests that a given concentration x of A in B should

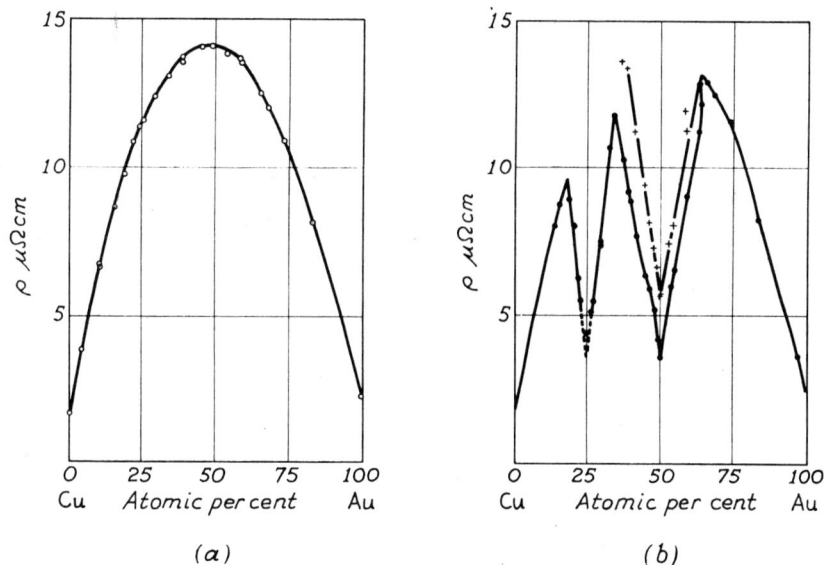

(a)   (b)

FIG. 4.6. (a) The resistivity of disordered copper-gold alloys which demonstrates the validity of (4.13). (b) The effect of ordering on the resistivity of copper-gold alloys (Johansson and Linde, 1936).

give the same resistivity increase as does the same concentration x, of B in A. Mott (1934) has shown that this holds for metals which have similar atomic volumes and electron configurations, e.g. Ag and Au, Mg and Cd, Pd and Pt.

4.9. *Alloys containing the transition elements*

The resistivity of the noble metals, when they are alloyed with those of the transition groups, does not fall into such a straightforward pattern as the alloys which we have so far considered. Neither the $(\Delta Z)^2$ rule nor the Nordheim rule (4.13) holds. The reason for this is due to the more complicated electronic states of the transition metals. As we have already indicated in connexion with the specific heats of these metals (section 1.14), the electrons lie partly in s states and partly in d states but in the noble metals the d levels are completely filled. They therefore take no part in the conduction process and only the s levels have to be considered. When we start adding a transition element to a noble metal, besides scattering from one s state to another, the possibility arises of scattering from s to d states and this will invalidate the simple story which we have outlined in the previous paragraphs. In addition to this, many such alloys show a resistance minimum and also sometimes a maximum at low temperatures, which makes their behaviour even more complicated (see section 4.15).

4.10. *Effect of ordering and phase transformation*

We have been concerned so far with disordered alloys, i.e. those in which the atoms of both constituents are distributed in a random manner over the lattice sites. For some special compositions of certain alloys, ordering can take place under appropriate thermal treatment in which the atoms of the two elements are arranged in a definite pattern with respect to one another. A well-known example is that of Cu_3–Au. When this occurs the periodicity of the lattice is improved and the resistivity decreases. In many cases the change in resistivity is so large that it can be observed very easily at room temperature. Fig. 4.6 (*b*) shows the resistivity of gold-copper alloys in the disordered state and also after heat treatment which will produce ordering. The compositions for which ordering occurs are clearly indicated by the resistivity decrease. A corollary to this behaviour is observed if phase changes occur as the concentration of the constituents is changed. The curve of resistivity against composition, instead of being smooth as is shown in Figs. 4.5 and 4.6 (*a*), will have discontinuities at the composition where the alloy changes phase.

4.11. *Determination of purity from residual resistance measurements*

We should emphasize the ease and utility of measurements of ρ_0 for checking the purity and purification of metals. As we have already seen, whilst ρ_0 (which can usually be obtained by measuring the resistivity at $4{\cdot}2°$ K) is very sensitive to the presence of impurity atoms, the room-temperature resistivity is determined mainly by thermal vibrational scattering and is in general independent of the purity (this applies only for fairly pure metals—in alloys the room-temperature resistivity will, of course, be different from that of the pure metal). A check on the purity of a metal can therefore be obtained by measuring the ratio of room-temperature to helium-temperature resistivity. The higher the purity, the smaller will be ρ_0 and the ratio will be greater. For commercially obtained high-purity samples (such as the Johnson Matthey Spectro-graphically Standardized materials), the resistivity ratio for metals such as copper, silver, and aluminium will be about 200. The impurities in these samples probably amount to several parts per million. Further purification, however, can increase the ratio very considerably. Experiments have been made on a sample of copper with a resistivity ratio as large as 8,600 (Langenberg, 1959). This was a piece of naturally occurring copper, which shows that on occasion nature can provide us with very pure raw materials. The purity of such specimens of copper is so high that it would be almost impossible to detect differences between two samples by any standard analytical technique, and yet the electrical resistivity would still enable them to be compared. Sodium always tends to have a high resistivity ratio. It is usually of the order of a thousand or more, even in specimens which have not been purified with great care. This is presumably due to the fact that hardly any other element is soluble in it and the usual process of distillation gets rid of most of the impurities. Certain metals always seem to have a low resistivity ratio—titanium, zirconium, tantalum, uranium, etc., usually have ratios which are between ten and twenty. Such metals are difficult to purify and some are liable to contain large amounts of gaseous impurity which is very difficult to remove. Even specimens which have been purified with very great care do not have very large resistance ratios—which presumably indicates that the purification techniques still have scope for improvement. There is no doubt that resistivity measurements are the easiest way of checking on the progress of purification. Their limitation is that they do not give any indication of the kind of impurity that is present.

Another use for residual-resistivity measurements is in the analysis of simple alloys of two metals, particularly when there is only a few per cent of one of the metals present. As we have already seen (4.13) the addition of another metal with concentration x gives a smooth curve for the relation between the residual resistivity and x which is proportional to $x(1-x)$. Thus after one has measured the resistivity of a few samples of known composition in order to define this curve, it is a comparatively simple matter to use it to find the composition of other samples.

4.12. *Resistivity of lattice imperfections*

In addition to the scattering of electrons by foreign atoms in the lattice we must still take account of the resistivity due to lattice defects, i.e. vacancies, interstitials, and dislocations. Some of these defects are produced in an otherwise pure metal during deformation or irradiation; resistivity measurements have very often been used as the main means of detection of defects in such experiments.

In order to make an estimate of the resistivity of these defects one might start by assuming that a vacancy would behave to a certain extent like an impurity atom of valency zero and that we could apply our $(\Delta Z)^2$ rule (section 4.7) to determine the resistivity which it produces. This will be an underestimate because the crystal lattice will relax around the vacancy to a much greater extent than would be possible around an impurity atom. Thus whilst for 1 per cent vacancies in copper (for which nearly all the calculations and experiments have been made) ΔZ will be unity, the resistivity should be somewhat larger than that which we obtain for 1 per cent zinc in copper, which is about $0 \cdot 2 \times 10^{-6}$ ohm cm. Calculations (e.g. see Blatt, 1955) give the resistivity for 1 per cent vacancies as about $1 \cdot 5 \times 10^{-6}$ ohm cm (i.e. about $1 \cdot 5 \times 10^{-27}$ ohm cm per vacancy) and this value appears to be generally accepted.

Experimental determinations of the resistivity of vacancies may be made by first finding the vacancy concentration at high temperatures, with precise X-ray techniques, and then measuring the actual resistivity after those vacancies have been quenched-in at low temperatures. Such experiments yield the following resistivities for 1 atomic per cent vacancies: Al, 3×10^{-6} ohm cm (Balluffi and Simmons, 1960); Ag, $1 \cdot 3 \times 10^{-6}$ ohm cm (Doyama and Koehler, 1960); Au, $1 \cdot 5 \times 10^{-6}$ ohm cm (Balluffi and Simmons, 1961).

The resistivity of an interstitial atom must be of a similar order of magnitude to that of a vacancy, although there has been much greater uncertainty about the value. Whilst ΔZ is now zero, this is offset by the fact that the lattice distortion is more severe. Blatt's calculations suggest that the resistivity of an interstitial is, in fact, about the same as that of a vacancy.

4.13. *Resistivity of dislocations and stacking faults*

The resistivity which we should associate with dislocations is not yet well established. We may very naïvely consider an edge dislocation as a row of defects each with a resistivity which is about the same as that of a vacancy or an interstitial. If there are about 10^{23} atoms cm^{-3} then a density of one dislocation cm^{-2} contains about 4×10^{7} atoms along the dislocation core and hence its resistivity should be of the order of 6×10^{-21} ohm cm. This is smaller than the value calculated by Harrison (1958), 5×10^{-20} ohm cm, for the resistivity due to electrons being scattered by the dislocation core. In other calculations (Hunter and Nabarro, 1953) only the scattering by the strain field around the dislocation has been considered and these give a lower value—about 10^{-20} ohm cm per cm of dislocation.

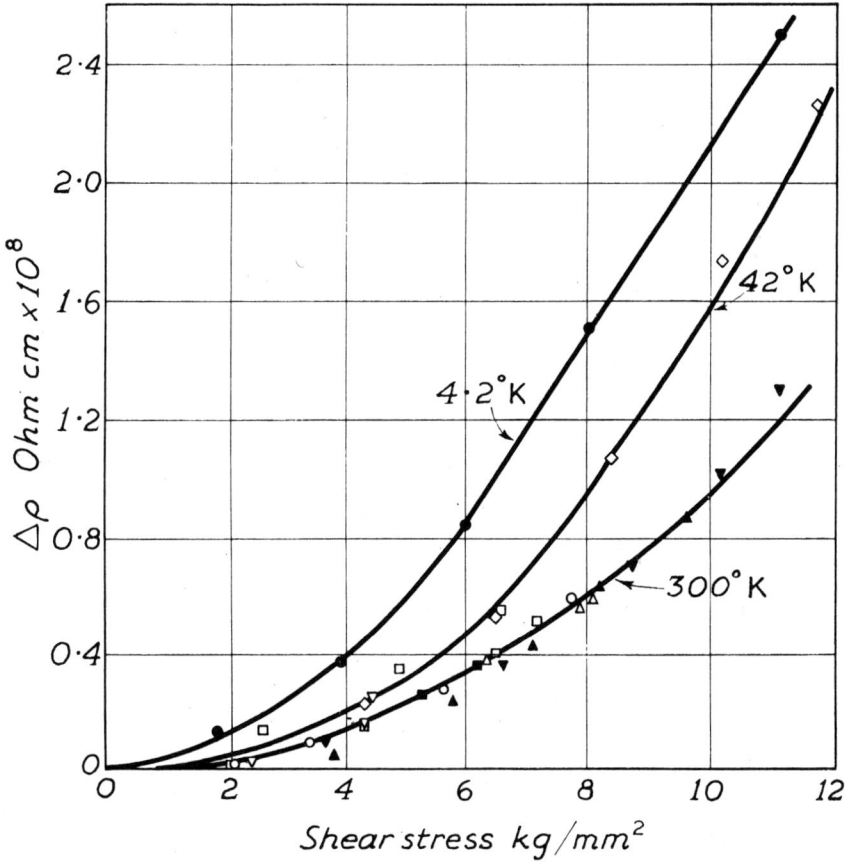

FIG. 4.7. The change in ρ_0 (at $4 \cdot 2^\circ$ K) of copper single crystals after being deformed at the temperatures indicated (after Blewitt, Coltman, and Redman, 1955).

Experiments which have been made to measure the resistivity of dislocations are sometimes difficult to interpret because when metals are deformed, both point defects and dislocations are produced. Since each of these will provide a contribution to the resistivity it is necessary to be able to separate the individual effects. Results from a typical series of experiments on copper deformed at various temperatures are shown in Fig. 4.7. It will be noted that the resistivity increase is greater when the deformation is carried out at low temperatures. One reason for this is that it is only at low temperatures that point defects are immobile so that they, in addition to the dislocations, are able to contribute to the resistivity. At higher temperatures the point defects can migrate to sinks (such as dislocations, or to the surface) or they can annihilate one another, and hence their effect on the resistivity will be less. In a metal such as copper it is generally assumed that most of the resistivity which is introduced after room-temperature deformation is that due to the dislocations themselves, because they are unable to anneal out until much higher temperatures. Nevertheless, if one makes reasonable assumptions about the number of dislocations which are introduced by deformation (in thin metallic films their number can now be determined by direct electron microscope observation, Hirsch, Horne, and Whelan, 1956) it is found that their resistivity should be much greater (about 40 or 50 times) than the values indicated in the preceding paragraph.

No satisfactory reason has been given as to why the experimental value for the resistivity of a dislocation should be so high. The discrepancy between theory and experiment is now generally ascribed to the fact that in a face centred cubic lattice (for which nearly all the experiments have been made) a dislocation can lower its energy by dissociating into two partial or half dislocations which are connected by a region of faulty lattice stacking.† The suggestion was first made by Broom (1952) that the electrons were scattered at these stacking faults and that these, rather than the dislocations themselves, were the cause of the large increases in resistivity.

Experiments have tended to confirm this suggestion. These have very often been made on a series of copper-zinc alloys in the alpha phase. As zinc is added to copper the stacking fault probability (as shown by the broadening of X-ray lines) is known to increase. Fig. 4.8 shows both the relative increase in the electrical resistivity on deformation and also the stacking-fault probability. The strong correlation between the two quantities is obvious. Other experiments (Lomer and Rosenberg, 1959) also give evidence in favour of a fairly high stacking-fault resistivity. Theoretical calculations by Seeger (1956), Ziman (1960), p. 243, and Howie (1960) suggest that the stacking fault resistivity in copper might be of the order of 30 or 40 times the dislocation resistivity and this would be in reasonable agreement with the experiments.

† For an explanation of partial dislocations and stacking faults see chapter 7 of *Dislocations in Crystals* by W. T. Read, Jr., McGraw-Hill, 1953.

4.14. *Recovery of the resistivity after irradiation or deformation*

We should mention another type of experiment on the residual resistivity caused by defects which has become very popular. In these, the defects are introduced at liquid-helium temperatures either by deformation or by bombardment with high-energy particles—electrons,

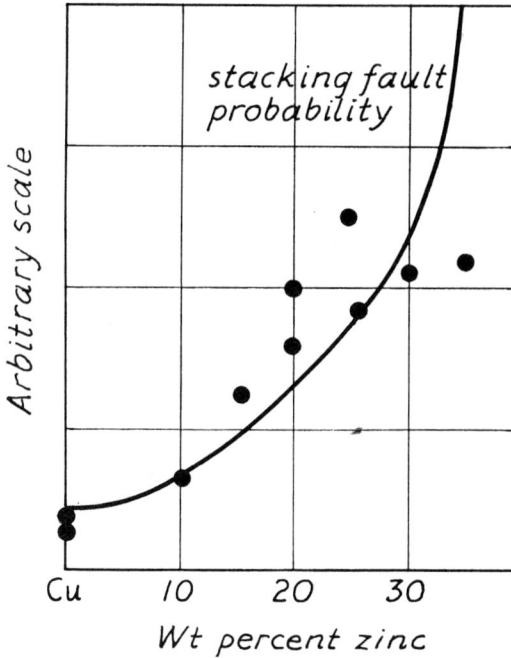

FIG. 4.8. The stacking fault probability (solid curve) and the electrical resistance increase (points) of deformed copper-zinc alloys. Arbitrary scale. The correlation suggests that the resistivity introduced by deformation is caused by stacking faults rather than by dislocations (after Christian and Spreadborough, 1956).

deuterons, or pile neutrons. The specimen is then warmed up to a slightly higher temperature for a certain time and subsequently recooled so that its residual resistance may be remeasured. The heating is repeated at progressively higher temperatures and between each stage ρ_0 is remeasured. These experiments show that, particularly after irradiation, a considerable fraction of the resistivity can anneal out at quite low temperatures. Reviews are given by van Bueren (1960), chapter 25, Cottrell (1958), and Broom and Ham (1958). Fig. 4.9 shows a typical series of results on neutron irradiated copper which indicates that a large amount of defect migration and annihilation occurs at about $40°$K. In order to obtain further information about the type of recovery process which such measurements of ρ_0 are able to detect, it is usual to measure ρ_0 as a function of time during a set of isothermal annealing stages. These are made in the region where the recovery occurs at temperatures which are fairly close together. The results of one series of measurements is shown in Fig. 4.10. If one assumes simple annealing kinetics with a single energy of activation, U, it is possible to calculate

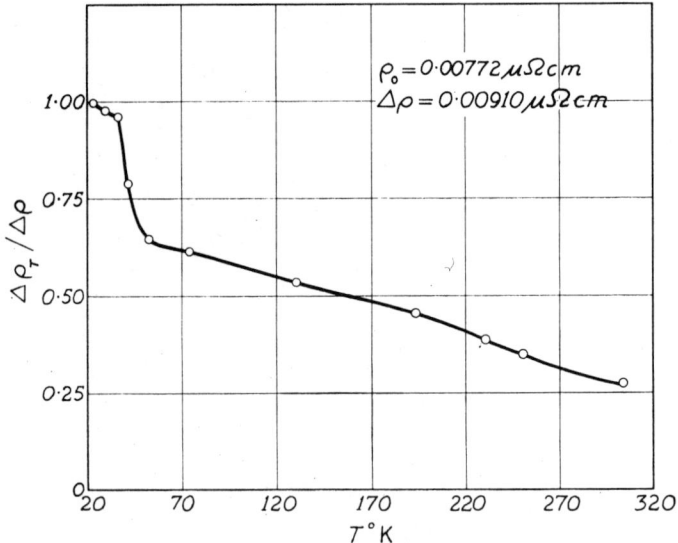

FIG. 4.9. An annealing curve for zone refined copper, bombarded with 8×10^{17} nvt fast neutrons at 16·5–18·5° K. The ordinate is the ratio of the extra resistivity after annealing at the various temperatures to the extra resistivity originally introduced by the bombardment. The sharp recovery at about 40° K should be noted (Blewitt, Coltman, Holmes, and Noggle, 1957).

this energy in the following manner. Let N_{def} be the number of defects which are present at any time t; then

$$-\frac{dN_{\text{def}}}{dt} = \text{constant} \times N_{\text{def}}^{y} \exp(-U/kT), \qquad (4.14)$$

where y is the order of the recovery mechanism, i.e. the number of defects which take part in each recovery process. If we assume that the resistivity increase, $\Delta\rho$, is proportional to N_{def} then from (4.14)

$$-\frac{d(\Delta\rho)}{dt} = \text{constant} \times (\Delta\rho)^{y} \exp(-U/kT). \qquad (4.15)$$

At a point on the annealing curve where the temperature is suddenly increased from T_1 to T_2, $\Delta\rho$ must be the same and so

$$\log_e \frac{d(\Delta\rho)}{dt}\bigg|_{T_2} - \log_e \frac{d(\Delta\rho)}{dt}\bigg|_{T_1} = \frac{U}{k}\left(\frac{1}{T_1} - \frac{1}{T_2}\right). \qquad (4.16)$$

Thus, using (4.16), U may be determined from the slopes of the annealing curve at the point where the temperature is changed. For the $40°$ K annealing region in Fig. 4.9, U appears to have a value of about $0·1$ eV and it is suggested that this is the energy for the migration of interstitial atoms in the lattice. These migrate and annihilate with vacancies and hence the resistivity is decreased. The interpretation of these experi-

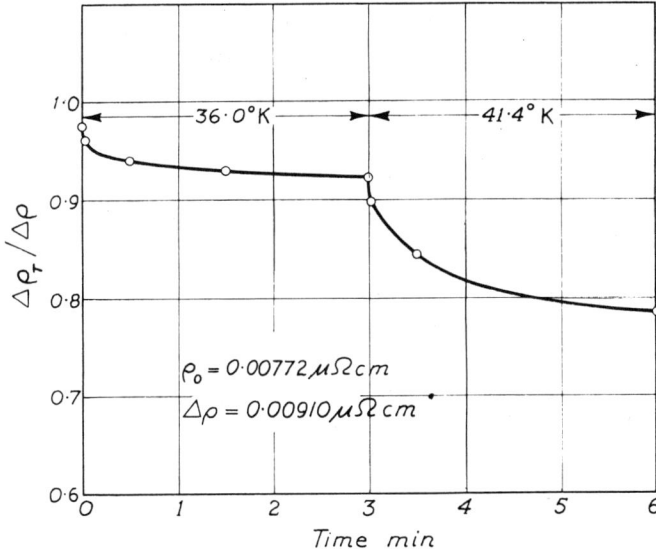

FIG. 4.10. Isothermal annealing curves for the same specimen as that of Fig. 4.9. The activation energy of the recovery process may be calculated with (4.16) from the slopes of the curve at the point where the temperature has been changed (Blewitt, Coltman, Holmes, and Noggle, 1957).

ments is usually complicated by the presence of further regions of recovery at higher temperatures which are due to other annealing mechanisms. This is particularly so for metals which have been deformed. These, except for the alkali metals, do not usually show recovery below about $100°$ K—see Fig. 4.11, but there is still considerable uncertainty about the detailed interpretation of these curves.

4.15. *Resistivity minima and maxima*

Experiments as early as 1930 (Meissner and Voigt) showed that for certain metals the simple explanation of the residual resistivity did not appear to be entirely correct because the resistance at low temperatures did not become constant. In some cases, particularly in gold, the resistivity passed through a minimum and, as the temperature was reduced further, it started to increase again (de Haas, de Boer, and van den Berg, 1934). Later experiments have shown that the effect tends to be most pronounced in certain samples of copper, silver, gold, and magnesium and that it is generally caused by the addition of the

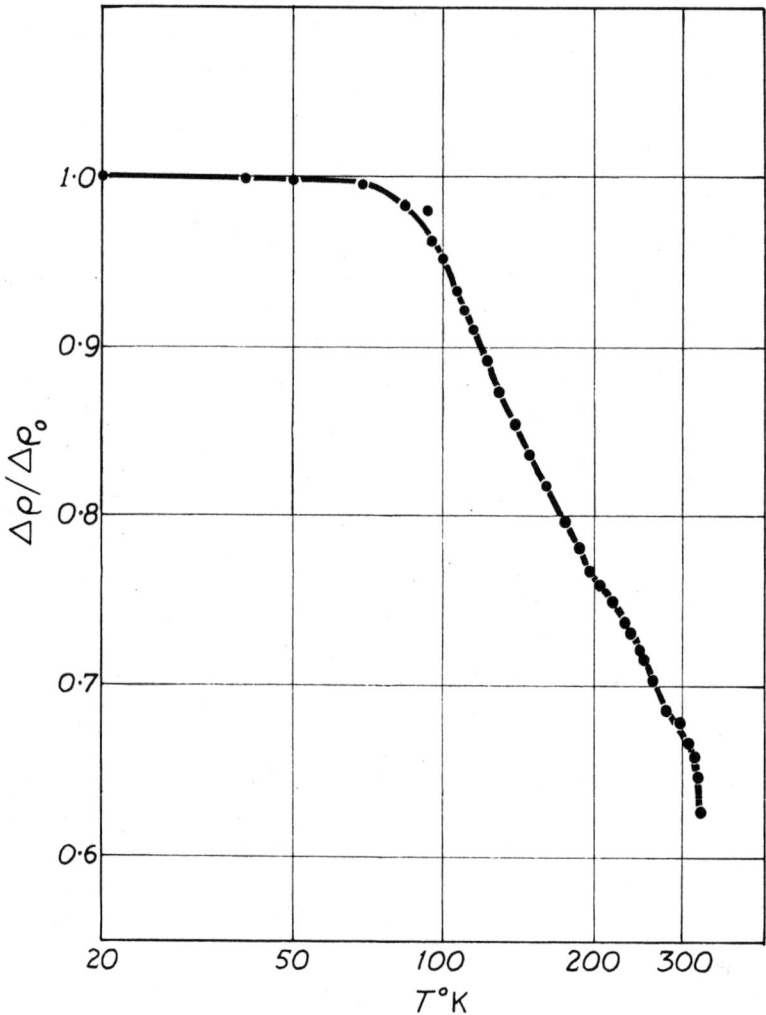

Fig. 4.11. The recovery of the extra resistivity introduced by the deformation of copper at 20° K (McCammon, 1957).

transition elements, Fe, Mn, Cr, Co, and sometimes Ni. In copper, minima have also been produced by Ga, In, C, Ge, Sn, Pb, and Bi. Surveys of the literature are given by MacDonald (1956), section 24, and Ziman (1960), p. 344. An example is shown in Fig. 4.12. Certain alloys, particularly those of the noble metals with chromium or manganese, sometimes exhibit a local maximum as well as, or instead of, a minimum (Fig. 4.13). The minimum is never a very deep one. At temperatures below the minimum the resistivity only rises up to values which are about 10 per cent higher than the minimum value. The lower the resistivity at the minimum, ρ_{min}, the lower is the temperature T_{min}, at which the minimum occurs. For copper-iron alloys Pearson (1955) has suggested

$$\rho_{min} = \text{constant} \times T^5_{min}. \tag{4.17}$$

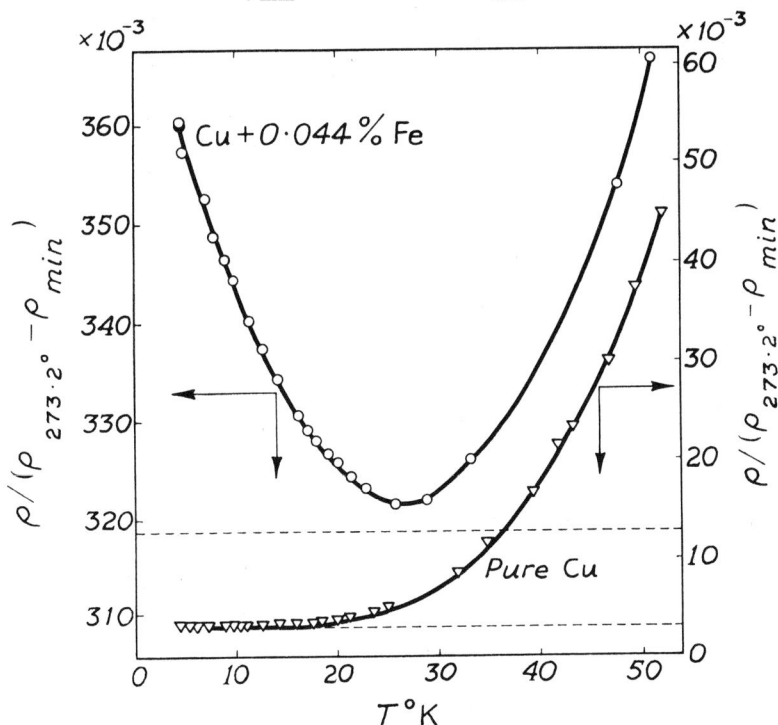

Fig. 4.12. The resistance minimum in a copper-iron alloy (Pearson, 1955).

An interesting point arises as to whether at very low temperatures the resistance continues to rise or whether it flattens off to a constant value. It appears that either can happen. In the case of dilute copper alloys White (1955) has shown that the resistance has become constant by about 2° K whereas Mendoza and Thomas (1951) for Cu, Ag, and Au, and Croft, Faulkner, Hatton, and Seymour (1953) for Au showed that the resistivity was still increasing in the demagnetization region.

Associated with the minimum are other anomalous effects. Mac-Donald and Pearson (1953) have shown that the absolute thermopower of alloys (section 8.8) which exhibit a resistivity minimum is large and negative instead of being small and positive (Fig. 8.7) and there are also anomalies in the electron and nuclear spin resonance and in the magnetic susceptibility of copper-manganese alloys which show resistance minima (Owen, Browne, Knight, and Kittel, 1956).

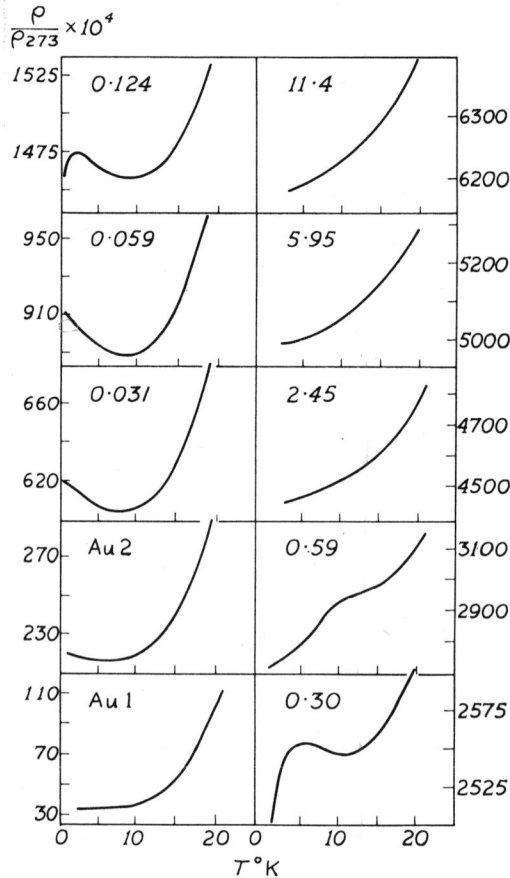

FIG. 4.13. The resistance of pure gold and of gold-manganese alloys, showing minima and maxima. The number on each graph gives the atomic percentage of Mn (Gerritsen and Linde, 1952).

There is still no satisfactory explanation for these phenomena. It is possible that they might be caused by interactions between the conduction electrons and the magnetic dipoles of some of the impurity atoms (e.g. see Franck, Manchester, and Martin, 1961) or they might be associated with the proximity of the Fermi surface to the zone boundary and the consequent large changes in electrical properties which can occur when impurities are added. For further (inconclusive) discussions the reader is referred to the reviews by MacDonald and Ziman already quoted.

The reader may think that, compared with other sections of this book, the treatment of the residual electrical resistivity has been unduly detailed. There is a good reason for this. There is no other type of low-temperature measurement which is so easy to make and yet which is so structure sensitive as is ρ_0. It has its disadvantages; as we have already emphasized, an increase in ρ_0 by itself does not give any information about the *nature* of the defect which is producing the change. Additional investigations (e.g. the annealing experiments following the irradiation and deformation measurements which have already been described) are sometimes necessary before a more complete interpretation is possible. Nevertheless, the measurement of the resistivity and particularly the residual resistivity, is one of the most sensitive and simple methods which we have available for studying impurities, defects, and structural changes in metals.

4.16. The temperature-dependent resistivity

Whilst a full treatment of the scattering of electrons by the thermal vibrations of the lattice is rather long and complicated (the most readable account is given by Ziman, 1960, chapter 5), it is quite straight-forward to describe the nature of the process in a simplified qualitative discussion. Electrons are scattered by phonons because the lattice vibrations set up perturbations of the regular lattice potential in various parts of the crystal. These will tend to change the electron state, designated by its wave vector, $\mathbf{k}$, into another state, say $\mathbf{k}'$. Putting this in other words we can say that an electron in state $\mathbf{k}$ interacts with a phonon of wave number $\mathbf{q}$ and so enters state $\mathbf{k}'$. More generally we could write

$$\text{electron } (\mathbf{k}) + \text{phonon } (\mathbf{q}) \rightleftharpoons \text{electron } (\mathbf{k}') \qquad (4.18)$$

since there is no reason why the process should not proceed in the reverse direction as well. Similar processes involving more than one phonon might be possible but (4.18) is the simplest type. If such an interaction occurs we must conserve both wave number and energy and these requirements give relations which are analogous to those used in phonon–phonon interaction (3.7), (3.8), and (3.9). It is necessary that

$$\mathbf{k} + \mathbf{q} = \mathbf{k}' + \mathbf{k}_0 \qquad (4.19)$$

and

$$\mathscr{E}(\mathbf{k}) + \mathscr{E}(\mathbf{q}) = \mathscr{E}(\mathbf{k}'), \qquad (4.20)$$

where $\mathscr{E}(\mathbf{q})$ is the energy of the phonon and $\mathbf{k}_0$ is either zero (i.e. a normal process) or is a basic vector of the reciprocal lattice which will bring $\mathbf{k}'$ back into the first zone (i.e. an umklapp-process, section 3.6). We should note the restrictions which these equations put on the possible processes. As has already been mentioned in section 4.2, the energy of a phonon $\mathscr{E}(\mathbf{q})$ is very much less than that of the electrons at the top of the energy distribution and since electrons can only undergo transitions into unoccupied states (4.20) can only be satisfied for electrons which are very near to the Fermi surface. In fact, because $\mathscr{E}(\mathbf{q})$ is so small it can be neglected in (4.20) and hence $\mathscr{E}(\mathbf{k})$ is approximately equal to $\mathscr{E}(\mathbf{k}')$. Since $\mathscr{E}(\mathbf{k})$ is on or very near the Fermi surface, which it will be recalled is a surface of constant energy, this means that electrons may only be scattered from one part of the Fermi surface to another. Taken in conjunction with (4.19) this implies that an electron in a state $\mathbf{k}$ can only interact with phonons which have a special range of $\mathbf{q}$ values.

4.17. Small-angle scattering

In order to determine the temperature dependence of ρ_i we first consider only the normal processes ($\mathbf{k}_0 = 0$). The most effective kind

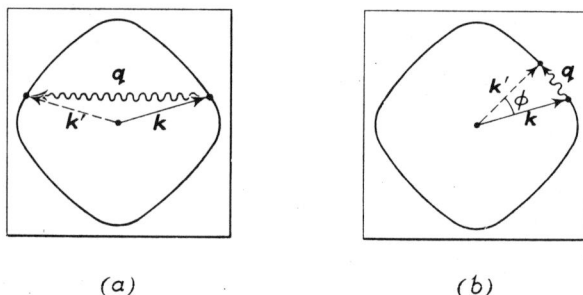

(a)   (b)

Fig. 4.14. (a) Large-angle and (b) small angle scattering of an electron by a phonon. At low temperatures only (b) can occur and the angle of scatter is $\sim T/\theta$.

of scattering will occur when an electron travelling in the direction of the applied field interacts with a phonon which has a large value of $\mathbf{q}$ as is shown in Fig. 4.14 (a). Whilst this type of process can occur at high temperatures, it becomes very rare at low temperatures because the abundance of such phonons decreases. Thus the angle ϕ, between $\mathbf{k}$ and $\mathbf{k}'$, will in general be small as is shown in Fig. 4.14 (b). We have already seen (1.16) that the dominant lattice wavelength is of the order of $a\theta/T$. This corresponds to a value of q given by

$$q \approx 2\pi T/(\theta a). \tag{4.21}$$

Since for a scattering from $\mathbf{k}$ to $\mathbf{k}'$ Fig. 4.14 (b) shows that

$$q \approx |2\mathbf{k}\sin(\phi/2)|, \tag{4.22}$$

then

$$2\pi T/(\theta a) \approx |\mathbf{k}\phi|. \tag{4.23}$$

If we assume that each atom has one free electron, the first zone will be half full and hence the value of $|\mathbf{k}|$ for an electron at the Fermi surface will be approximately π/a. Putting this into (4.23) we see that at low temperatures the angle of scattering is

$$\phi \approx T/\theta. \tag{4.24}$$

4.18. *The T^5 law*

If the electric field is in the x direction, the electric current will be dependent on the component of the velocity in that direction, i.e. it will depend on k_x. If after scattering $k_x \to k_x'$, then the resistivity associated with this scattering will be proportional to $(1-\cos\phi)$. This, however, is not the only factor involved. The probability that such a transition will actually occur depends on the scattering cross-section of an atom which is vibrating thermally and hence it should be proportional to the square of the amplitude of its oscillation, i.e. to its energy/elastic modulus. The energy is $\sim kT$ and the elastic modulus is proportional to the square of the phonon velocity (3.3). But we see from (1.8) that the velocity is proportional to ν_{max}, i.e. to θ (1.10). Thus the scattering cross-section is proportional to T/θ^2. Hence, when we integrate over the surface of the Fermi sphere up to a scattering angle of T/θ (4.24) we find that

$$\rho_i \propto \frac{T}{\theta^2} \int_0^{T/\theta} (1-\cos\phi)\sin\phi \, d\phi \tag{4.25}$$

and since ϕ is small this reduces to

$$\rho_i \propto \frac{T}{\theta^2} \int_0^{T/\theta} \phi^3 \, d\phi.$$

Hence

$$\rho_i \propto T^5/\theta^6. \tag{4.26}$$

This is not the only way to derive the T^5 law by simple arguments. An alternative explanation is given by Ziman (1960), p. 365.

4.19. *The Bloch–Grüneisen function and θ_R*

The most well-known expression which has been used for ρ_i is the Bloch–Grüneisen function, $\mathscr{G}$:

$$\mathscr{G} = \rho_i = \text{constant} \times \frac{T^5}{M\theta^6} \int_0^{\theta/T} \frac{z^5 \, dz}{(e^z-1)(1-e^{-z})}, \tag{4.27}$$

where M is the atomic weight and the constant is characteristic of the metal. This neglects u-processes and assumes a spherical Fermi surface and a Debye-type lattice spectrum. At high temperatures $(T > \theta/2)$ it reduces to

$$\rho_i = \text{constant} \times \frac{T}{4M\theta^2} \tag{4.28}$$

and at low temperatures $(T < \theta/10)$ the upper limit of the integral in (4.27) can be taken as infinity and we may obtain

$$\rho_i = \text{constant} \times \frac{124 \cdot 4 T^5}{M\theta^6}. \tag{4.29}$$

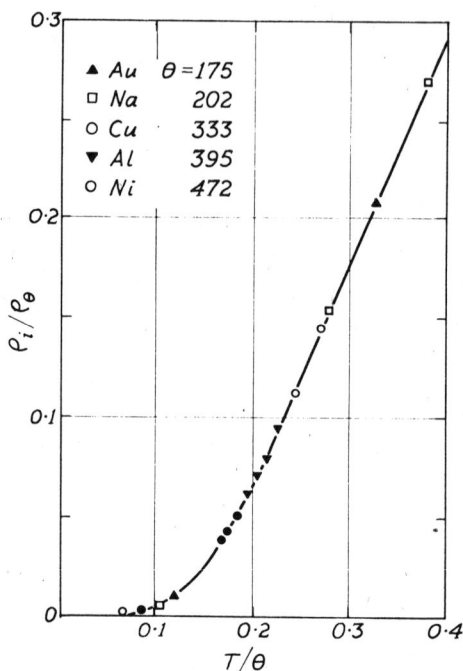

Fɪɢ. 4.15. A plot of the Bloch–Grüneisen function for the electrical resistivity together with points from experimental data (after Meissner).

Thus if ρ_i has the value ρ_1 at a low temperature T_1 $(< \theta/10)$, and ρ_2 at a high temperature T_2 $(> \theta/2)$, then

$$\frac{\rho_1}{\rho_2} = \frac{497 \cdot 6 T_1^5}{\theta^4 T_2}. \tag{4.30}$$

A plot of the reduced resistivity (ρ_i/ρ_θ) against the reduced temperature (T/θ) should give the same curve for all materials (ρ_θ is the resistivity at $\theta°$ K). Fig. 4.15 shows the theoretical curve for ρ_i/ρ_θ from (4.27) (it is tabulated by MacDonald, 1956, section 17) together with some experimental points and the agreement is quite good. These expressions would suggest that one can obtain a characteristic temperature, θ_R, from electrical resistivity data. Values of θ_R are given in Table 4.1 for the

TABLE 4.1

		Li	Na	K	Rb	Cs	Cu	Ag	Au
θ_R (From MacDonald, 1956, table 5)	°K	330	180	114	65	45	320	200	200
θ (from specific heats)	°K	369	158	89	55	40	348	225	164

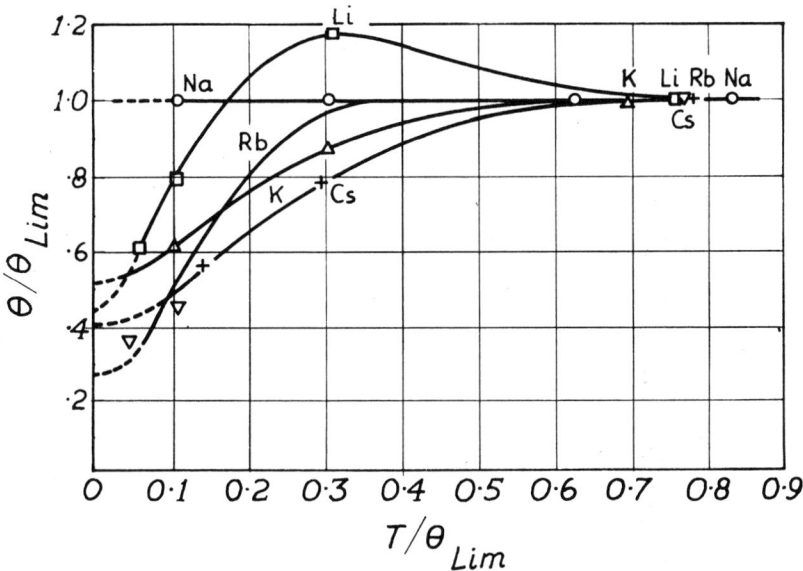

FIG. 4.16. The relative variation of θ_R for the alkali metals. θ_{Lim} is the apparent value of θ_R for $T/\theta \approx 1$ (Macdonald, 1956, section 18).

group I metals which shows that these values are reasonably close to the θ obtained from specific-heat measurements although they are certainly not exactly the same. However, from a detailed examination of the data for the alkali metals it is clear that, except for sodium, θ_R is not a constant, but that it tends to lower values as the temperature is reduced (Fig. 4.16). Hence the Grüneisen function, whilst it is very useful, is not a completely accurate representation of ρ_i. The variation in θ_R tends to be greater than the variation in θ which is obtained from specific heat measurements (section 1.8).

4.20. *Umklapp-processes*

So far we have not considered the contribution of u-processes towards the electrical resistivity. These are most easily visualized on a repeated zone diagram as in Fig. 4.17. If the phonon interaction induces a transition from OA to OB, such that OB crosses the zone boundary, then this is a u-process produced by a phonon with a wave vector AB. Note that B in the right-hand zone is equivalent to point C in the left-hand zone. If we look at the left-hand zone we have the relation

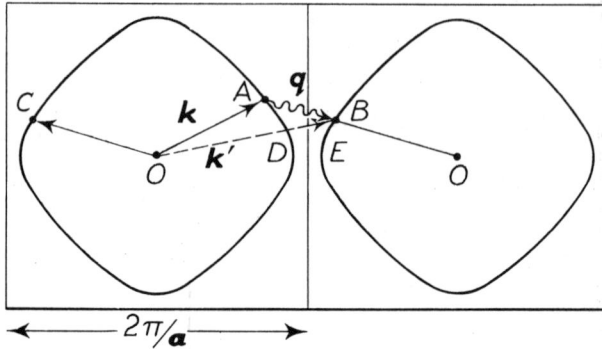

FIG. 4.17. An umklapp-process. Interaction of the electron which has a wave vector OA with the phonon AB, yields a final wave vector OB which lies over the zone boundary. OB, however, is equivalent to $O'B$ (and also to OC) and hence the direction of the incident electron has been changed. DE is the closest distance of approach of the Fermi surfaces and represents the smallest phonon wave vector which can produce a u-process.

$\mathbf{k}' \equiv OC = \mathbf{k} + \mathbf{q} - 2\pi/\mathbf{a}$ which is the expression for a u-process in (4.19). We see that for this type of scattering to occur phonons must be available with values of $\mathbf{q}$ equal to at least DE, i.e. the closest distance of approach of the Fermi surfaces in the two zones. Even if we have a metal with a spherical Fermi surface and one electron per atom, this value of $\mathbf{q}$ is not very high, but the surface area of the sphere which is sufficiently close to the zone boundary is small and so few of the electrons will be scattered in this manner at low temperatures. It is for this reason that many workers have neglected the effect of u-processes on ρ_i. If, however, the Fermi surface is not spherical, so that more of it is close to the zone boundary, then u-processes can occur with quite low values of $\mathbf{q}$. In such cases they might yield an important contribution to the electrical resistivity even at low temperatures.

The influence of u-processes was first taken into account by Bardeen (1937) and his calculations for sodium, together with those of later workers (see Ziman, 1960, p. 372), are in good agreement with experiment. Unfortunately the theoretical expressions contain integrals which have to be calculated numerically and one cannot give an explicit formula for ρ_i. The temperature dependence at high and at low temperatures remains the same as in the Bloch–Grüneisen formulae. At high temperatures u-processes can account for over 70 per cent of the resistivity.

4.21. *Resistivity of the transition metals*

Room-temperature measurements show that the resistivity of the transition metals is considerably higher than that of the alkali and the noble metals. Mott (1936 a) has explained this using the model involving overlapping s and d bands, which has already been described in section 1.14 in connexion with the high electronic specific heat of these metals. This model assumes that the d electrons occupy part of a narrow energy band with a high density of states whereas the s electrons are in a band which is more nearly like that for free electrons. Now a high density of states implies a small value of $d\mathscr{E}/dk$ and from (4.1) this means that the velocity of the electrons in the d band will be small. Thus we should not expect the d electrons to contribute very much to the electric current. Because of the high density of states in the d band, however, there will be a good probability that the s electrons (which, since they have a higher velocity, will be carrying most of the current) will be scattered into vacant d states. Their velocity is then reduced and they will not be very effective in determining the conductivity. This, of course, will be in addition to any scattering into other s states by the phonon interactions which we considered in the previous section. This extra s–d interaction will therefore lead to a higher resistivity.

4.22. *Electron-electron scattering*

At low temperatures it is also possible that electron-electron scattering (without any phonon interaction) might become an important process. Theoretically this resistivity should be proportional to T^2. This temperature dependence can be understood very simply. The conduction electrons which can be scattered will be those whose energy lies within kT of the Fermi energy $\mathscr{E}_F$ and thus the number which can be scattered is $\sim kT\, F(\mathscr{E}_F)$, where $F(\mathscr{E}_F)$ is the density of states at $\mathscr{E}_F$. These electrons can only be scattered into vacant states and the number of vacant states is also $\sim kT\, F(\mathscr{E}_F)$. Hence the total probability of scattering, i.e. the resistivity, will be proportional to $\{kT\, F(\mathscr{E}_F)\}^2$. In most metals this type of scattering has not been detected experimentally, but for the transition metals, due to the high density of states at the top of the d band, there is a much higher probability for the s-electrons (i.e. those that carry the current) to be scattered and in many instances a resistivity approaching a T^2 dependence has been measured (Olsen-Bär, 1956). Fig. 4.18 shows how the resistivity of Fe, Co, Ni, Pd, W, Nb approaches a T^2 dependence at very low temperatures. The decrease in the power of T from the usual T^5 is quite marked. de Haas and de Boer (1934) suggested that

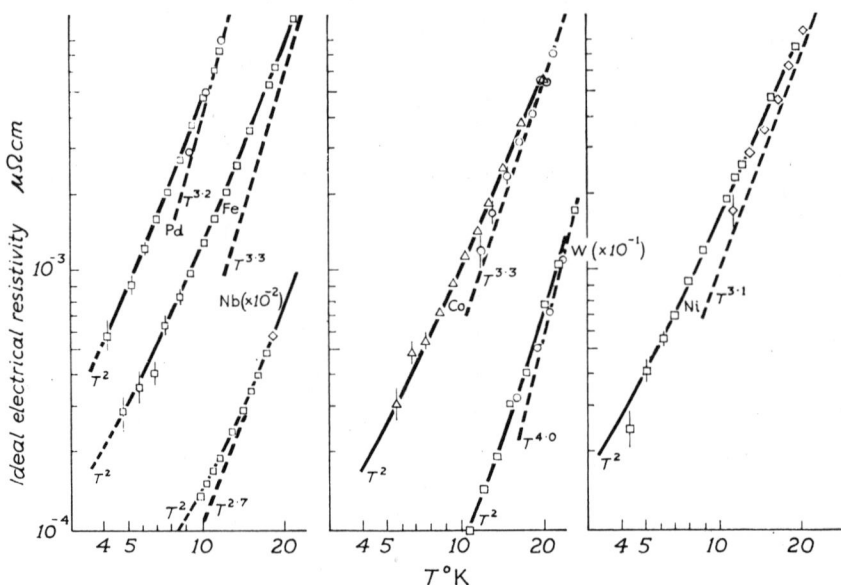

FIG. 4.18. The ideal electrical resistivities for some metals of the transition groups showing the trend towards a T^2 behaviour at low temperatures (White and Woods, 1959)

the resistivity of the transition metals could be represented by a combination of the Grüneisen function $\mathscr{G}$ and a T^2 term

$$\rho_i = A\mathscr{G} + BT^2, \tag{4.31}$$

where A and B are constants. Whilst a suitable choice of A and B does give a reasonable fit with the experimental results, the agreement, as is shown in Fig. 4.19, is by no means perfect.

4.23. Resistivity of thin films and rods

The mean free path of an electron may be roughly calculated by estimating v_F from (4.1) and then using the kinetic theory equation (4.3). It will then be found that, even at room temperature, l is about 100 atomic distances, i.e. $\sim 10^{-6}$ cm. Since the resistivity of pure metals can decrease by a factor of several hundreds or even thousands at low temperatures, this means that l might become as large as 10^{-3} cm, or perhaps even more in favourable cases. It is a fairly straightforward matter to prepare specimens in the form of films, or thin wires whose dimensions are comparable with these long mean free paths. In such specimens we shall have a situation which is analogous to that with which we have already dealt in our description of phonon conduction (section 3.7), where at the lowest temperatures the phonons are scattered

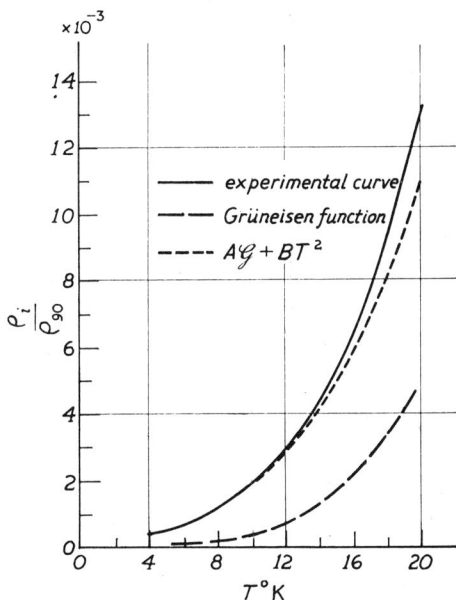

FIG. 4.19. The relative value of ρ_i, compared with its value at 90° K, for palladium showing an attempt to fit (4.31) to the experimental data (Olsen-Bär, 1956).

by the boundaries of the specimen. In metals the electrons will be similarly scattered; the thinner the specimen, the shorter will be the effective mean free path and hence the higher will be the resistivity. If D is the diameter of a rod and l_b is the mean free path in the bulk specimen, the effective mean free path l will be given by

$$\frac{1}{l} \approx \frac{1}{l_b} + \frac{1}{D} = \frac{l_b + D}{l_b D}. \tag{4.32}$$

Since the resistivity is inversely proportional to l, the ratio of the resistivity (ρ_b) in the bulk to that in the thin specimen will be

$$\frac{\rho_b}{\rho} = \frac{l_b D}{(l_b + D)l_b} = 1 - \frac{l_b}{D} \quad \text{for } D \gg l_b. \tag{4.33}$$

More precise calculations give

$$\frac{\rho_b}{\rho} = 1 - \frac{3}{4}\left(\frac{l_b}{D}\right) \quad \text{for } D \gg l_b, \tag{4.34}$$

and one can modify this by including a term which takes account of any possible specular reflection of the electrons at the surface of the metal. Since the experimental evidence suggests that the scattering is entirely diffuse, we have omitted this term.

For a film of thickness t, the mean free path is only limited in one dimension and it is not surprising to find that the correction to the resistivity ratio is one-half of that for a thin wire, i.e.

$$\frac{\rho_b}{\rho} = 1 - \frac{3}{8}\frac{l_b}{t} \quad \text{for } t \gg l_b. \tag{4.35}$$

For very thin specimens, where the smallest dimension is much less than l_b, we see that (4.33) reduces to

$$\frac{\rho_b}{\rho} \approx \frac{D}{l_b} \quad (D \ll l_b), \tag{4.36}$$

which is also the solution obtained from a more rigorous calculation. For a film, the corresponding formula is

$$\frac{\rho_b}{\rho} \approx \frac{3}{4}\frac{t}{l_b}\log_e\left(\frac{l_b}{t}\right) \quad (t \ll l_b). \tag{4.37}$$

Further details of the theory are given by MacDonald (1956), section 21.

Experimental work bears out the above relationships quite well. In order to test them it is necessary to use specimens which have a large value of l_b, i.e. they must have a very small resistivity. The lowest resistivity which can be attained is the residual resistivity at low temperatures and as we have already seen in this chapter this will have its smallest values for specimens of very high purity. Fig. 4.20 shows the result of Andrew's experiments (1949) on the resistivity of thin foils of tin measured at 3·8° K. It will be noted that the points agree quite well with the theoretical curve both for specimens which are. thicker and thinner than l_b. Experiments on very fine sodium wires (MacDonald and Sarginson, 1950) also show a considerable size effect.

4.24. *The magneto-resistance size effect*

It is possible to increase the electron mean free path which has been limited by the dimensions of a very thin sample by the application of a magnetic field. This may be understood quite easily by referring to Fig. 4.21. Let us apply a magnetic field, H, parallel to the length of the wire in which an electric current is flowing. This will cause an electron, which was previously travelling in a straight line, at an angle ϕ to H, to be diverted into a helical path parallel to H and of radius r, where

$$r = \frac{m_e vc}{eH}\sin\phi. \tag{4.38}$$

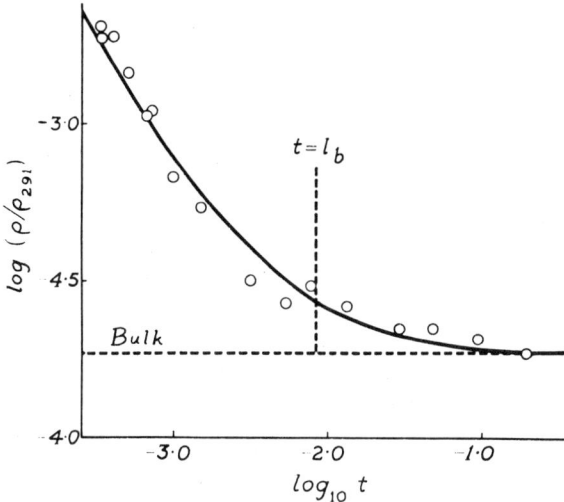

FIG. 4.20. The resistivity of thin tin foils at $3 \cdot 8°$ K as a function of their thickness, t. The solid line is the theoretical curve (Andrew, 1949).

c is the velocity of light *in vacuo*. An electron which, before the magnetic field was applied, would have been scattered at the specimen boundary will now traverse a greater distance before it reaches the boundary. The mean free path is increased and hence the resistivity is *reduced*. When H is increased to such an extent that r becomes equal to or less than the radius of the specimen, then most of the electrons will not reach the boundaries of the specimen at all. They will then have the ordinary mean free path of the bulk material and so the resistivity will also tend to approach the bulk value. This effect is generally

FIG. 4.21. The effect of a magnetic field, **H**, on an electron trajectory.

masked in the experimental results by the ordinary magneto-resistance effect (see section 4.27) in which the resistance tends to *increase* in a magnetic field. Nevertheless, the results on sodium by White and Woods

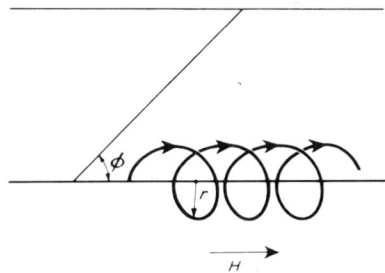

(1956) shown in Fig. 4.22 indicate the initial drop in the resistivity quite convincingly. The curve for high fields, where the resistivity is increasing, extrapolates at $H = 0$, to the bulk resistivity of the metal in zero field Similar results are found for the thermal resistivity (White and Woods, 1956).

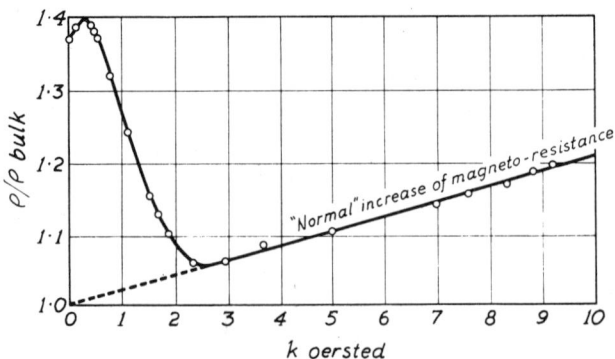

Fig. 4.22. The magneto-resistive size effect at 4·2° K for a thin sodium rod of diameter 0·13 mm. The drop in resistance as the field is increased is caused by the increase in the electron path as shown in Fig. 4.21 (White and Woods, 1956).

4.25. The anomalous skin effect

Whilst in the size effects which have been dealt with in the preceding two sections, the mean free path of the electrons has been restricted by using very thin specimens, a similar effect can be produced in large specimens by other means. As is well known, a high-frequency electric field does not penetrate very far into a metal and all the current flows in a 'skin' at the surface. The higher the frequency of the applied field, the thinner is the conducting layer. H. London (1940) first suggested that if the frequency could be increased to a value sufficient for this layer to be thinner than the mean free path of the electrons, there should be a change in the behaviour of the skin resistivity. This change in the properties, which was first investigated in detail by Pippard (1947), has been called the anomalous skin effect. Detailed calculations have been made by Reuter and Sondheimer (1948), Dingle (1953), and Sondheimer (1954). A general review of the theoretical work is presented by Pippard (1954) and by Ziman (1960), p. 474. Following Pippard (1947) we give a simple physical derivation of the main formula.

Classical electromagnetic theory shows that an alternating electric field E of angular frequency ω penetrates into a conductor of d.c. conductivity, σ_b, according to the relation

$$E = E(0)\exp[-(1+i)z/d],   (4.39)$$

where the z-axis is normal to the surface of the conductor and $E(0)$ is the field at the surface. d is the classical skin depth, and its value is given by

$$d = (2\pi\sigma_b\,\omega)^{-\frac{1}{2}}, \quad \text{where } \sigma_b \text{ is in e.m.u.} \tag{4.40}$$

The surface resistance of the metal $R(\omega)$ is then

$$R(\omega) = (2\pi\omega/\sigma_b)^{\frac{1}{2}}. \tag{4.41}$$

If the mean free path of the electrons is now increased (by lowering the temperature) so that it is greater than d, then E is found to fall off very much less rapidly within the conductor than (4.39) would suggest;

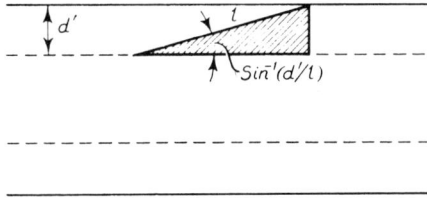

FIG. 4.23. The anomalous skin effect. When the electron mean free path, l, becomes greater than the skin depth, d', the only electrons which contribute to the current flow are those which are travelling at an angle less than $\sin^{-1}(d'/l)$ to the surface.

i.e. the effective skin depth is much increased and the current is transported by the electrons in a deeper layer of thickness d' below the surface. A full treatment of the problem shows that the only electrons which contribute to the electric current are those which have their velocity approximately parallel to the specimen surface. These electrons will remain in the surface layer for the longest time and hence will be the ones most influenced by the applied field. Pippard suggested that one should take as a criterion for the 'effective' electrons, those which are able to complete a mean free path within the surface layer without striking the surface first. These electrons will be those which are moving in directions which make an angle of less than $\sin^{-1}(d'/l)$ with the surface (Fig. 4.23). Very roughly, these effective electrons will comprise a fraction of the order of d'/l of the total number of conduction electrons. Let us call this fraction $\beta d'/l$, where β is a factor of the order of unity. Having taken account of the long mean free path by considering the effective number of electrons, rather than the total number, we can calculate the skin depth by the ordinary formula (4.40), except that, instead of using the d.c. conductivity σ_b, we use the conductivity σ_e associated with the effective electrons, i.e.

$$\sigma_e = \sigma_b\,\beta d'/l. \tag{4.42}$$

Hence from (4.40) and (4.42) we obtain

$$d' = (2\pi\sigma_b\,\omega\beta d'/l)^{-\frac{1}{2}}, \tag{4.43}$$

i.e.
$$d' = (2\pi\sigma_b\,\omega\beta/l)^{-\frac{1}{3}}. \tag{4.44}$$

The surface resistance in the anomalous skin effect region, $R'(\omega)$, will be given by (4.41) where again we use σ_e (4.42) instead of σ_b.

Hence
$$R'(\omega) = (2\pi\omega l/\beta d'\sigma_b)^{\frac{1}{2}}. \tag{4.45}$$

With the aid of (4.44) for d' we then obtain

$$R'(\omega) = \left(\frac{4\pi^2\omega^2 l}{\beta\sigma_b}\right)^{\frac{1}{3}}. \tag{4.46}$$

A rigorous mathematical treatment for the case when the metal surface is rough (100 per cent diffuse scattering) gives

$$R'(\omega) = \{\sqrt{3}\,\pi\omega^2 l/\sigma_b\}^{\frac{1}{3}}. \tag{4.47}$$

The important thing to notice about this result is that, since the bulk conductivity σ_b is proportional to the mean free path l, then R' is independent of σ_b, whereas in the region where the classical skin effect is operative, (4.41) shows that R is proportional to $\sigma_b^{-\frac{1}{2}}$. Experimentally σ_b can be altered by changing the temperature (until the residual resistance range is reached). The value of $R(\omega)$ is actually determined by measuring the Q of a resonant cavity of which the specimen forms a part. Fig. 4.24 shows some results from Chambers's work (1952) on lead, silver, and gold using a frequency of 3,600 Mc/s. The proportionality of $R(\omega)$ to $\sigma^{-\frac{1}{2}}$ is quite good but as σ increases, $R(\omega)$ becomes constant showing the onset of the region of the anomalous skin effect.

4.26. *Determination of the shape of the Fermi surface*

Once we have measured this constant value R', (4.46) gives the value for σ_b/l. This is a very important quantity. The kinetic formula for the resistivity (4.3) shows that

$$\sigma_b/l = ne^2/mv_F. \tag{4.48}$$

Since n determines the volume of the Fermi 'sphere' and mv_F is the momentum of the electrons at the Fermi surface and is equal to $\hbar k_F$, where k_F is the radius of the sphere, we see that R' can give us information about the area of the Fermi surface. In actual fact much more may be obtained. Further mathematical development (Pippard, 1954a) which is outside the scope of this book shows that by using values of R' one can obtain an expression involving the radius of curvature of the Fermi surface. By making measurements of R' on copper single crystals of several orientations Pippard (1957) was able to calculate the shape of

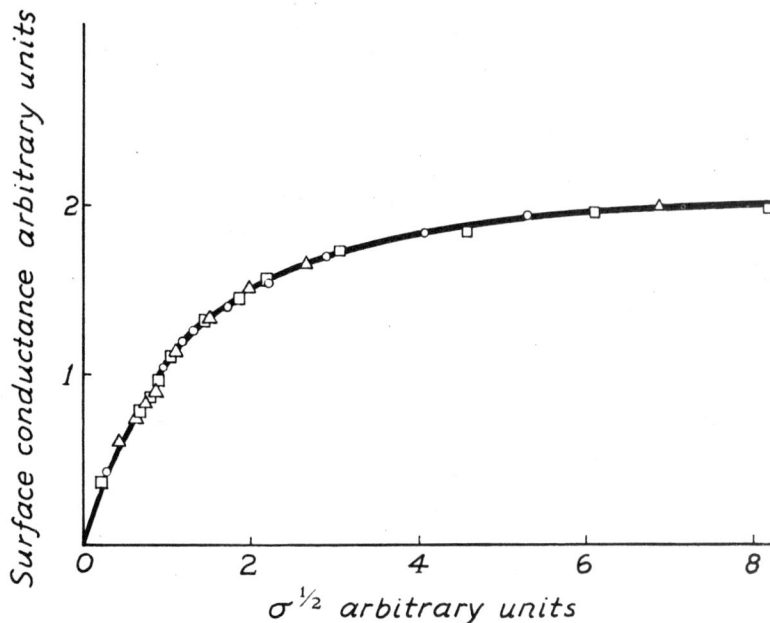

Fig. 4.24. The anomalous skin effect in silver ○, gold □, and lead △. The dependence of the a.c. surface conductance on the d.c. conductivity σ. For small values of σ, the conductance is proportional to $\sigma^{\frac{1}{2}}$ in accord with (4.41), but for higher values of σ, i.e. for a longer free path, the region of the anomalous skin effect is reached, where the conductance is constant (4.47) (Chambers, 1952).

the Fermi surface. His results suggested that the Fermi surface must touch the zone boundary along the [111] axes as is shown in Fig. 4.25.

The computation involved in this work is very tedious and the results are not always capable of a unique interpretation. Nevertheless, measurements of the anomalous skin effect are a powerful tool for investigating the electron distribution in metals.

4.27. Magneto-resistivity

In addition to the rather sophisticated effects of a magnetic field which have been dealt with in the preceding two sections we must also consider the increase in the d.c. resistivity which is observed when a magnetic field is applied to a bulk specimen.

Whilst there is a very large amount of

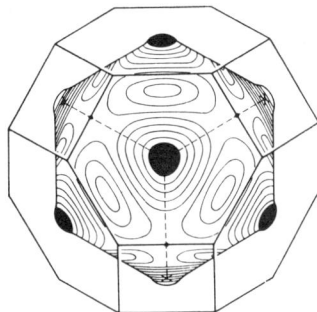

Fig. 4.25. The Fermi surface of copper deduced from measurements of the anomalous skin effect. Contact is made with the hexagonal faces of the Brillouin zone (Pippard, 1957, 1960).

experimental data available, the theory of the effect is by no means fully developed and it certainly does not permit of a simple treatment. Experiments show that for all metals the resistivity increases when a magnetic field is applied to a specimen. The magneto-resistance is usually considerably greater when the field is transverse to the direction of the electric current than when it is parallel to it and the effect becomes more pronounced as the temperature is reduced. In general the magneto-resistivity of the monovalent metals is much less than that for metals with more complicated electronic structures. For most metals the increase in resistivity is proportional to H^2 although in a few cases it is more nearly directly proportional to H.

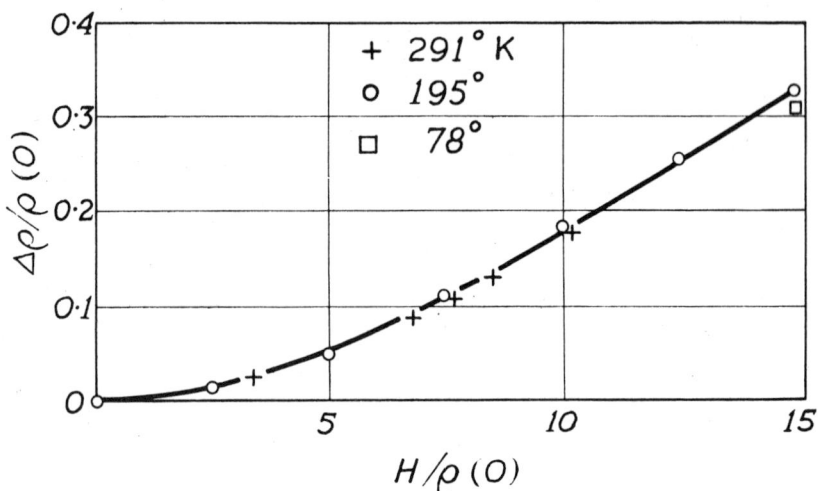

Fig. 4.26. Magneto-resistance data for Mg at 78, 195, and 291° K plotted to show the validity of Kohler's rule (4.49) (Kohler, 1938).

4.28. *Kohler's rule*

The one simple rule in a subject of disturbing complexity is that the change in resistivity $\Delta\rho$, in a field H is of the form

$$\Delta\rho/\rho(0) = \text{function}\{H/\rho(0)\}, \qquad (4.49)$$

where $\rho(0)$ is the resistivity in zero field. This is the well-known Kohler rule (1938). $\rho(0)$ may be changed by altering the specimen temperature but in the residual resistance range it can only be altered by using a specimen of different purity. Fig. 4.26 shows data for Mg taken at various temperatures which show the very good agreement with Kohler's rule. Measurements on many other metals are shown in a reduced Kohler diagram in Fig. 4.27. This figure shows the plots of $\Delta\rho/\rho(0)$

against $H\rho_\theta/\rho(0)$, where ρ_θ is the resistivity at the Debye temperature. This type of diagram is useful because it enables the magneto-resistive effect in different metals to be compared conveniently. The enormous magneto-resistance of bismuth (and to a lesser extent of antimony and

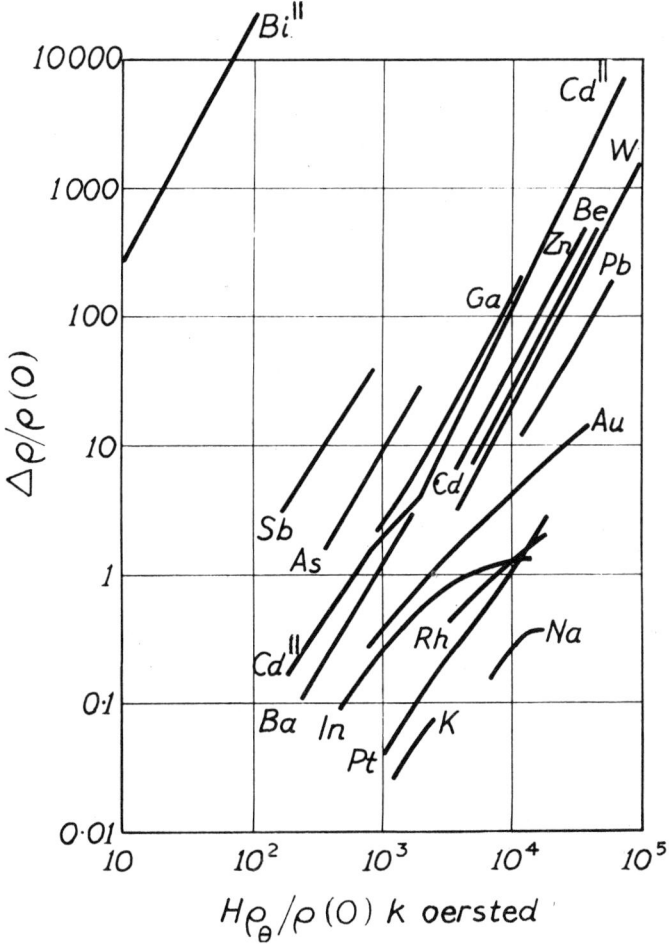

FIG. 4.27. A reduced Kohler plot for several metals (Kohler, 1949).

arsenic) should be noted. It has been utilized to measure magnetic fields, particularly when rapid variations over short distances are encountered, e.g. see section 6.20. Large effects are also observed in metals which have a more complicated electronic structure than the monovalent metals. Except in very high fields the magneto-resistance is approximately proportional to H^2. It is clear from (4.49) that the largest magneto-resistive effect will occur when $\rho(0)$ is very small, i.e. at low temperatures.

4.29. *Magneto-resistance of a free-electron metal*

Using the free-electron theory the magneto-resistive effect should be zero. This can be understood very easily for a longitudinal magnetic field (parallel to the direction of the current) because since the force $\mathbf{F}_H$ on an electron moving with velocity $\mathbf{v}$ is

$$\mathbf{F}_H = e(\mathbf{v} \times \mathbf{H})/c, \tag{4.50}$$

it is clear that the component of $\mathbf{v}$ parallel to $\mathbf{H}$ (which is the part of the velocity which determines the current) will not be affected by $\mathbf{H}$. If a transverse field is applied there will be a non-zero value of $\mathbf{F}_H$ for the electrons which have a component of $\mathbf{v}$ which is perpendicular to $\mathbf{H}$. These electrons will be diverted from their paths along the specimen and they will accumulate along the specimen sides. In so doing they create a space charge (this is the origin of the Hall effect, section 7.5) whose repulsion just cancels out the effect of $\mathbf{F}_H$ on the other electrons, which therefore continue to travel down the specimen along their original paths. Thus again the resistivity should not be affected. The explanation of the magneto-resistance must be therefore found in a study of the real as opposed to the ideal electronic structure of metals, and the theoretical work (Ziman, 1960, chapter 12; Chambers, 1960) shows that if the Fermi surface is not spherical, then a magneto-resistive effect should be observed. This is essentially because the behaviour of the electrons now depends on their position on the Fermi surface. Ziman has shown that reasonable agreement with the experimental data for copper can be obtained. Another argument has been used by Sondheimer and Wilson (1947) to account for the rather large magneto-resistivity of the divalent metals. They assumed an energy scheme consisting of two overlapping bands of electrons and holes. Since the effective masses and mobilities of the holes and of the electrons will not be the same, the precise balancing effect of the magnetic field by the Hall field, which has just been described, cannot take place. The electrons and the holes would each require a different field in order to maintain zero deflexion and hence each will be deviated to some extent in the compromise field which is actually set up.

4.30. *Fermi surface investigations by magneto-resistance measurements*

In a sufficiently high magnetic field the radius of curvature of an orbit will become so small that an electron will be able to complete an entire circuit before it is scattered. It seems reasonable to suppose that once this condition has been reached, the magneto-resistance behaviour will be modified. The theory does indeed show that for high fields $\Delta\rho/\rho(0)$ saturates to a constant value.

There are cases, however, in which the electrons do not execute closed

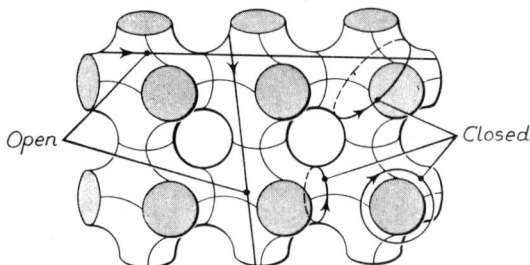

FIG. 4.28. A repeated zone scheme for a simple cubic lattice in which the Fermi surface makes contact with the zone boundary (cf. Fig. 4.25). Open and closed orbits are indicated.

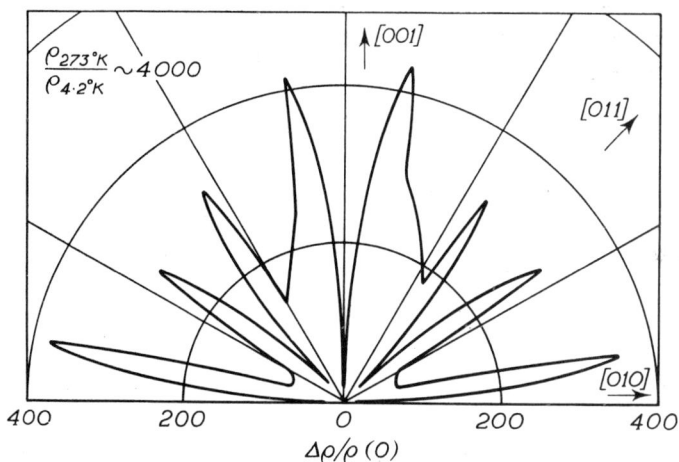

FIG. 4.29. A polar diagram showing the variation of the magneto-resistance of a high-purity copper single crystal as a transverse field is rotated about the specimen axis. $T = 4\cdot2°$ K, $H = 18$ k oersted, specimen axis approximately [100] (Klauder and Kunzler, 1960).

orbits. This can arise when the Fermi surface touches the zone boundary. We have already mentioned Pippard's experiments which showed that this was the case for copper (Fig. 4.25). If contact does occur, then in the repeated zone scheme of Fig. 4.1 (b), the Fermi surface forms a multiply connected surface of the type shown in Fig. 4.28. Recalling that a magnetic field does not alter the electron energy, an electron whose energy state is on the Fermi surface will continue its motion so that it remains on the surface. It can therefore execute closed orbits as shown in the figure, but it can also travel on the 'open' orbits which are indicated, if the magnetic field is in the correct direction for these paths.

When open orbits are permitted the magneto-resistance does not saturate; instead it continues to increase proportional to H^2. Thus measurements of $\Delta\rho/\rho(0)$ as a function of the direction of the magnetic field show the saturation behaviour when only closed orbits are allowed but there will be a sharp rise in the resistivity when the field is in a suitable direction for open orbits to be permitted. Single crystals must, of course, be used for these experiments. The fields required are usually of the order of 15–20 k oersted. A typical polar diagram is shown in Fig. 4.29. From the analysis of such data it is possible to identify the orbits and hence one can get more information about the shape and the connectivity of the Fermi surface (see Chambers, 1960). In addition to these Fermi surface studies and the ones already described using the anomalous skin effect (section 4.26), we should also mention another method of investigation, that using the de Haas–van Alphen effect. This is described in Chapter 10.

5

THE THERMAL CONDUCTIVITY OF METALS, ALLOYS, AND SEMICONDUCTORS

5.1. Introduction

WHEN we described the mechanism of thermal conduction for non-metals in Chapter 3 we only needed to consider the heat which was transported by the phonons. For metals we must take account of the fact that the conduction electrons can also carry thermal energy. This is very important because experiments show that in reasonably pure metals most of the heat is, in fact, carried by these electrons and it is only in semiconductors, impure metals, and alloys and that an appreciable proportion of the thermal conductivity is due to lattice conduction. The total thermal conductivity, κ, of a metal will be due to the sum of the two separate conductivities, i.e.

$$\kappa = \kappa_e + \kappa_g \tag{5.1}$$

where κ_e and κ_g are the electronic and the phonon conductivities, respectively. Both κ_e and κ_g will be limited by various scattering mechanisms which will each give rise to a thermal resistance. In this chapter we shall first consider the behaviour of κ_e and in later sections we shall deal with those aspects of κ_g which were not described in Chapter 3.

5.2. The electronic thermal resistivity

We have already discussed the various mechanisms by which electrons can be scattered in a metal, and how such scattering affects the electrical resistivity, in Chapter 4. Since it is these same electrons which also carry the thermal energy, it is clear that we must once again consider the same interaction mechanisms. It will be recalled that these are the scattering of the electrons by static defects in the crystal and the scattering by the thermal vibrations of an otherwise perfect crystal lattice. Once again we shall only attempt a simplified theoretical explanation and the reader is referred to Ziman (1960), chapter 9, for a full review and exposition.

The static defects in the metal—impurity atoms, vacancies, dislocations, etc.—will introduce a thermal resistivity W_0, which is analogous to the residual electrical resistivity ρ_0 (section 4.5). These defects give rise to a mean free path, l, which is temperature-independent and we can calculate the form of the conductivity by using the kinetic formula which was introduced in section 3.3; i.e.

$$\kappa = c_v \, vl/3. \tag{5.2}$$

For electrons the specific heat, c_v, is proportional to T (1.23) and we assume the velocity v to be constant. Thus the thermal conductivity, when limited by static defects, is proportional to T; i.e. W_0 is of the form

$$W_0 = \beta/T. \tag{5.3}$$

5.3. The Wiedemann–Franz law

It is clear that there must be a mathematical relationship between the electrical and thermal conductivities. This may be obtained very simply from expressions which we have already presented. If we calculate the value of the quantity $\kappa/(\sigma T)$ using (5.2) for κ and the kinetic expression (4.3) for the electrical conductivity, σ, we obtain

$$\kappa/(\sigma T) = \frac{m_e v^2 c_v}{3ne^2 T}, \tag{5.4}$$

and using (1.22) for c_v this becomes

$$\frac{\pi^2 k^2}{3e^2}\left(\frac{m_e v^2}{2\mathscr{E}_0}\right), \tag{5.5}$$

but the fraction in brackets is equal to unity, since $m_e v^2$ is equal to twice the kinetic energy of the electrons at the Fermi surface, i.e. $2\mathscr{E}_0$. Thus we obtain

$$\rho/(WT) = \kappa/(\sigma T) = \pi^2 k^2/(3e^2) = 2\cdot 45 \times 10^{-8} \equiv L, \tag{5.6}$$

where ρ is expressed in ohm cm and W in watt^{-1} cm deg. This relationship, which can also be derived more generally, and is valid even if a metal does not approximate to the simple free electron model, is commonly referred to as the Wiedemann–Franz law. L is usually called the Lorenz constant.

We should emphasize that this simple expression is obtained because we have been able to cancel the mean free path, l, from the two equations (5.2) and (4.3), assuming its value for electrical and thermal conduction to be the same. As we shall see in section 5.8, it is only legitimate to do this when impurity scattering is the dominant resistive mechanism, i.e. at very low temperatures, and also at high temperatures. At intermediate temperatures the Wiedemann–Franz law does not hold.

5.4. *The impurity resistivity*

We are now able to return to the problem of calculating W_0. Using (5.6) we see that

$$W_0 T = \beta = \rho_0/L \qquad (5.7)$$

and since we have already derived the values of ρ_0 for the most important types of defect (section 4.6 ff.) it is quite straightforward to transform these into their effective thermal resistivity. Thus for 1 atomic per cent of vacancies or of interstitial atoms, $\beta \approx 40$ watt^{-1} cm deg^2. For dislocations we have the same uncertainty as has already been discussed in the case of electrical resistivity (section 4.13). The values will range from $\beta \approx 2 \times 10^{-13}$ watt^{-1} cm deg^2 per unit density of dislocations in which the core scattering is neglected, to a value about five times greater in which the effect of the core is possibly overestimated. Stacking faults will presumably play a role similar to that which they play in electrical resistivity and hence should give a value of β which is about forty times that for dislocations. Impurity atoms, too, will increase W_0. The $(\Delta Z)^2$ rule (section 4.7) should apply and β will be approximately proportional to $x(1-x)$, where x is the concentration of a given element (section 4.8).

It should be pointed out that no critical test of these rules has been made for thermal resistivity. The experiments are much more difficult to carry out than are those for electrical resistivity and the precision of measurement which can be attained is very much less (1 per cent is a standard which is quite difficult to achieve).

5.5. *Resistivity due to phonon scattering*

The other type of mechanism of electronic thermal resistivity which must be discussed is that which is due to the scattering of the electrons by the thermal vibrations of the lattice. By analogy with the electrical case it is sometimes called the ideal thermal resistivity and it is designated by W_i.

There is a very important difference between the scattering which determines W_i and that which is responsible for the electrical resistivity ρ_i. As we have already seen (section 4.3), electrical conduction comes about because when an electric field is applied, the $\mathbf{k}$ values of all the electrons are changed by the same amount in the direction of the field. The whole electronic distribution in $\mathbf{k}$ space is shifted bodily (Fig. 5.1 (a)) and hence more electrons now travel in one direction than in the other, i.e. a current flows. The amount the distribution is shifted for a given field is determined by the various scattering processes which will tend to restore electrons from the right-hand side of the figure to the left. These have been termed 'horizontal' transitions by Klemens (1956), section 13. We have already noted (section 4.17) that at low temperatures, due to the lack of phonons of high wave number, the electrons cannot be transferred from one side of the diagram to the other, but they can only be scattered through small angles of the order of T/θ and

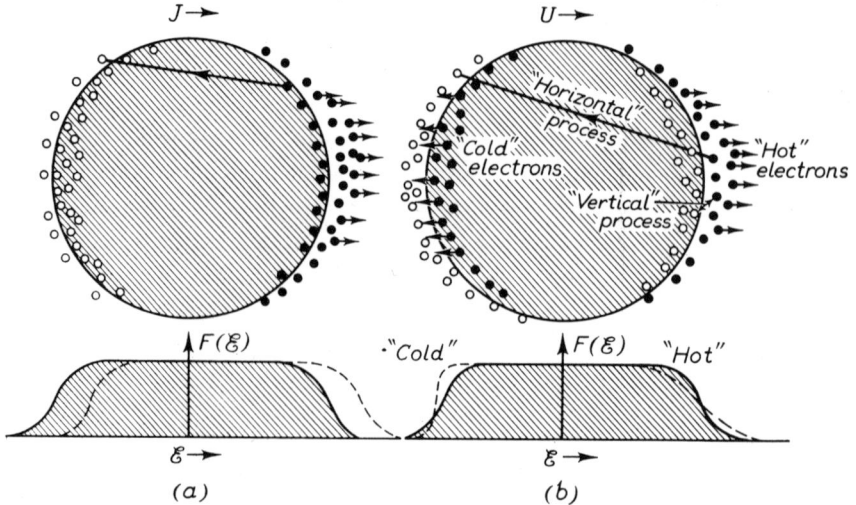

FIG. 5.1. The electron distribution for (a) electrical and (b) thermal conduction. In (a) there is an excess of electrons at the right-hand side of the Fermi surface, whilst the left hand is depleted. This gives a net transport of charge, J. In (b) there are equal numbers in the neighbourhood of the Fermi surface on both sides, but the detailed distribution is such that there are more electrons slightly above the Fermi surface on the right and more just below the surface on the left. Thus there is a flow of heat, U. The lower part of the figure shows the energy distribution function in the two cases (after Ziman, 1960).

this leads to the rapid T^5 decrease in ρ_i as the temperature is reduced. For thermal conductivity the effectiveness of this type of scattering is somewhat different. In the first place, under the normal conditions of a thermal conductivity experiment, no electric current is allowed to flow and so the distribution in $\mathbf{k}$ space is still more or less symmetrically arranged about $\mathbf{k} = 0$. Since, however, there is a flow of heat, this means that at any point in the metal the electrons which are flowing in one direction will on average be hotter than those flowing in the opposite direction, so that the Fermi surface will be modified as is shown in Fig. 5.1 (b). This Fermi surface is not a well-defined line but is, of course, a diffuse band with a thickness (in energy units) of $\sim kT$. Thus when there is heat transport occurring the right-hand side of Fig. 5.1 (b) will have more electrons at the outer edge of this band, whereas on the left-hand side more of them are nearer the inner edge. There is thus a surplus of hot electrons above the normal equilibrium value on the right-hand side of the surface, whilst on the left-hand side there are more cold electrons than normal within the surface. In this way we get a flow of heat. The lower sections of Fig. 5.1 show the distribution functions plotted as a function of the energy for electrical and thermal conduction.

When an electron interacts with a phonon at low temperatures, energy of the order of kT is either emitted or absorbed by the electron in accordance with (4.20). This energy is sufficient to transfer an electron from the inside of the surface layer to the outside, or vice versa. Hence even though the scattering angle is small, the distribution of hot and cold electrons can still be materially affected by these 'vertical' transitions (Klemens, 1956, section 13). Thus when we consider the transport of heat each interaction with a phonon remains an effective scattering mechanism. In order to estimate the temperature dependence of the thermal resistivity we should note that the number of phonons which are present at any temperature is proportional† to T^3, and so the mean free path, l, of the electrons will be inversely proportional to the number of phonons, i.e. to T^{-3}. Thus using the kinetic formula (5.2), together with the assumption that c_v is proportional to T and that v is constant, we find that κ is proportional to T^{-2}, i.e. W_i is of the form

$$W_i = \alpha T^2. \tag{5.8}$$

If we had assumed the validity of the Wiedemann–Franz law we should have found that as a consequence of ρ_i being proportional to T^5, W_i should have been proportional to T^4. Thus the thermal resistivity does not decrease as rapidly as the behaviour of the electrical resistivity would lead us to believe.

For completeness we should point out that at high temperatures $(T > \theta)$ the 'band' at the Fermi surface is still of thickness kT, but the phonon energy cannot become greater than $k\theta$. Thus 'vertical' transitions through the band are not possible, but the phonons now have wave numbers which are sufficiently large to induce 'horizontal' transitions from right to left in Fig. 5.1 (b). These are the same types of transition that are effective for electrical resistivity and hence in this temperature region the Wiedemann–Franz law is valid. Thus since at these temperatures ρ_i is proportional to T, it follows that W_i is constant. Further

† The total energy of the crystal at low temperatures is proportional to T^4 in the T^3 specific heat region, and the average energy of a phonon is kT, hence the number of phonons is proportional to T^3.

discussion on the validity of the Wiedemann–Franz law is given in section 5.8.

A quantitative calculation of W_i is outside the scope of this book. A full solution can only be obtained by solving the Boltzmann transport equation and this leads to a complicated integral equation which has only been solved for the case of quasi-free electrons. For details and reviews the reader is referred to articles by Wilson (1953), Klemens (1956), and Ziman (1960).

The general result of these detailed calculations is that at low temperatures the αT^2 dependence of W_i (5.8) is confirmed, with a value of α given by

$$\alpha = \frac{G n_0^{\frac{2}{3}}}{\kappa_\infty \theta^2},\qquad(5.9)$$

where n_0 is the number of free electrons per atom, κ_∞ is the limiting constant thermal conductivity at high temperatures and G is a numerical factor which is about 70, but whose precise value depends on which approximation is used for solving the Boltzmann equation.

5.6. *Modifications to the theory*

It is unfortunate, as (5.9) shows, that α can only be calculated with the aid of other experimental data—κ_∞ and θ. However, if a constant value of θ is used, this implies that a simple Debye spectrum is valid, but as we have already seen in section 1.8, it is just at low temperatures that the deviations from this simple model are most apparent. κ_∞ appears in (5.9) because it is not possible to calculate the true interaction function of the electrons and the lattice vibrations. To overcome this difficulty, a substitution involving κ_∞ is made for this function, since κ_∞ will also be dependent on the interaction between the electrons and the phonons. If, however, n_0 is taken as unity and experimental values of κ_∞ and θ are used, Hulm (1950) showed that the calculated value of α appears to be about four times greater than that obtained from actual measurements of the thermal conductivity at low temperatures.

This substitution for the interaction function assumes that it has the same form both at high and at low temperatures. It is very unlikely that this is correct. At high temperatures we have already seen (section 4.20) that there is a much greater probability of an umklapp-process occurring. Such processes will enhance the scattering of the electrons and will increase the value of the interaction function at high temperatures. Thus the use at low temperatures of an expression which involves this function will result in an overestimate of the low-temperature scattering, i.e. the calculated value of α in (5.9) will be too large, and this is what has been observed. It has also been suggested by Blackman (1951) that (5.9) should be modified so that θ refers to the longitudinal lattice vibrations only, because according to the Bloch theory electrons will not be scattered by transverse vibrations. Hence a special value of θ should be used, say θ_L. The specific heat θ takes account of lattice vibrations of all polarizations. Since the value of θ_L is about 1·5 times greater than the specific heat θ, this would help to reduce the value of α.

Ziman (1954) has developed the theory, taking into account both the u-processes and also the fact that only longitudinal phonons should be considered. His numerical calculations for sodium are in quite good agreement with the experimental results of Berman and MacDonald (1952). The calculations also show that a maximum in W_i which the earlier theories had suggested ought to exist at about $\theta/5$ and which had never been observed, was now almost entirely removed. Unfortunately he is not able to give any explicit expression for the variation of the thermal conductivity with temperature. Nevertheless, the T^2 dependence of W_i still holds at low temperatures although as we have already mentioned the expression (5.9) for α should be reduced by a factor of about 4 to be in accord with experiments on many metals.

5.7. *Comparison with experiment*

The total electronic thermal resistivity, W_e, is therefore of the form

$$W_e = W_i + W_0 = \alpha T^2 + \beta/T. \qquad (5.10)$$

This is shown graphically in Fig. 5.2 (a). This expression will only apply in the region where W_i is proportional to T^2 and for a Debye lattice spectrum this is taken to be below $\theta/10$. For many metals, however, as we have already seen in section 1.8, the Debye theory is not accurate until temperatures of $\theta/50$ to $\theta/100$ have been reached and therefore W_i will not be strictly proportional to T^2 until such temperatures. Fig. 5.2 (b) shows the behaviour of the thermal conductivity for specimens of differing purity.

The preceding discussion shows that α is the more fundamental parameter in (5.10) because it should be a constant for a certain metal whereas β, the impurity coefficient, will depend on the particular sample of metal which is being measured. Provided that β is fairly small it can be seen that there will be a minimum in the value of W_e at low temperatures, i.e. there will be a maximum in the electronic thermal conductivity κ_e. The purer the sample of the metal the smaller will be β and so the maximum in κ_e will be higher and it will occur at a lower temperature. This maximum can be very high—100 watt cm^{-1} deg^{-1} or more for very pure and perfect specimens. Some typical conductivity curves are shown in Fig. 5.3.

A check on the validity of (5.10) can be made by noting that

$$W_e T = \alpha T^3 + \beta. \qquad (5.11)$$

Thus a graph of $W_e T$ against T^3 should be a straight line with a slope of α and an intercept on the $W_e T$ axis of β. In this way α may be deter-

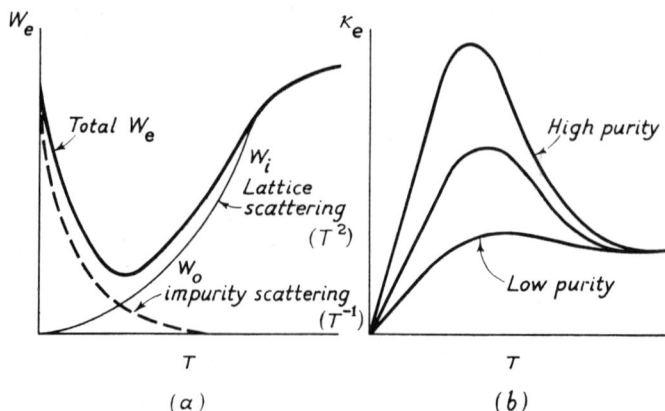

Fig. 5.2. (a) The electronic thermal resistivity as a function of temperature showing the individual contributions due to impurity and to lattice scattering. (b) The electronic thermal conductivity as a function of temperature, showing the behaviour to be expected from specimens of various purities.

mined. It is found that (5.11) is obeyed by many metals, particularly those which have a relatively simple quasi-free electron structure, such as the alkalis. Nevertheless, all metals have W_i proportional to T^x where x is ~ 2, although there is a tendency for x to be a little larger than 2 (see White and Woods, 1959, table 4). Even the transition metals with their complicated overlapping bands are in rough agreement with the simple theory, although for tungsten, rhenium, osmium, rhodium, and iridium, x is greater than 2·5. It is not yet clear whether these deviations from $n = 2$ are (a) the effect of a more complicated Fermi surface, (b) a deviation of the phonon distribution from the Debye spectrum, or (c) the decreasing probability of u-processes as the temperature is reduced.

We have suggested that in pure metals all the heat transport can be accounted for by electronic conduction. Nevertheless, there are some metals for which the curve of κ against T is apparently linear at low temperatures (i.e. W_0 should be dominant) and yet the value of W_0, calculated from the residual electrical resistance, ρ_0, using the Wiedemann–Franz law (5.6) is higher than the experimental value, i.e. the measured conductivity is more than it ought to be. The metals in

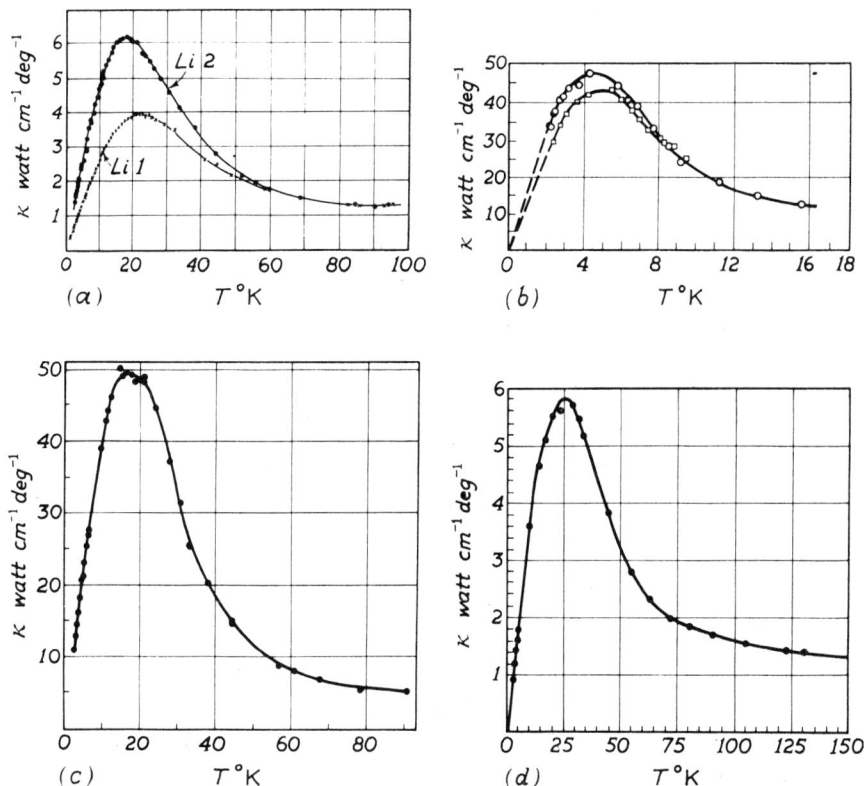

FIG. 5.3. Typical thermal conductivity curves (a) lithium (Rosenberg, 1956); (b) sodium (MacDonald, White, and Woods, 1956); (c) copper (Berman and MacDonald, 1952); (d) chromium (Harper, Kemp, Klemens, Tainsh, and White, 1957).

which this occurs all have a low value of κ—titanium, zirconium, hafnium, niobium, chromium, vanadium, and manganese (Rosenberg, 1955, White and Woods, 1959)—they are all metals which are very difficult to obtain in a high state of purity and many of them, even for the most carefully prepared samples, are physically very hard, which indicates that some impurities might still be present. In view of the reliability of the Wiedemann–Franz law in calculating W_0, it seems reasonable to suppose that the small amount of extra conduction is due to the phonons. As we shall see in section 5.9, the effect of impurities is to decrease the electronic conduction at low temperatures, but not the phonon conduction. Thus it is very likely that the discrepancy between the observed and the calculated values of W_0 is due to a small contribution to κ which is caused by lattice conduction.

5.8. *The temperature variation of the Lorenz number*

In section 5.5 we have discussed the fact that when the electrons are scattered through small angles by phonon interaction, the

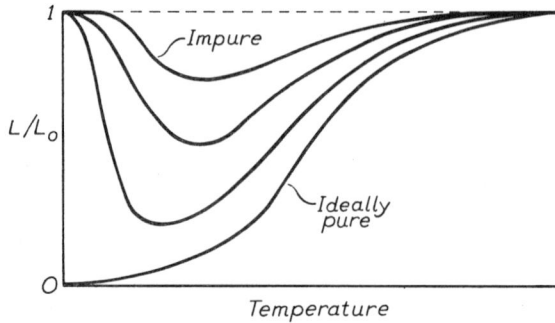

Fig. 5.4. The temperature variation of the Lorenz number as a function of purity and temperature. L_0 is the limiting value of L at low and at high temperatures as given by (5.6). The top curve relates to the most impure specimen.

scattering is much more effective in producing a thermal than an electrical resistance. This means that, except at very low or at high temperatures, the Wiedemann–Franz law (5.6) is not valid and the value of L is smaller than the value quoted. The higher the purity, the smaller will be the value of L, since the impurity scattering (for which (5.6) still holds) is then a smaller fraction of the total scattering. Fig. 5.4 shows the manner in which L might be expected to vary for specimens of different purity.

5.9. Lattice thermal resistivity of metals

In Chapter 3 we dealt at some length with the conduction of heat by phonons and, in principle, all the scattering mechanisms which were discussed there will be operative in a metal crystal. However, there is one mechanism in metals which tends to dominate all others. This is the scattering of the phonons by the conduction electrons and it gives rise to a phonon resistivity W_{ge}, which can be shown (Ziman, 1960, chapter 8) to be of the form

$$W_{ge} = ET^{-2}. \tag{5.12}$$

This temperature dependence can be derived by a simple argument. The mean free path, l, of the phonons will be proportional to the number of electrons with which they can interact. The only electrons with which this is possible are those whose energy lies within kT of the Fermi energy $\mathscr{E}_0$, and the proportion of these electrons to the total number present will be of the order $kT/\mathscr{E}_0$. Hence the number of electrons available for scattering will be proportional to T and therefore l will vary as T^{-1}. If one assumes that the phonon velocity, v, is constant and that c_v is proportional to T^3, then, from the kinetic theory (5.2) W_{ge} is proportional to T^{-2}.

The determination of the constant of proportionality, E, in (5.12) involves the same problem that we have already encountered in the calculation of W_i—that is, the uncertainty as to the form of the interaction constant between the electrons and the phonons. Makinson (1938) tried to overcome this by using the limiting *high* temperature value of the electronic thermal conductivity κ_∞ in his expression for E, but as has already been remarked (section 5.6), such a substitution ignores the fact that u-processes are more likely to be operative at high than at low temperatures. Klemens (1954) avoided this source of error by using the *low*-temperature electronic thermal resistivity, W_i, as the experimental quantity which contained the interaction constant. His expression for W_{ge} is

$$W_{ge} = 3\cdot2 \times 10^{-3} \alpha \theta^4 n_0^{\frac{4}{3}} T^{-2} \equiv E T^{-2}, \tag{5.13}$$

where α is defined by $W_i = \alpha T^2$ as in (5.8).

Since for most metals α is of the order of 10^{-4}, we see that at 4° K, W_{ge} is about 10^2 watt^{-1} cm deg (putting $n_0 = 1$ and $\theta = 300^\circ$ K). This is in general much greater than the phonon resistances which one would expect at 4° K from most of the other scattering mechanisms described in Chapter 3.† Hence in metals the main limitation to κ_g is the scattering of the phonons by the conduction electrons.

The only other contribution to the phonon resistivity which can become important in metals (particularly those which have been cold worked) is that produced by dislocations. This, as shown by (3.22) and (3.23), is directly proportional to the dislocation density and varies as T^{-2}, i.e. it has the same temperature dependence as W_{ge}. Thus a T^{-2} phonon resistivity might be due to the combined scattering effect of both dislocations and electrons and this must be borne in mind when results are analysed.

It should be noted once again that the reason why impurity atoms and other point defects do not reduce the low-temperature *phonon*

† Except in superconductors, see section (5.13).

conductivity is that they are so much smaller than the phonon wavelength that the phonons are not scattered. This is in direct contrast to the behaviour of the *electronic* thermal resistivity W_0, which is inversely proportional to T and therefore becomes much more pronounced at low temperatures. The situation can and does arise, therefore, that in an impure metal or an alloy, W_0 might become of the same order of magnitude, or perhaps even greater than, W_g. Under these circumstances a large proportion of the heat transport in the metal will be by the phonons. It is this kind of situation which is thought to be operating for metals such as titanium and zirconium whose behaviour has been described in section 5.7.

5.10. *Separation of the lattice and the electronic conductivities*

If, as has been described in the previous section, a reasonable amount of phonon conduction is present, then it is obviously of interest to separate the total measured conductivity, κ, into its two components, κ_e and κ_g. From the practical point of view it is very difficult to measure the thermal conductivity with an accuracy greater than 1 per cent. It is clear, therefore, that if either κ_e or κ_g is very small compared with the total conductivity, then it will not be possible to determine the smaller component. In practice this means that it is not possible to determine κ_g for a pure metal, since nearly all the heat is transported by the electrons. The exception to this is for a metal in the superconducting state where, as will be described in section 5.13, at very low temperatures nearly all the heat is carried by the phonons and hence the heat transport which is measured is due to lattice conduction. It should be noted, however, that this value of κ_g is peculiar to the superconducting state and is not the same as that appertaining to the normal state when the electrons can interact with the phonons and so reduce κ_g.

In order to determine κ_g for an impure metal or an alloy we make the very reasonable assumption that at low temperatures the electronic conductivity, κ_e, is limited by impurity scattering and is therefore of the form $\kappa_e = 1/W_0 = T/\beta$. This can be checked by ensuring that the specimen has a constant electrical resistivity (i.e. the residual resistivity, ρ_0) in the temperature region we are considering. From the Wiedemann–Franz law (5.6) we can then calculate W_0 and hence κ_e. κ_g is then equal to $\kappa - \kappa_e$. A variant of this method is to assume that in an annealed metal, κ_g is limited by the scattering of phonons by electrons. Thus from (5.12) and (5.3) the total conductivity should be of the form

$$\kappa = T^2/E + T/\beta. \tag{5.14}$$

Hence a graph of κ/T against T should be a straight line with a slope of $1/E$. Fig. 5.14 shows such a plot for a copper-zinc alloy. It should be noted that the value of κ_g which is obtained in this way only relates to the impure sample which has been measured. It need not necessarily be equal to κ_g for the pure metal. There are two main reasons for this.

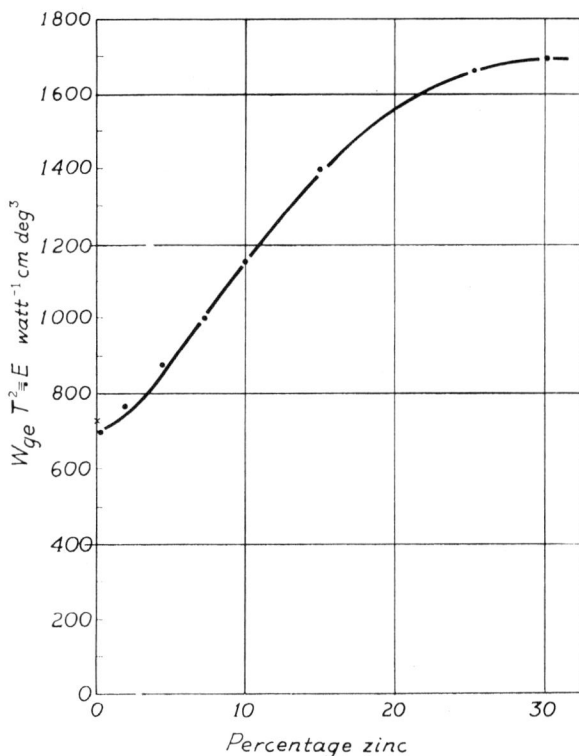

FIG. 5.5. The variation of the lattice conductivity, characterized by the coefficient E in (5.13), as a function of zinc content for Cu-Zn alloys (Lomer, 1958).

Firstly the addition of impurities might affect the electron density and energy distribution, and hence the amount of phonon scattering could be changed. Secondly, the addition of impurity atoms will alter the phonon spectrum (i.e. the θ will be different); this effect can be reduced by using as an impurity an atom whose atomic weight and radius are similar to those of the parent metal. By measuring E as a function of impurity content and then extrapolating to zero impurity, E for the pure metal can be estimated. This is shown in Fig. 5.5.

If we wish to find κ_g at higher temperatures the simple expression for κ_e cannot be used. We must also take into account the αT^2 term (i.e. W_i)

of (5.10). This can only be done by using for α the value found for the pure metal since values of α for alloys cannot be measured. Provided the temperature is not too high, however, this term is rather small and hence any inaccuracies in α will not be very important. A fuller discussion on the separation of κ_g and κ_e is given by Klemens (1956) section 22.

5.11. *The thermal conductivity of alloys*

Analysis of experimental data on the heat conductivity of alloys confirms the results of the discussion in the preceding two sections. The presence of the impurity atoms reduces the electronic heat transport to such an extent that the phonon conductivity becomes important. It is difficult to give any guide as to when κ_g must be taken into account in the interpretation of κ, since the reduction of κ_e which is necessary for this to occur is dependent on the type of impurity atom which has been introduced. To give an example, however, it has been found (Lomer, 1958) that when zinc is added to copper the effect of κ_g becomes evident (although κ_g cannot be estimated with much accuracy) at an impurity concentration of about 1 per cent zinc. At 5 per cent zinc about one-half of the total conductivity can be ascribed to lattice conduction, but if the valency difference between solute and solvent is greater (section 4.7), then κ_g should be observable at lower impurity concentrations. Results for the heat conduction of some Cu-Zn alloys are shown in Fig. 5.6. The reduction in κ when greater impurity is present is quite evident.

At low temperatures the total conductivity of alloys agrees quite well with (5.14). When the temperature is increased, however, other scattering mechanisms take over; in particular κ_g will tend to be dominated by point defect scattering. This will tend to stop the original T^2 increase in κ_g and it will make it pass through a maximum; thereafter it will decrease. It is difficult to determine κ_g accurately at these higher temperatures, however, although since κ_e can be estimated roughly from the behaviour of the electrical resistivity, an approximate value of κ_g can be found by using (5.1).

5.12. *Low-conductivity alloys*

Before concluding this description of the heat conduction of alloys we should mention those materials which have an exceedingly low thermal conductivity at low temperatures. These are very important technically since they are invariably used for tubing and the supports of cryostats, when heat influx to the low-temperature chamber must be reduced to a minimum. All the alloys contain nickel and the ones used most commonly are stainless steel, Inconel, Monel, German silver, and, for wires, constantan. Conductivity curves for some of these alloys are shown in Fig. 5.7, together with curves for solder and some non-

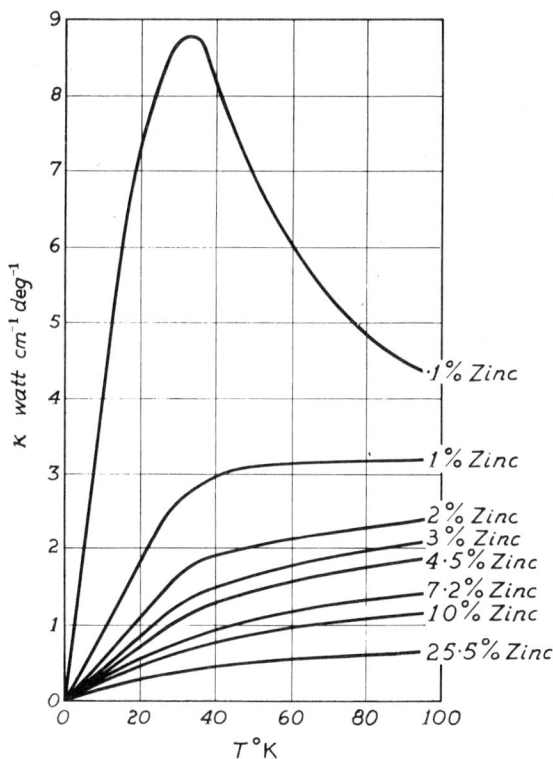

FIG. 5.6. The thermal conductivity of some Cu-Zn alloys (Lomer, 1958).

metallic thermal insulators. In order to compute the heat influx, however, it is useful to have values of the mean thermal conductivity between certain pairs of temperatures which are commonly used in cryogenic apparatus. Some useful values of the heat flow down rods 10 cm long and 1 mm² cross-section are given in Table 5.1. Although stainless steel is the worst conductor it is not always used because the copper-nickel alloys are more easily soldered. If strength is an important requirement, however, then stainless steel is the most satisfactory material.

5.13. The thermal conductivity of superconductors

Soon after the phenomenon of superconductivity had been established, experiments were made to see whether, at the temperature at which the electrical resistance dropped to zero, there was any corresponding change in the thermal conductivity. Early experiments by Onnes and Holst (1914) showed that at the superconducting transition temperature, T_c, no discontinuous change in the thermal conductivity accompanied the

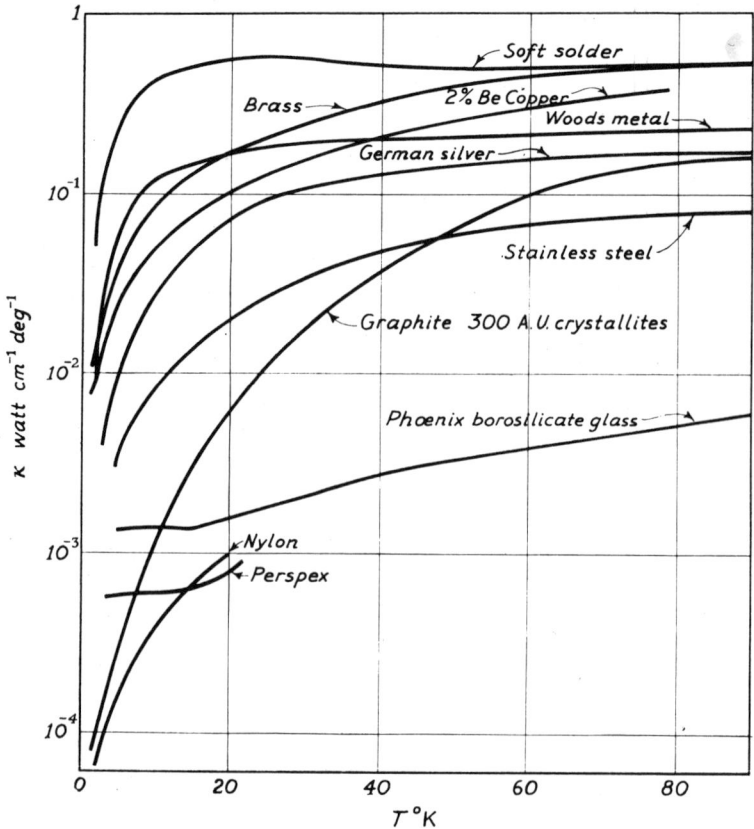

Fig. 5.7. The thermal conductivity of some technical alloys, solders, nylon, perspex, glass, and micro-crystalline graphite.

disappearance of the electrical resistivity. It was not until the 1930's, however, that further investigations were made—mainly by de Haas and Bremmer (1931, 1936) and also by Mendelssohn and Pontius (1937). They showed that when a metal became superconducting, the thermal conductivity had a lower value than when a magnetic field was applied to bring it back to the 'normal', non-superconducting state. Whilst most of the properties of superconductors are described in Chapter 6, it is more convenient for us to discuss the heat conductivity at this juncture.

TABLE 5.1

Heat flow in watts down rods 10 cm long and 1 mm² cross-section when the two ends are at the temperatures indicated

(After Berman, 1951, and White, 1959)

	300 to 77° K	300 to 20° K or 300 to 4° K	77 to 20° K	77 to 4° K	20 to 4° K	4 to 2° K
Pyrex glass	$1{\cdot}8\times10^{-3}$	$2{\cdot}0\times10^{-3}$	$1{\cdot}6\times10^{-4}$	$1{\cdot}8\times10^{-4}$	$1{\cdot}9\times10^{-5}$	$1{\cdot}4\times10^{-6}$
Stainless steel†	$2{\cdot}7\times10^{-2}$	$3{\cdot}0\times10^{-2}$	$3{\cdot}1\times10^{-3}$	$3{\cdot}3\times10^{-3}$	$1{\cdot}6\times10^{-4}$	4×10^{-6}
Inconel (c. 72 Ni, 14–17 Cr, 6–10 Fe, 0·1 C) hard-drawn	$2{\cdot}8\times10^{-2}$	$3{\cdot}1\times10^{-2}$	$3{\cdot}5\times10^{-3}$	$3{\cdot}7\times10^{-3}$	$1{\cdot}9\times10^{-4}$	6×10^{-6}
Monel (c. 66 Ni, 2 Fe, 2 Mn, 30 Cu) annealed	$4{\cdot}6\times10^{-2}$	$5{\cdot}4\times10^{-2}$	$7{\cdot}6\times10^{-3}$	$8{\cdot}3\times10^{-3}$	$6{\cdot}4\times10^{-4}$	$1{\cdot}4\times10^{-5}$
German silver (47 Cu, 41 Zn, 9 Ni, 2 Pb) as received	$4{\cdot}5\times10^{-2}$	$5{\cdot}3\times10^{-2}$	$8{\cdot}0\times10^{-3}$	$8{\cdot}8\times10^{-3}$	$6{\cdot}2\times10^{-4}$	1×10^{-5}
Constantan (60 Cu, 40 Ni) wire, as received	$4{\cdot}9\times10^{-2}$	$5{\cdot}9\times10^{-2}$	$9{\cdot}1\times10^{-3}$	$1{\cdot}0\times10^{-2}$	$7{\cdot}4\times10^{-4}$	$1{\cdot}2\times10^{-5}$
Copper (electrolytic tough pitch), as received	$0{\cdot}9$	$1{\cdot}6$	$0{\cdot}56$	$0{\cdot}71$	$0{\cdot}15$	8×10^{-3}

† The stainless-steel data are for alloys which approximate to the composition 18 per cent Cr, 9 per cent Ni, with traces of Mn, Nb, Si, Ti, totalling 2·3 per cent.

A simple éxplanation of the thermal conductivity of superconductors may be given in terms of the phenomenological 'two fluid' model of superconductivity (section 6.7) which was developed by Gorter and Casimir (1934). According to this theory the electrons form two groups or fluids, the 'normal' electrons, which occupy higher energy levels and the 'superconducting' electrons which exist in a lower set of levels. Below the superconducting transition temperature a certain fraction, $1-x$, of the electrons remain in the normal state, whilst x go into the superconducting state. As the temperature is reduced, more electrons become superconducting, and at the absolute zero $x = 1$. The properties of the superconducting electrons are such that they are not scattered by the lattice waves or by impurities. Below T_c, even though x might be quite small there will be a thread or a fine network of superconducting electrons which will 'short circuit' the resistance due to the normal ones, and hence the metal will have zero electrical resistivity. The experiments of Daunt and Mendelssohn (1938, 1946) led to the conclusion that the specific heat of these superconducting electrons is zero, and hence from (5.2) they cannot contribute to the heat transport. Thus, as the temperature is reduced below T_c, fewer and fewer electrons càn carry the heat, and so the thermal conductivity becomes much less than it is when the metal is in the normal state. In pure metals this is indeed the general observation, but the behaviour has been shown to fall into two groups. When the transition temperature is on the high side of the thermal conductivity maximum (i.e. where the electrons are being scattered by the lattice vibrations) the thermal conductivity for the superconducting state meets the curve for the normal state at quite a large angle, as in Fig. 5.8. Such behaviour is typical of metals with a small Debye θ and a fairly high transition temperature, e.g. lead and mercury. If, on the other hand, T_c is on the low-temperature side of the thermal conductivity maximum (where impurity scattering is dominant), then the superconducting curve gradually drops below the curve for the normal state, as is shown in Fig. 5.9. Most superconductors fall into this latter category.

By making certain assumptions regarding the mean free path of the normal electrons when the metal is in the superconducting state, Koppe (1947) derived a function which describes the thermal conductivity when limited by impurity scattering. A similar function has been derived from the more recent Bardeen, Cooper, and Schrieffer (1957) theory of superconductivity (see section 6.28 ff.). This function is shown in Fig. 5.10 as a graph of the ratio of the superconducting to the normal

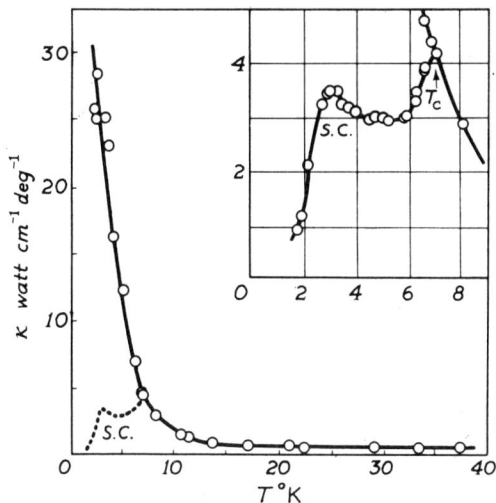

FIG. 5.8. The thermal conductivity of lead in the normal and the superconducting states showing the sharp breakaway of the superconducting curve when the transition tempera-ture, T_c, lies on the high-temperature side of the maximum conductivity. The inset shows the superconducting curve on a larger scale (Rosenberg, 1955).

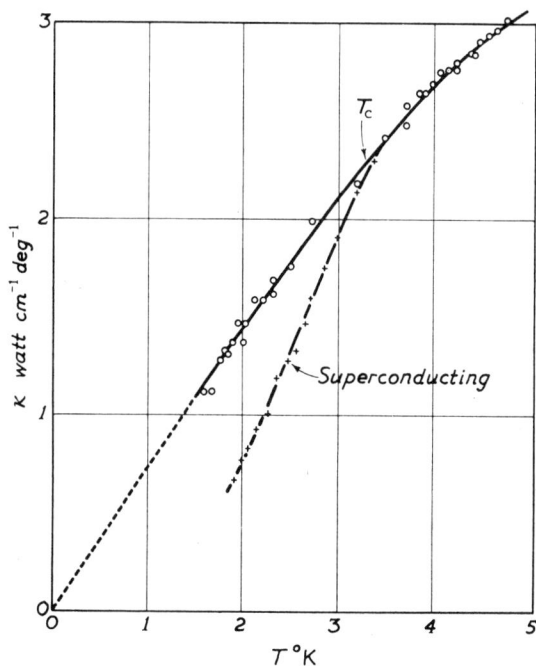

FIG. 5.9. The thermal conductivity of indium in the normal and the superconducting states showing the gradual shift of the superconducting curve from the normal one when the transition temperature, T_c, is on the low-temperature side of the conductivity maximum (Hulm, 1950).

thermal conductivities, κ_s/κ_n, plotted against the reduced temperature, T/T_c. The experimental points for aluminium are seen to fit the theoretical curve quite well, down to about $T/T_c = 0.4$. Up to the present, however, there is no satisfactory calculation of the behaviour of the

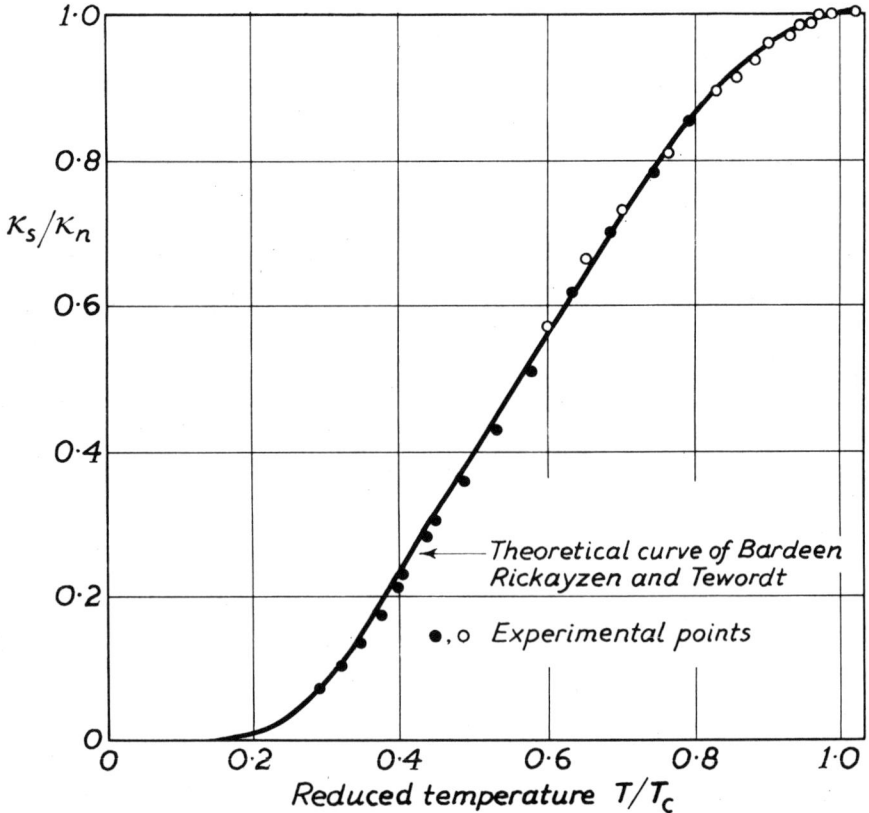

FIG. 5.10. The theoretical function for κ_s/κ_n from the work of Bardeen, Rickayzen, and Tewordt (1959). The points are from the experiments on aluminium by Satterthwaite (1962). The excellent agreement with the theory is to some extent fortuitous (see Satterthwaite).

conductivity when, as in Fig. 5.8, it is limited by lattice scattering (see latter part of section 6.35).

So far we have assumed that the conductivity in the superconducting state is entirely electronic, and at temperatures just below T_c this is probably true. If, however, as the electrons fall into the superconducting state they cannot be scattered by the phonons, then conversely, it seems reasonable to assume that the phonons themselves will not be scattered by these electrons (Hulm, 1950). As the temperature is reduced, more and more electrons leave the normal state and become superconducting (according to the Gorter–Casimir theory the fraction of normal electrons is $(T/T_c)^4$), and hence the lattice waves are no longer scattered so much. This means that the lattice conductivity will start to increase appreciably below, say, $0 \cdot 3T_c$, and at lower temperatures one might expect the conductivity to be almost entirely due to phonons. To check this, experiments must be made in the rather difficult temperature region below 1° K. This was, until recently, an unknown region for thermal-conductivity measurements, and new thermometric techniques had to be devised before satisfactory results were obtained. The earliest experiments were made by Heer and Daunt (1949) and later by Goodman (1953), but more precise work has since been published by Olsen, Renton, and Mendelssohn (1952, 1955), Laredo (1955), and Graham (1958). These workers show that at the lowest temperatures the thermal conductivity in the superconducting state is proportional to T^3 (Fig. 5.11). This is the same temperature dependence as is obtained for the heat conduction of non-metals when the phonons are scattered by the boundaries of the specimen or the crystallites (section 3.7). It will be recalled that in the case of non-metals very good agreement could be obtained between the smallest dimension of the specimen and the mean free path, l, of the phonons, when it was calculated by putting the measured conductivity into the kinetic formula (3.4). In the case of superconductors, however, the agreement between l and the diameter of the specimen is not quite so satisfactory, although there is no doubt that boundary scattering is operative. This is demonstrated in Fig. 5.11 for superconducting tin. The upper curve is for a specimen with a good surface as cast in a glass tube, whilst the lower curve is for the same specimen after its surface had been etched. The decrease in conductivity is due to the greater amount of diffuse scattering of phonons at the etched surface and it confirms the importance of boundary scattering. Nevertheless, the specimen diameter was $2 \cdot 82$ mm whereas from the experimental values of κ_s, l was $2 \cdot 26$ and $1 \cdot 32$ mm before and after the etching, respectively. Thus in both cases there is some other type of scattering in addition to that produced by the specimen boundary. All workers find similar discrepancies between l and the diameter of the specimen.

5.14. *The thermal conductivity of superconductors in the intermediate state*

When a superconductor is placed in a gradually increasing magnetic field, the field suddenly starts to penetrate the specimen, at a strength dependent on the metal being used, the shape of the specimen, and the direction of the field. The electrical resistance also starts to increase. When the field reaches a critical value, H_c, all the metal becomes normal

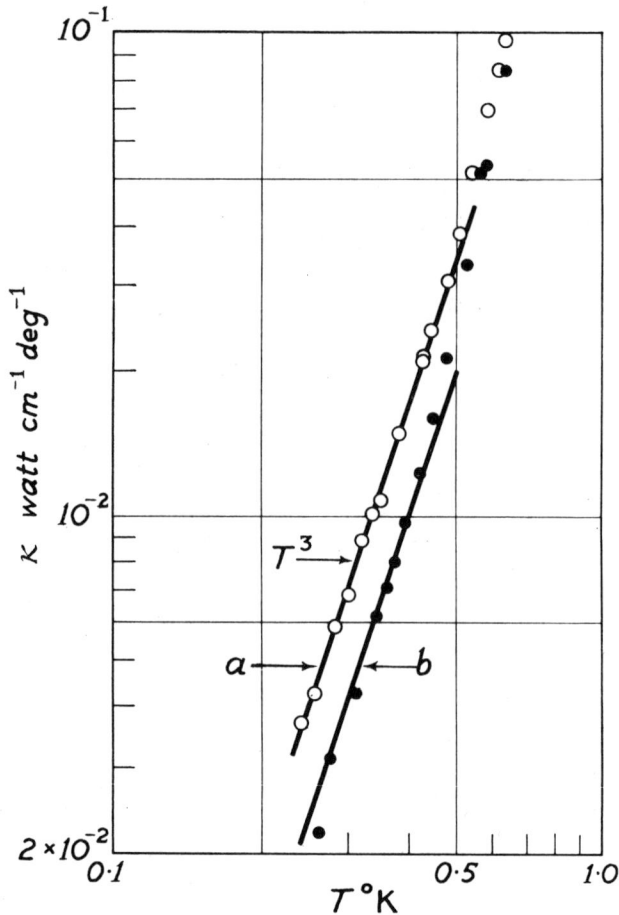

FIG. 5.11. The thermal conductivity of tin in the superconducting state at temperatures below 1° K. The line corresponds to a T^3 dependence. (a) refers to the specimen as cast whilst (b) is for the same specimen after the surface had been etched, thereby demonstrating the effect of diffuse boundary scattering (Graham, 1958).

(H_c depends on the metal and the temperature). The electrical resistance is completely restored and the electronic properties are those of an ordinary metal. This is described in more detail in section 6.17. The range between the point of initial penetration of the field and H_c is called the intermediate state. If now the field is once more reduced to zero, the specimen does not always return to its original superconducting state although its electrical resistance again drops to zero. Magnetic-susceptibility measurements, however, quite often show that some parts of the specimen are still normal in zero field and there remains a so-called frozen-in flux. It will be seen at once that this type of behaviour is very suitable for investigation by thermal-conductivity measurements, since it should be possible to get an idea of the amount of metal which, trapped by this frozen-in flux, remains normal. Accordingly, many experiments have been made on these lines. The thermal conductivity has been measured at various values of the field as it has been increased up to H_c and then decreased to zero once more. Usually two sets of measurements are taken, in fields which are transverse and parallel to the specimen axis. It is unfortunate that such experiments are not so easy to interpret as might have, at first, been supposed, although at higher temperatures the results are what might have been expected from the susceptibility experiments. In general the conductivity as the field is increased is not exactly the same as when it is decreased. Some results for a lead-bismuth alloy at 5·4° K are shown in Fig. 5.12 (a). At the point where the magnetic field begins to penetrate the specimen (for a cylindrical sample in a transverse field this should occur at $\frac{1}{2}H_c$) the thermal conductivity starts to increase as parts of the specimen become normal, and with increasing field it gradually rises until, at H_c, the full normal conductivity is attained. As the field is reduced this curve is more or less retraced, although there might be some hysteresis. In a longitudinal field (Fig. 5.12 (b)) the field penetration should occur at H_c, in a much more discontinuous fashion and this is borne out by the heat-conduction results.

At lower temperatures, particularly with superconducting alloys, much more complicated behaviour has been observed (Fig. 5.13). As the transverse field first penetrates the specimen the conductivity drops, passes through a minimum, and then rises to its normal value of H_c. On reducing the field this behaviour is not repeated—no minimum is observed and the conductivity in zero field is usually lower than it had been in the original superconducting state. Such behaviour was first shown in niobium and lead-bismuth alloys by Mendelssohn and Olsen (1950), but it has since been found in pure metals, e.g. tin, lead, and indium. The minimum in increasing field is probably due to *lattice* conduction which is reduced because the lattice waves are being scattered within the normal regions (Laredo and Pippard, 1955). There is also the possibility of a scattering of either electrons or phonons at the actual boundaries of the normal/superconducting phases, but this would appear to be a very much smaller effect.

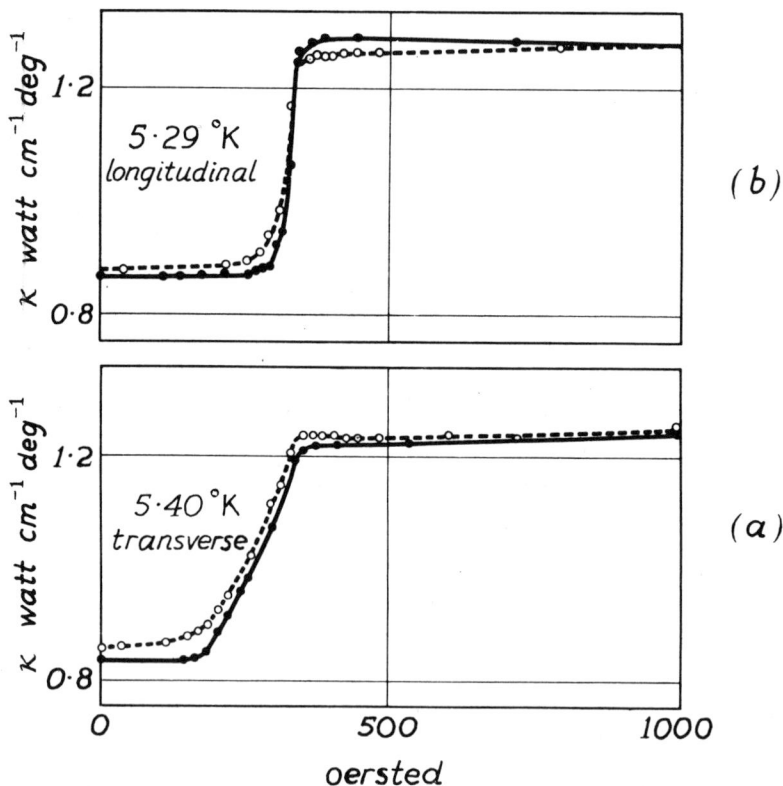

FIG. 5.12. The change in thermal conductivity of a lead-bismuth alloy at ∼5·4° K when superconductivity is destroyed by a magnetic field. Full line—field increasing; broken line—field decreasing (Mendelssohn and Olsen, 1950).

5.15. *The thermal switch*

The small values of κ_s/κ_n which are attained at very low temperatures have been utilized to make a thermal contact which can be made or broken without direct mechanical contact being required. This switch operates when a magnetic field which is greater than H_c is applied to or removed from a superconductor. Since the most effective switch will be that for which κ_s/κ_n is as low as possible, it is an advantage to work at a temperature T such that T/T_c is very small. For this reason Pb, which has the relatively high value of $T_c = 7·2°$ K, has always been used. Pb also has the advantages that H_c is not very high (it approaches ∼ 800 oersted at 0° K), and it can be made into thin foils very easily; as we have seen, this helps to reduce κ_s. Such a thermal contact is particularly useful in the region attained by adiabatic demagnetization (below $\sim 0.1°$ K) where the vibrations which would be transmitted to the specimen by operating a mechanical contact would be sufficient to cause considerable

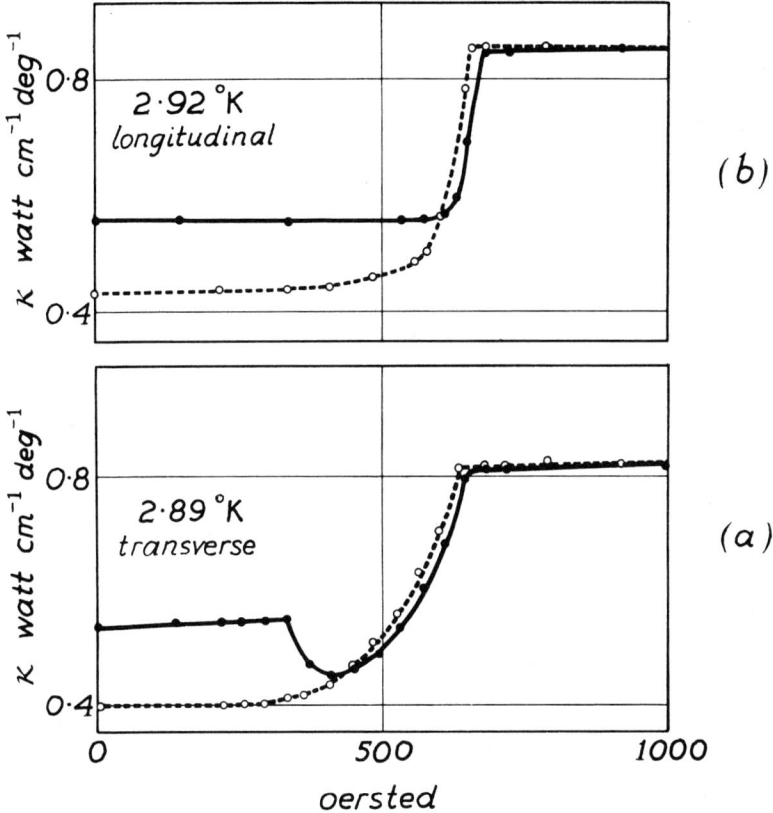

FIG. 5.13. The change in the thermal conductivity of a lead-bismuth alloy at $\sim 2.9°$K when superconductivity is destroyed by a magnetic field, showing the minimum which appears as the transverse field first penetrates the specimen. Full line—field increasing; broken line—field decreasing (Mendelssohn and Olsen, 1950).

heating in the specimen. It is just in this very low-temperature range that κ_s/κ_n will be exceedingly small and where the superconducting switch will be most effective. These devices have been used successfully in several experiments.

The first was used in the two-stage demagnetization apparatus of Darby, Hatton, Rollin, Seymour, and Silsbee (1951). In this experiment a paramagnetic salt which had been demagnetized to $0.25°$ K was used to cool a second salt via a Pb contact. This second salt could be demagnetized further, after it had been isolated by making the Pb super-conducting. By starting the demagnetization of the second salt from $0.25°$ K it was possible to reach approximately 10^{-3} °K with the relatively small magnetic field of 9,000 oersted. The switch was so efficient that the heat influx to the salt was only one erg/min and the temperature was kept below 10^{-2} °K for 40 min.

5.16. The study of imperfections in metals by thermal conduction experiments

The study of imperfections in metals has today achieved very great importance. In theories of work-hardening, information concerning the multiplication of dislocations and point defects during deformation is necessary in order that mechanisms of plastic deformation can be formulated. In studies of irradiation damage it is obviously essential that we should know as much as possible about the type of damage that is introduced. As we have seen in Chapter 4, both of these fields can be investigated by means of electrical-resistivity measurements, but the disadvantage of such experiments is that, for a certain resistivity change, it is very difficult to separate out the contributions which are due to specific types of damage. In particular there is considerable uncertainty in separating the effects of dislocations from those of point defects and in estimating the actual numbers of each defect which are present, particularly since there is still some doubt about the relative importance of dislocation and stacking fault scattering (section 4.13).

In order to obtain additional information, heat-conductivity measurements can be very useful. We should emphasize that these have not been concerned with the electronic thermal conductivity, κ_e, which will, of course, imitate the behaviour of the electrical resistivity and will be liable to the same uncertainties of interpretation. It is from the measurement of the lattice conductivity, κ_g, that additional information can be obtained. There are two reasons for this. Firstly, as has been shown in sections 3.9 and 3.8 respectively, the thermal resistance due to the scattering of phonons by dislocations has a T^{-2} dependence whereas that for the scattering by point defects is proportional to $T^{\frac{3}{2}}$. This means that any extra thermal resistance which might be detected at low temperatures will tend to be due to dislocation scattering. Secondly, the calculations for the scattering of phonons by defects can be made with far more confidence than can those for the scattering of electrons because the detailed configuration, for example at the core of a dislocation, is not required since at low temperatures only long phonon wavelengths are involved. Thirdly, one would not expect stacking faults to scatter phonons as much as they scatter electrons (section 4.13) since their lattice disarrangement is only of atomic dimensions, i.e. much smaller than a phonon wavelength at low temperatures.

The problem which remains is to measure the lattice conductivity of a metal. There are two possible methods; either the conductivity of an alloy can be measured and then κ_g can be estimated as in section 5.10, or the conductivity κ_s of a pure or impure metal in the superconducting state can be determined at very low temperatures. As has been described in section 5.13, κ_s is then due entirely to phonon conduction. We shall describe briefly some experiments which have relied on each of these methods.

5.17. *Experiments on alloys*

The method of determining κ_g from the total heat conductivity has been described in section 5.10, where it was assumed that at low temperatures κ_g was proportional to T^2. Now this dependence can be produced by the scattering of phonons both by electrons or by dislocations (5.12) and (3.22). This was confirmed in some experiments on copper-10 per cent nickel and on silver-cadmium and silver-palladium alloys which showed that at low temperatures κ_g was proportional to T^2 both when the specimens had been cold-drawn and also after they had been annealed, although in the latter case the absolute value of κ_g was appreciably higher. The smaller conductivity in the cold-drawn state was ascribed by Klemens (1955) to scattering by the dislocations which were introduced during the deformation. Similar experiments (Kemp, Klemens, and Tainsh, 1959) on copper-zinc alloys gave similar results, although in all these experiments the actual dislocation density due to the cold-drawing which was estimated from the decrease in conductivity appeared to be higher than would have been expected from other types of investigation.

A later series of experiments was made on copper-zinc alloys (Lomer and Rosenberg, 1959) in which both single and polycrystals containing 7, 15, and 30 per cent zinc were extended in stages by small amounts, the heat conduction in the range 2 to $4\cdot2°$ K being measured after each successive deformation. After the specimens broke they were deformed further by drawing one of the broken halves through dies, the conductivity again being measured at various stages. κ_g was found to decrease after each deformation, although it remained proportional to T^2, thereby showing that the additional low-temperature thermal resistivity was indeed due to dislocations. Fig. 5.14 illustrates the type of change which is observed. It shows plots of κ/T against T, the slope of which should be a measure of κ_g (5.14), for an annealed specimen and then after successive deformations. The decrease in slope (i.e. of κ_g) is quite marked. From these changes in slope the dislocation density may be calculated

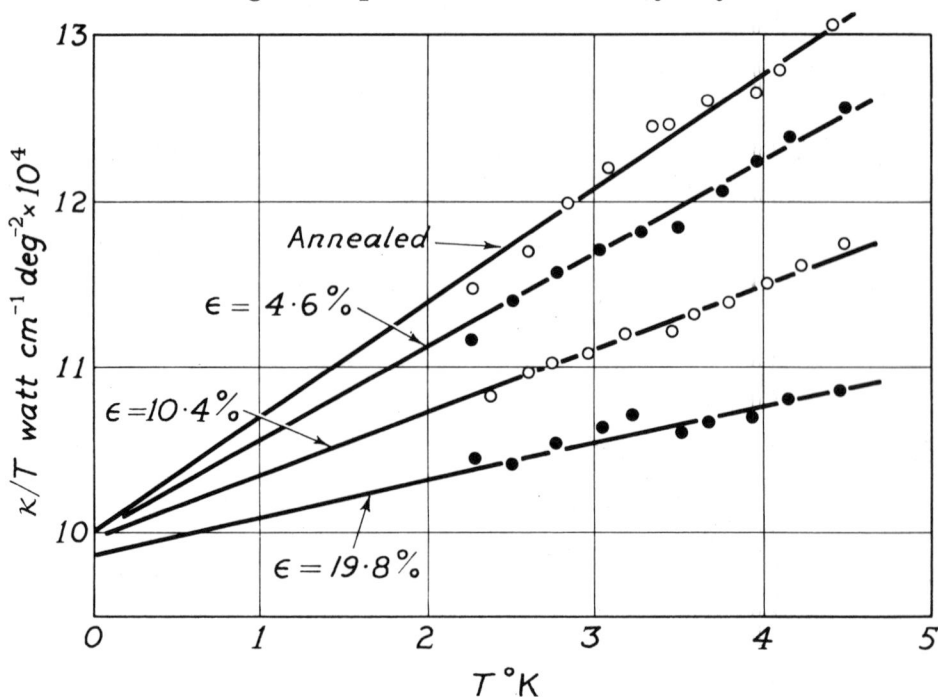

FIG. 5.14. Typical curves of κ/T against T for a specimen of Cu-Zn (15 per cent), after various amounts of strain, ϵ, showing the decrease in slope (i.e. of κ_g) due to the extra dislocation scattering (Lomer and Rosenberg, 1959).

with the aid of (3.23). The results of these experiments are shown in Figs. 5.15 and 5.16. For small to moderate extensions (Fig. 5.15) the dislocation density is independent of the zinc content, but it rises rapidly for polycrystals and very much less rapidly for single crystals oriented for single slip. For larger extensions (Fig. 5.16) the dislocations stop multiplying at a value which is very strongly dependent on the zinc content. It would be out of place in this book to give a detailed interpretation of these results but they are in good qualitative agreement with modern theories of work-hardening. It is clear that heat-conductivity measurements can provide a very useful means of detecting the multiplication of dislocations in a crystal.

It has already been noted earlier in this section that the numerical

value for the dislocation densities when determined by thermal conductivity measurements was higher than that which could be deduced
from other experiments (e.g. X-ray data). Some experiments were made
to check this (Lomer and Rosenberg, 1959) by an independent counting
method. The heat conduction of thin brass strips was measured before

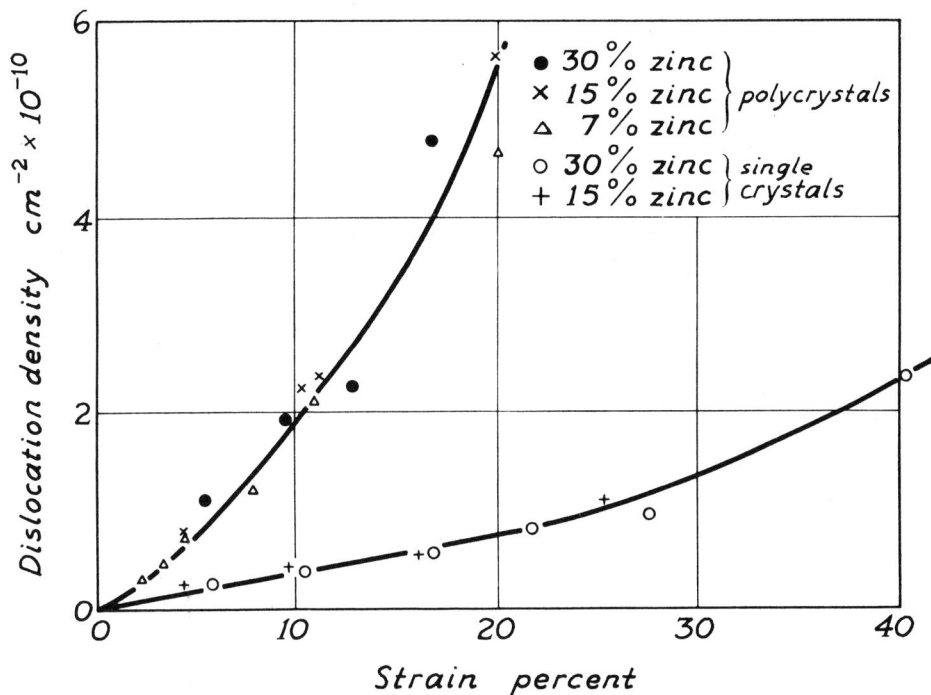

FIG. 5.15. The dependence of the dislocation density (as determined from heat conduction measurements) on the strain, for small and moderate strains on Cu-Zn specimens.
It will be noted that there is little difference in behaviour between specimens of different
zinc content (Lomer and Rosenberg, 1959).

and after deformation and the increase in the dislocation density which
was introduced was calculated by the decrease in κ_g. The strips were
then electro-polished until they were very thin and they were then
examined by transmission electron microscopy (Hirsch, Horne, and
Whelan, 1956). This technique shows up the dislocations directly and
they can be counted. It was found that (3.23) overestimated the dislocation density by a factor of about 6. For this reason the ordinates
in Figs. 5.15 and 5.16 have been reduced by this factor so that they are
more in accord with the actual number of dislocations which are present.
This does not mean that these numerical values are now accurate,

because the electron-microscope technique only samples a very small part of the specimen and this might not be typical of the whole. Other independent experiments, however, also suggest that (3.23) overestimates the dislocation density and it seems probable that the amended values are a good approximation.

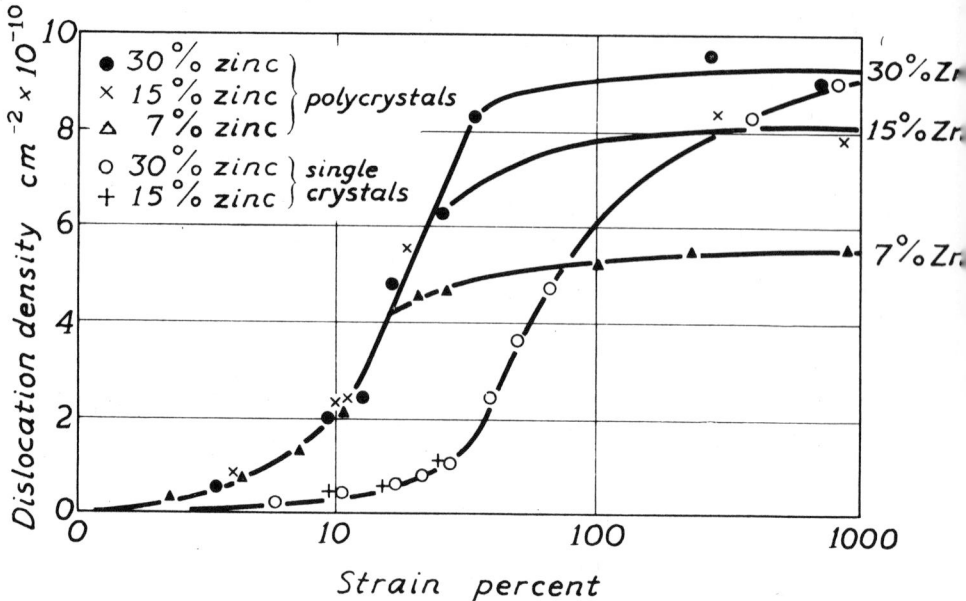

FIG. 5.16. The dependence of the dislocation density (as determined from heat conduction measurements) for large strains, on Cu-Zn specimens. It will be noted that the dislocation density at which the curves flatten off is dependent on the zinc content (Lomer and Rosenberg, 1959).

5.18. Experiments on superconductors

We have already seen (section 5.13) that the thermal conductivity in the superconducting state at temperatures well below the transition temperature is due almost entirely to lattice conduction. If we wish to study the effects of imperfections on κ_g it is therefore not necessary to make any special analysis of the results in order to separate κ_e and κ_g, since κ_e is approximately zero. There are disadvantages in this type of investigation. One is limited to superconductors and these do not in general have the simple structure which one wishes to study in the already complicated world of imperfections; but of course the experiments on alloys described in the previous section are very often no better in this respect. An advantage of experiments on superconductors is that by applying a field greater than H_c, the heat conduction in the normal

state can be measured and used to detect the point defects in the specimen. Fig. 5.17 shows some results by Rowell (1960) on the deformation
of lead at liquid-helium temperatures. The decrease in κ_s is very marked.

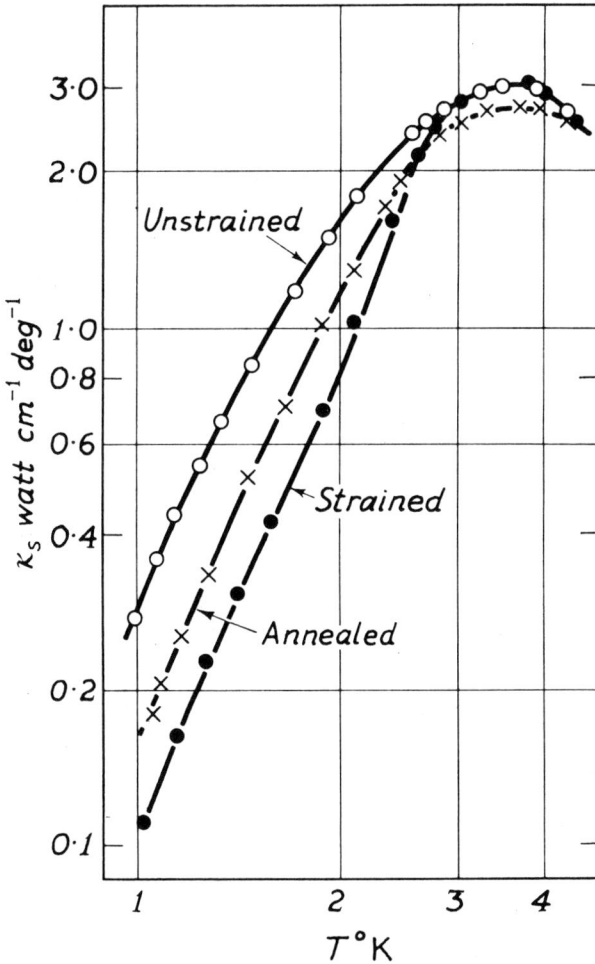

FIG. 5.17. The effect of deformation and annealing on the thermal conductivity of lead
in the superconducting state (P. M. Rowell, 1960).

Experiments have also been made on superconductors which have
been irradiated with neutrons in a reactor. Some work on niobium is
shown in Fig. 5.18, where it is seen that both κ_n and κ_s are reduced after
irradiation. The decrease in κ_n has been ascribed to the production of
vacancies and it is suggested that the decrease in κ_s might be due to
small dislocation loops which are able to form from the general debris

of irradiation damage. A review of this type of work is given by Mendelssohn and Rosenberg (1961).

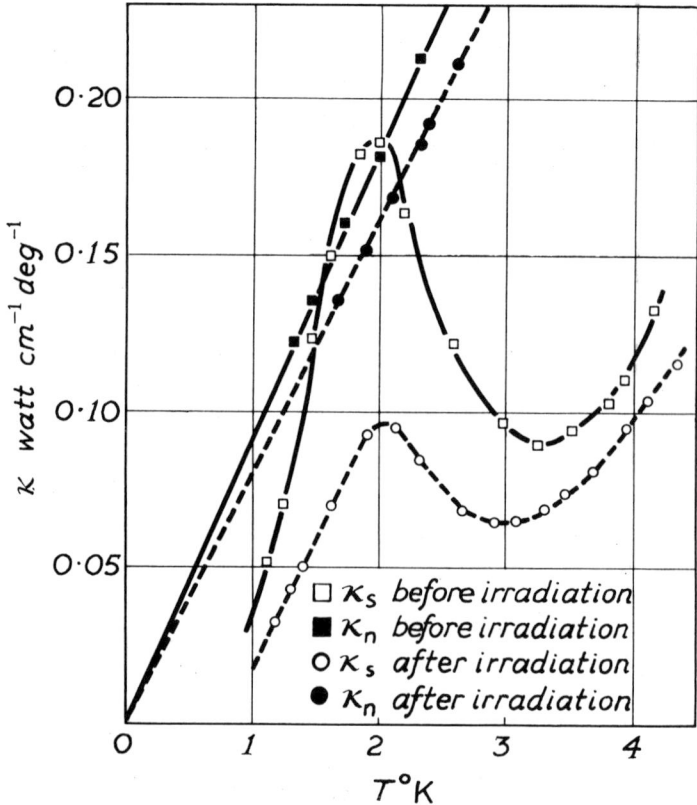

FIG. 5.18. The effect of neutron irradiation (10^{18} fast neutrons cm^{-2}) on the heat conductivity of a niobium single crystal in the normal and the superconducting states
(Chaudhuri, Mendelssohn, and Thompson, 1960).

5.19. The thermal conductivity of semi-metals

Some elements, such as bismuth and antimony, which have very few
conduction electrons, still have metallic properties. Nevertheless, this
small number of electrons does not produce much phonon scattering
and hence the lattice conductivity is not reduced to the degree that it
is in a metal. Thus even in a pure sample κ_g plays an important role
in the heat transport. Fig. 5.19 shows κ_g and κ_e for antimony in which
κ_e was estimated by measuring the electrical resistivity and then assuming the validity of the Wiedemann–Franz law (5.6). It will be seen that
by far the larger part of the conductivity is due to the phonons. For

temperatures up to about 6° K, κ_g is approximately proportional to T^2 which, as we have already seen (section 5.9), is the behaviour which one would expect if the main resistive process was the scattering of phonons by electrons. At higher temperatures κ_g decreases due to the scattering of the phonons by point defects (section 3.8). For bismuth the heat conductivity below liquid-air temperature is due almost entirely to the phonons. In the liquid-helium range κ is approximately proportional

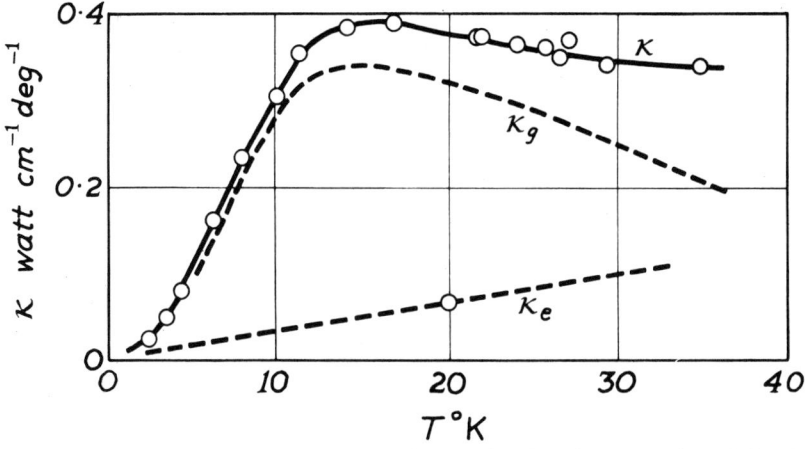

FIG. 5.19. The thermal conductivity of antimony showing the separation of the total conductivity into the electronic contribution κ_e, and the lattice conductivity κ_g (Rosenberg, 1955).

to T^3, which is typical of the conductivity of a non-metal in the boundary-scattering range (section 3.7).

5.20. *The thermal conductivity of semiconductors*

Experiments on single crystals of high-purity germanium and silicon show that phonons are responsible for all the heat conduction. The results which are obtained (e.g. the top curve of Fig. 5.20) are of the same type as those for good single crystals of dielectric substances. At low temperatures there is the T^3 boundary-scattering region, and above the conductivity maximum the conductivity is proportional to $\sim T^{-1\cdot5}$, which in view of the very high purity of the samples must be ascribed to isotope scattering (section 3.8). Experiments on germanium which is enriched in one isotope (Geballe and G. W. Hull, 1958) do indeed show a higher conductivity and a more rapid temperature dependence than $T^{-1\cdot5}$, indicating that the isotope scattering has been reduced (see Fig. 3.10).

The thermal conductivity of germanium samples which have been doped with p-type impurity (e.g. gallium) is not so easy to interpret. Some results are shown in Fig. 5.20, where it will be noted that not only is the conductivity reduced by the doping (which is not surprising) but that for a sample, Ge 7, containing $2\cdot3\times10^{16}$ carriers cm^{-3} at room

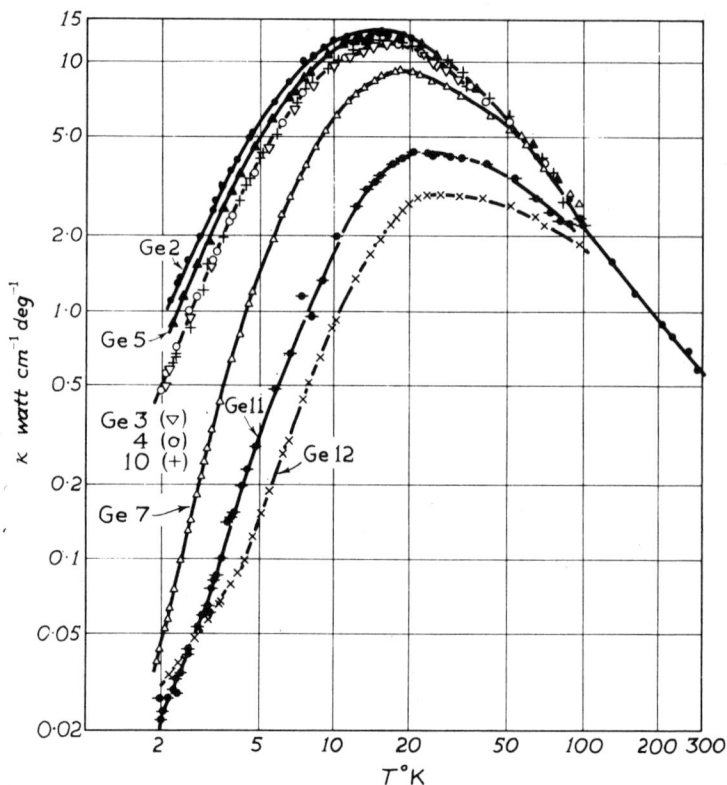

FIG. 5.20. The thermal conductivity of germanium single crystals plotted on a logarithmic scale. Ge 2 and Ge 5 are high-purity n-type samples. The others contain the following densities of p-type impurity: Ge 3, 10^{14}; Ge 4, $1\cdot9\times10^{14}$; Ge 10, 10^{15}; Ge 7, $2\cdot3\times10^{16}$; Ge 11, 2×10^{18}; Ge 12, 10^{19} cm^{-3} (Carruthers, Geballe, Rosenberg, and Ziman, 1957).

temperature κ was proportional to T^4 at low temperatures. This in terms of the kinetic theory (3.4) means that the mean free path of the phonons was *decreasing* as the temperature was reduced which is a rather unusual state of affairs. Ziman (1956) has suggested that this behaviour is due to the scattering of the phonons by an electron band which is non-degenerate. The density of occupied states at the top of this band will then vary with temperature and might give the effects which are observed. For very large amounts of doping (e.g. Ge 12) there are always a large number of current carriers present and the conductivity of germanium tends to become proportional to T^2—the temperature dependence which is due to the scattering of phonons by electrons (section 5.9).

5.21. Other thermal conductivity experiments

For completeness we mention some other types of experiments which have been made which involve the heat conductivity, but we give no

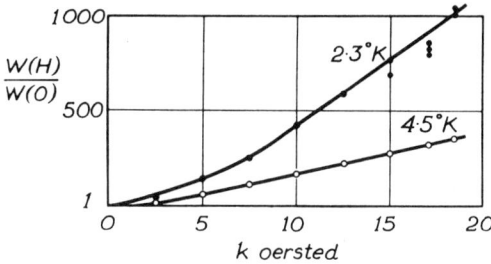

FIG. 5.21. The variation of thermal resistance $W(H)$ in a transverse magnetic field H for a cadmium single crystal. $W(0)$ is the thermal resistance in zero field (Mendelssohn and Rosenberg, 1953).

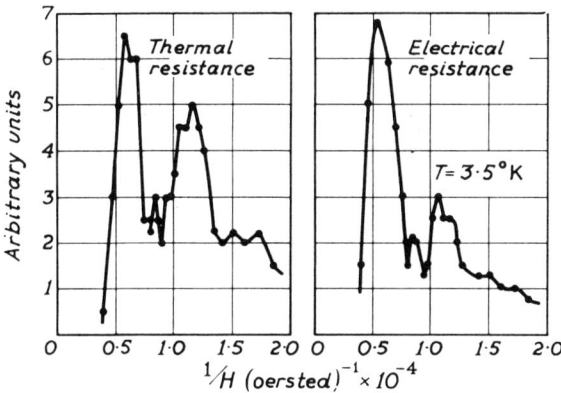

FIG. 5.22. The oscillatory nature of the thermal and the electrical resistance of zinc in high magnetic fields. The extra anomalous resistance is plotted against $1/H$ (Alers, 1956).

details because whilst the results are interesting, no satisfactory interpretation has been made in view of the theoretical complexity.

These experiments have been concerned with observing whether anomalies which have been detected in the electrical conductivity are reflected in the thermal conductivity. In general, as might be expected, similar effects are found. Thus the extreme anisotropy of gallium is shown in the thermal as well as in the electrical conductivity (Rosenberg, 1955). Several workers have found deviations in the $W_e T \sim T^3$ plot (5.11) which are associated with the minimum in the electrical resistivity which is found in some metals at low temperatures (section 4.15). Others have measured the decrease in the thermal conductivity when a magnetic field is applied to a specimen, which is analogous to the ordinary magneto-resistive effect (section 4.27). In some cases, e.g. cadmium, this change in thermal conductivity can be very large. As Fig. 5.21 shows, a decrease by a factor of over 1000 is obtained in a field of 18 k oersted (since only κ_e and not κ_g is reduced by a magnetic field this type of measurement can be used to provide an upper estimate of the lattice conductivity). In very high fields, oscillatory effects akin to the de Haas–van Alphen effect (Chapter 10) have been observed (Fig. 5.22). For a review of these experiments which are all complicated functions of the convolutions of the Fermi surface, the reader is referred to Rosenberg (1958).

6

SUPERCONDUCTIVITY

6.1. Introduction

THE discovery by Onnes (1911) that the electrical resistivity of mercury fell practically discontinuously to zero at a temperature of about $4 \cdot 2°$ K has since unleashed a flood of experimental results and theoretical speculation which would require a large volume rather than a modest chapter to do them justice. We shall therefore content ourselves with a brief description of the important features of superconductivity together with an elementary phenomenological treatment of the effect and an introduction to the current theory. In conclusion we shall describe some of the technological uses of superconductivity. For more detail, particularly of the earlier work up to about 1951, the excellent monograph by Shoenberg (1952) is recommended. Thereafter there are reviews by Bardeen (1956) and Serin (1956) on the theoretical and the experimental aspects respectively, by Bardeen and Schrieffer (1961) which deals particularly with the many elegant experiments which have given confirmation to the modern theory, and a recent text by Lynton (1962).

6.2. The basic properties of superconductors

The most outstanding property of a superconductor is, of course, the complete disappearance of the electrical resistivity at some low temperature, T_c, which is characteristic of the material. So far twenty-four elements have been shown to be superconductors and they, together with T_c, are outlined in the periodic table shown in Fig. 6.1. Many alloys and compounds have also been found to be superconducting, but we delay a discussion of their properties until section 6.23. No clue as to the explanation of the phenomenon can be obtained from a study of this figure, although the metals which one expects to have a simple electronic structure, such as the alkalis and the noble metals do not appear to become superconducting. Elements representing most types of crystal structure can be superconductors. The only correlation appears to be that the atomic volumes of the superconducting elements are not very different from one another (Clusius, 1932). The highest value of T_c for an element is about $8°$ K for niobium, followed closely by lead with

$T_c = 7.22°$ K. The lowest value of T_c so far found is that for iridium at $0.14°$ K (Hein, Gibson, Matthias, Geballe, and Corenzwit, 1962).

The actual value of T_c depends very little on the particular sample of the metal which is measured, although the width of the temperature range over which the resistance drops to zero is very dependent on its physical and chemical state. The more perfect the sample, the sharper

Period	I	II	III	IV	V	VI	VII	VIII		
1	1 H							2 He		
2	3 Li	4 Be	5 B	6 C	7 N	8 O	9 F	10 Ne		
3	11 Na	12 Mg	13 Al 1.197°	14 Si	15 P	16 S	17 Cl	18 A		
4	19 K	20 Ca	21 Sc	22 Ti 0.39°	23 V 4.89°	24 Cr	25 Mn	26 Fe	27 Co	28 Ni
	29 Cu	30 Zn 0.93°	31 Ga 1.10°	32 Ge	33 As	34 Se	35 Br	36 Kr		
5	37 Rb	38 Sr	39 Y	40 Zr 0.55°	41 Nb 8.9°	42 Mo ~0.95	43 Tc 11.2°	44 Ru 0.47°	45 Rh 0.9°	46 Pd
	47 Ag	48 Cd 0.56°	49 In 3.40°	50 Sn 3.74°	51 Sb	52 Te	53 I	54 Xe		
6	55 Cs	56 Ba	57 La 48.58	72 Hf 0.37°	73 Ta 4.38°	74 W	75 Re 1.70°	76 Os 0.71°	77 Ir ~0.1°	78 Pt
	79 Au	80 Hg 4.16°	81 Tl 2.39°	82 Pb 7.22°	83 Bi	84 Po	85 At	86 Rn		
7	87 Fr	88 Ra	89 Ac	90 Th 1.37°	91 Pa	92 U 1.1°	93 Np	94 Pu	95 Am	96 Cm

FIG. 6.1. The periodic table showing the superconducting elements. The transition temperature for each superconductor is shown in the lower part of the box. (After *Heat and Thermodynamics*, 4th ed., by M. W. Zemansky, 1957, McGraw-Hill Book Company. Used by permission.)

is the transition. Thus in a high-purity single crystal of tin, the transition region is 10^{-3} degree. Specimens which are impure or are in a strained condition may have a transition spread over 10^{-1} degree or possibly more (Fig. 6.2).

The question is always raised as to whether the resistance of a super-conductor is really zero, or whether it is just very small. This has been effectively answered by several series of experiments in which a current has been started in a closed superconducting loop by magnetic induction. The presence of this current and its strength can be detected, for example, by bringing a small suspended magnet close to the apparatus. It is found that as long as the superconducting ring is kept below T_c the induced current persists and within the limits of measurement it does not diminish in value. Since in one apparatus a persistent current was made to run for over a year (Collins, 1955) the upper limit to the possible value of the resistance is exceedingly small and there is little doubt that it is in fact really zero.

6.3. *The critical field*

The other fundamental property of a superconductor, besides its zero resistance, is that its ordinary resistance may be restored on the application to the specimen of a magnetic field greater than a critical value H_c.

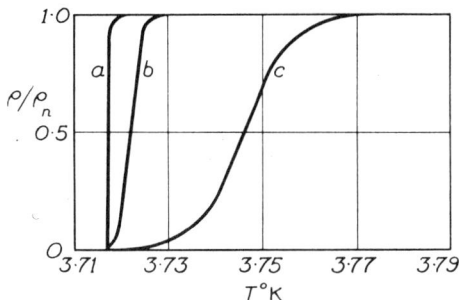

FIG. 6.2. The superconducting transition for tin showing the influence of the specimen quality on the sharpness of the transition to zero resistivity. (*a*) High-purity single crystal; (*b*) high-purity polycrystal; (*c*) less pure polycrystal. ρ_n is the resistance in the normal state (de Haas and Voogd, 1931).

H_c depends both on the material and on the temperature. It is zero at T_c and as the temperature is reduced it increases, following approximately a parabolic law of the form

$$H_c = H_0\{1-(T/T_c)^2\}, \tag{6.1}$$

thus tending to a constant value H_0 as $T \rightarrow 0°$ K (Fig. 6.3). The sharpness of the recovery of the normal electrical resistivity as the field is increased depends not only on the purity and perfection of the sample but also on the direction of the field relative to its length. If the field is parallel to the axis of a cylindrical specimen, the transition from the superconducting to the normal (i.e. resistive) state is quite sharp when H_c is reached. If the field is perpendicular to the sample, the recovery of the resistivity is much more gradual. It commences at $\sim \frac{1}{2}H_c$ and the full resistivity is finally attained at H_c (Fig. 6.4). For most of the pure metals H_c is never very high. The lower the transition temperature, the smaller is H_c. This can be seen for the curves shown in Fig. 6.3. If T_c is around or below $1°$ K, H_0 is a few tens of oersted, otherwise it is usually a few hundred oersted. Only in a few cases (e.g. Ta, V, and Nb) are transition

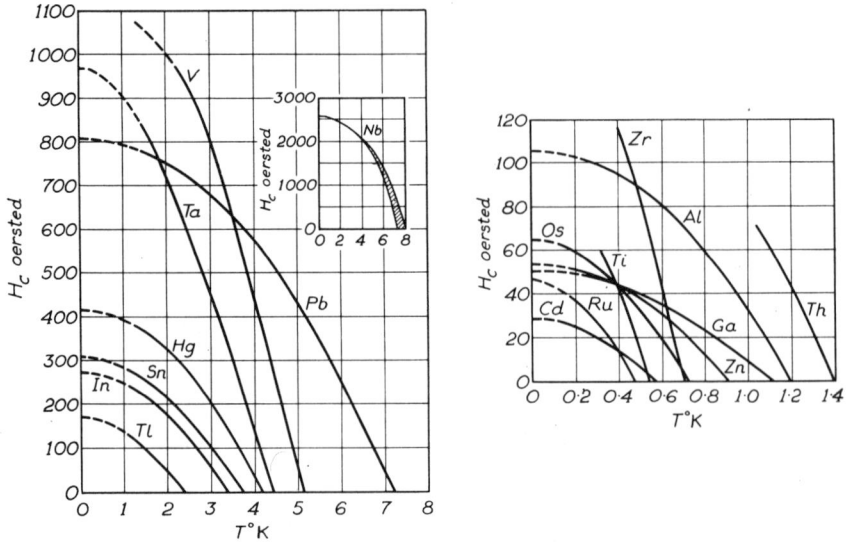

FIG. 6.3. Curves showing the dependence of the transition field, H_c, on the temperature for several superconductors (after Shoenberg, 1952).

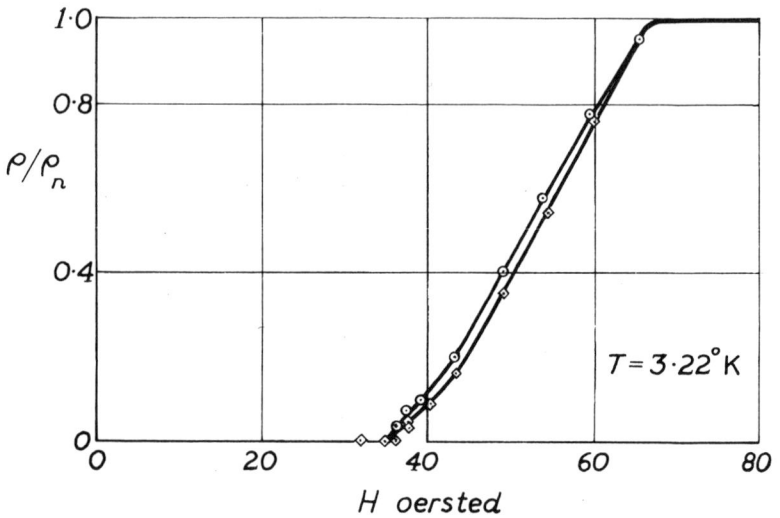

FIG. 6.4. The recovery of the electrical resistance for two tin single crystals when the external field is transverse to the axis of the specimens showing the broadened transition which begins at about $\frac{1}{2}H_c$ (de Haas, Voogd, and Jonker, 1934).

fields found which approach or are greater than 1000 oersted. These metals were at one time thought to form a distinct class known as 'hard' superconductors both from their mechanical properties and their high values of H_c. There is, however, no particular significance in this separation and modern methods of purification and preparation can produce 'softer' materials, both mechanically and with smaller values of H_c. For Nb in particular, H_c can be much higher than the values shown in Fig. 6.3 if the material is not carefully annealed. It should be noted

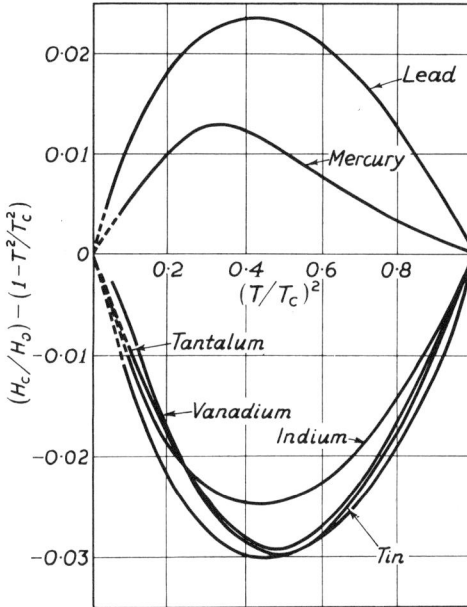

FIG. 6.5. The deviation of the critical magnetic field H_c from the parabolic relation $H_c = H_0\{1-(T/T_c)^2\}$ (Mapother, 1962).

that (6.1) is only an approximate relation. Deviations from the parabolic law are shown in Fig. 6.5.

Superconductivity can also be destroyed if too high an electric current is passed through the specimen. It was suggested by Silsbee (1916) that the transition was caused by the magnetic field produced by the current. In fact some of Onnes's very early work showed that the transition temperature was lowered if the current used for measuring the resistance was increased. This was ascribed to the fact that the material would now appear to be superconducting only if the field produced by the measuring current was less than H_c. Later experiments showed quite accurately that when the field at the surface of the wire due to this current became equal to H_c the resistance started to be restored, although the full normal state resistance was not attained until about double the initial restoration current was passed through the specimen.

6.4. *The Meissner effect*

For over twenty years after the discovery of superconductivity, the phenomenon was thought to be essentially the manifestation of zero resistance. In fact this is only one facet of the behaviour of super-conductors. In a perfect conductor there will, of course, be no electric field, $\mathbf{E}$, and therefore the application of Maxwell's equation

$$-\frac{1}{c}\frac{\partial \mathbf{B}}{\partial t} = \mathrm{curl}\,\mathbf{E} = 0 \tag{6.2}$$

might lead us to suppose that when a metal in an external field is cooled below T_c, the induction in the metal would remain constant at the value $\mathbf{B}$ which it had just at the time when the resistance vanished. On the other hand, a field which was applied after the metal had become superconducting would have been excluded from the body of the sample (until it reached a value $\geqslant H_c$). A series of experiments by Meissner and Ochsenfeld (1933) showed that the first of these predictions was incorrect. They measured the magnetic field around a superconducting tin cylinder and found that not only was the field unable to penetrate the specimen if it was applied at a temperature *below* T_c, but that if the field was first applied *above* T_c and the specimen was then cooled below that temperature, then $\mathbf{B}$ did *not* remain constant as one might have expected from (6.2). Instead, the field was expelled from the sample as soon as it became superconducting (Fig. 6.6), i.e. $\mathbf{B} = 0$ within the sample. Thus, independently of whether the field was applied above or below T_c, it cannot exist within the body of the superconductor. This expulsion of the magnetic field is commonly referred to as the Meissner effect. As we have seen, it cannot be predicted from the classical electromagnetic equations and it must be regarded as another novel property of a superconductor. One can regard the effect as being caused by induced surface currents which are just sufficient to create an opposing field of the correct value so that the resultant field in the sample is zero. If a hollow cylindrical specimen is cooled below T_c whilst in a magnetic field, then the flux within the cylinder is 'trapped' by the superconductor as is shown in Fig. 6.7 (*b*). Conversely if the field is applied whilst the

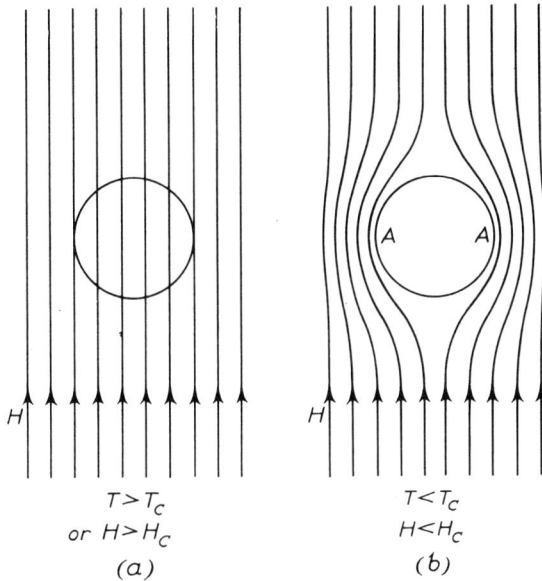

T > T_c
or H > H_c

(a)

T < T_c
H < H_c

(b)

FIG. 6.6. The Meissner effect. The magnetic lines of force around a superconductor (a) in the normal state; (b) in the superconducting state the field is expelled from the superconductor. Note the concentration of the field at the points A, A when the metal is superconducting.

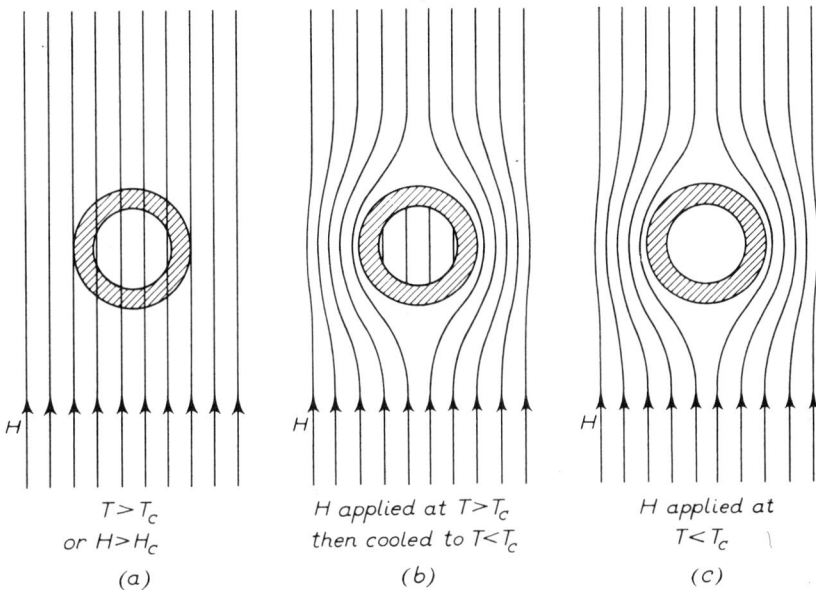

T > T_c
or H > H_c

(a)

H applied at T > T_c
then cooled to T < T_c

(b)

H applied at
T < T_c

(c)

FIG. 6.7. The Meissner effect for a torus or hollow cylinder. (a) In the normal state; (b) in the superconducting state if the field is applied *above* T_c and the metal is then cooled below T_c, showing the flux trapped in the hole; (c) in the superconducting state if the field is applied *after* the metal has been cooled below T_c, showing how the superconductor screens the central hole.

metal is superconducting, the flux cannot enter the cylinder and hence the tube acts as a perfect magnetic shield (Fig. 6.7 (c)).

The distortion of the lines of force around a cylindrical superconductor as is shown in Fig. 6.6 (b) increases the field strength at the points A, A, to a value which is twice the value of H in regions distant from the specimen. It is for this reason, as has already been remarked, that the destruction of superconductivity in a transverse field begins at $\frac{1}{2}H_c$ (for a sphere it is $2H_c/3$)—under such conditions the field at the specimen surface is actually H_c.

Some typical results are shown in Fig. 6.8. In a longitudinal field **B** remains zero until H_c is reached, and it then rises very rapidly to **B** = **H** (Fig. 6.8 (a)). If the field is transverse to the specimen axis, then, as in the case of the electrical resistance, **B** starts to increase from zero at $\frac{1}{2}H_c$ and finally becomes equal to **H** at H_c (Fig. 6.8 (b)).

If these measurements are first taken with H increasing it is very unusual for the initial curve to be retraced when the field is being reduced to zero. Some hysteresis is nearly always found (there is a small amount in Fig. 6.8 (b)). In general, even after the field has been reduced to zero once again, the specimen still has some magnetic induction remaining, i.e. the Meissner effect is not complete. The situation is usually described by saying that some flux is trapped or frozen in the specimen. A complete expulsion of the magnetic field after such an experiment can usually only be observed in carefully annealed, strain-free single crystals. Otherwise, in order to achieve zero induction again it is usually necessary to warm the specimen above T_c and then to re-cool it in the absence of a magnetic field.

The zero (or near zero) magnetic induction of a superconductor may be expressed in another way by saying that a superconductor is a perfect diamagnetic material. Using the relation

$$\mathbf{B} = \mathbf{H'} + 4\pi\mathbf{m}_v, \tag{6.3}$$

where **H'** is now the *actual* field at the surface of the specimen and $\mathbf{m}_v$ is the magnetic moment per unit volume, we see when this is equated to zero that

$$\mathbf{m}_v = -\mathbf{H'}/(4\pi). \tag{6.4}$$

In the modern theoretical approaches to the subject it is generally considered that the perfect diamagnetism is a more fundamental property than infinite conductivity.

From the aesthetic point of view the existence of the Meissner effect was a very satisfying discovery. The fact that, independent of the

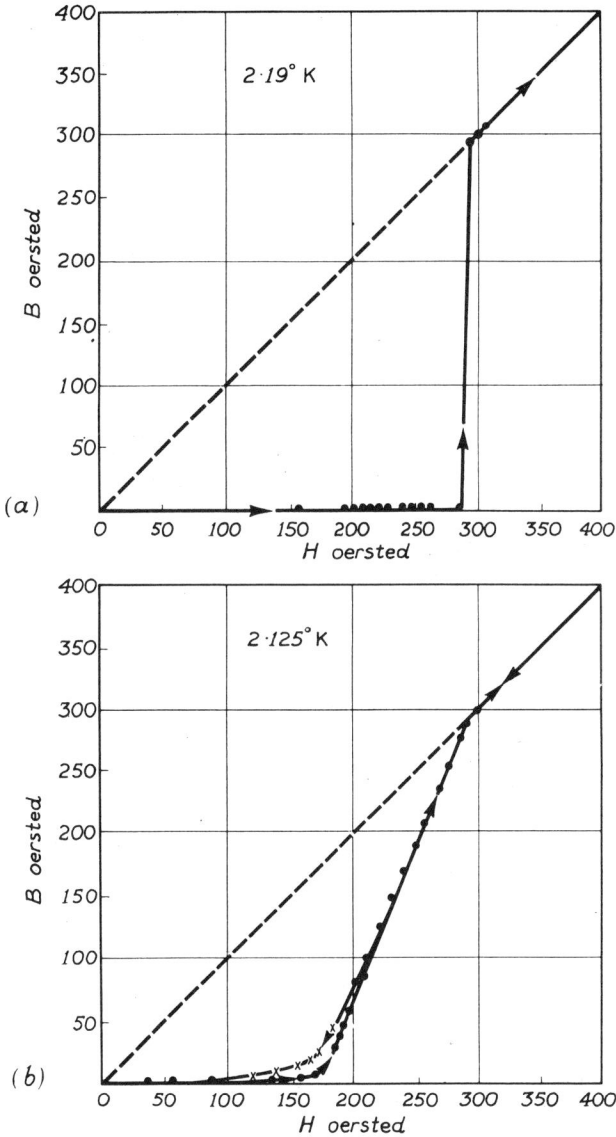

Fig. 6.8. The change in the magnetic induction, B, when a field is applied to a cylindrical mercury specimen when the field is (a) parallel to the specimen axis (after Mendelssohn, 1936), (b) transverse to the specimen axis (after Désirant and Schoenberg, 1948a).

previous history of the specimen, the magnetic induction inside it is zero (apart from the hysteresis effects which we have mentioned), means that the change from the superconducting to the normal state (or vice versa) by the application (or removal) of a field $\geqslant H_c$ is a reversible process. The properties of a perfect conductor might lead one to expect irreversible behaviour because, as we have already pointed out, the induction in the specimen could depend on whether the field was applied when the metal was still in the normal state (above T_c) and was then cooled down, or whether the field was applied after the metal had become superconducting. Since, however, the transition is a reversible one, it can be treated by the ordinary methods of thermodynamics.

6.5. Thermodynamics of a superconductor

The most important results which can be derived from a thermo-dynamical treatment of the transition to superconductivity are those which enable the entropy and specific-heat differences between the normal and the superconducting states to be determined from the temperature variation of the critical magnetic field. A detailed dis-cussion both of these and of other thermodynamical aspects is given by Shoenberg (1952), chapter 3, from which this treatment is taken.

In a reversible change the transition from one phase to another occurs when the Gibbs free energy, G, is the same in each phase. For a substance with magnetic moment $\mathbf{m}_v$ per unit volume, we may write the free energy of the superconducting phase in a field $\mathbf{H}$ as

$$G_s(H) = U - TS + pV - V \int_0^{H_c} \mathbf{m}_v \, d\mathbf{H} \tag{6.5}$$

where the energy of the magnetic field is included in the internal energy U, and the other symbols have their usual meaning. A derivation of this equation is given by Pippard (1957a), p. 133. If we consider a long thin specimen whose axis is parallel to H, then $H = H'$ and, as we have already described, the transition to the normal state occurs practically discontinuously at $H = H_c$. Hence for $H < H_c$ we have perfect dia-magnetism and so we may write

$$\mathbf{m}_v = -\mathbf{H}/(4\pi). \tag{6.6}$$

On integrating, therefore, we obtain from (6.5)

$$G_s(H_c) = G_s(0) + VH_c^2/(8\pi). \tag{6.7}$$

But $G_s(0)$ is the free energy of the superconducting phase in zero field. $G_s(H_c)$ is the free energy at the transition, and so is the same for both the superconducting and the normal states in the field H_c. Hence $G_s(H_c) = G_n(H_c)$, the free energy of the normal phase. Thus writing G_s

and G_n for $G_s(0)$ and $G_n(H_c)$ respectively, we have,† for the free energy difference,

$$G_n - G_s = VH_c^2/(8\pi). \tag{6.8}$$

Whilst we have derived (6.8) for the special case of an axial field, the result is clearly of general application since G_s cannot depend on H.

Since the entropy, S, may be obtained from the relation

$$S = -\partial G/\partial T \tag{6.9}$$

we obtain for the difference in entropies between the normal and the superconducting states

$$S_n - S_s = -\frac{VH_c}{4\pi}\frac{dH_c}{dT}. \tag{6.10}$$

From the curves in Fig. 6.3 we see that dH_c/dT is always negative and hence the entropy of the superconducting state is always lower than that of the normal state, i.e. the superconducting state is one of greater order. We also note that since, according to the third law of thermodynamics, $S_n - S_s$ must vanish at $0°$ K, that dH_c/dT must tend to zero at very low temperatures and a trend towards this behaviour is shown in the H_c curves.

From (6.10) we see that when superconductivity is destroyed by the application of a field $> H_c$, a latent heat Q equal to $T(S_n - S_c)$ is absorbed, which is given by

$$Q = -VT\frac{H_c}{4\pi}\frac{dH_c}{dT}. \tag{6.11}$$

There is, however, no latent heat associated with the onset of superconductivity at the transition temperature (where $H_c = 0$), because dH_c/dT at T_c is still finite.

Thus if superconductivity is destroyed adiabatically by the application of a magnetic field, the specimen will cool down. This effect can be observed and has been used as a means of producing temperatures below $1°$ K (Yaqub, 1960). It is of limited application, however, because the entropy difference between the two states is very small (of the order of $10^{-3}R$ J deg^{-1} mole^{-1}).

6.6. The specific heat

We may find the difference in the specific heats of the two phases by differentiating (6.10) with respect to the temperature and multiplying

---

† This neglects the contribution due to the very weak dia- or paramagnetism of the metal in the normal state.

by T. For unit volume we obtain

$$c_s - c_n = \frac{T}{4\pi}\left\{H_c\frac{d^2H_c}{dT^2} + \left(\frac{dH_c}{dT}\right)^2\right\}. \tag{6.12}$$

Thus in zero field at T_c there is a discontinuity in the specific heat given by

$$c_s - c_n = \frac{T_c}{4\pi}\left(\frac{dH_c}{dT}\right)^2_{T_c} \tag{6.13}$$

an essentially positive quantity (Fig. 6.9 (a)). At lower temperatures, however, the difference in the specific heats will become negative because, as can be seen from the form of the $H_c \sim T$ curves of Fig. 6.3, the first term in the bracket of (6.12), which is negative, will dominate as the H_c curve flattens off at lower temperatures. As T tends to $0°$ K the specific-heat difference will become zero. Fig. 6.9 (b) shows the temperature variation of $(c_s - c_n)$ for tin as calculated from the $H_c \sim T$ curve and also from specific heat measurements. Quite reasonable agreement is obtained between the value of the discontinuity at T_c and that calculated using the H_c data with (6.13).

6.7. *The two-fluid model*

Before discussing the results of the measurements of the specific heat of superconductors in more detail, it is illuminating at this stage to describe in terms of a simple model, why over a large part of the temperature range the specific heat is greater in the superconducting than in the normal state. The model is a very useful one to have in mind when many aspects of the behaviour of superconductors are being considered. In essence we assume that in the superconducting state the assembly of conduction electrons can be considered as being composed of two interpenetrating fluids which have different properties: a fluid of normal or n-electrons which behaves in the same way as the conduction electrons of the normal metal and a fluid of superconducting or super-electrons which has zero entropy and which experiences no resistance to flow. For the moment we offer no explanation as to how these two states arise or what is responsible for the special properties of the super-electrons. We shall return to this in section 6.35.

As the metal is cooled below T_c a certain fraction x of the electrons become super-electrons, the value of x depending on the temperature. Since these electrons have zero resistance they will, in effect, short circuit the n-electron stream and so the specimen as a whole exhibits zero resistance. The further the metal is cooled below T_c the larger does x become and it attains unity at $0°$ K. From the Gorter and Casimir

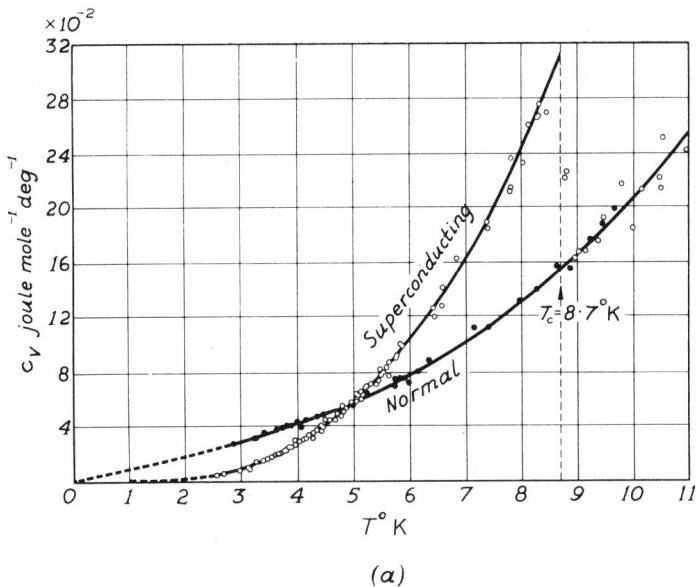

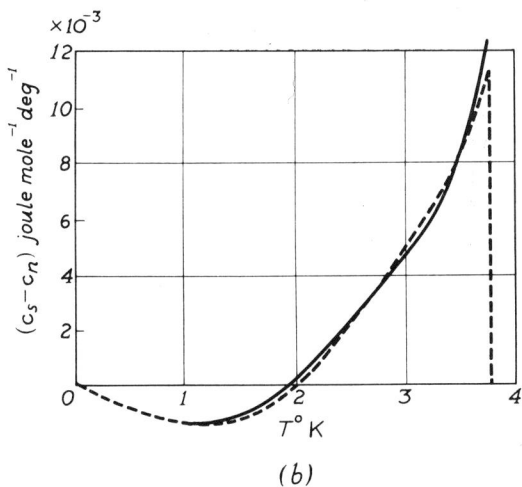

Fig. 6.9. (a) The specific heat of niobium in the normal and the superconducting states, showing the discontinuity in c_s at T_c (Brown, Zemansky, and Boorse, 1953). (b) $(c_s - c_n)$ for tin; full line, from the experimental data of Keesom and van Laer (1938), dotted line, calculated from (6.12) (after Shoenberg, 1952).

theory (1934) the value of x which fitted the experimental specific heat results most satisfactorily was

$$x = 1 - (T/T_c)^4. \qquad (6.14)$$

The explanation of the behaviour of $c_s - c_n$ is then based on the assumption that when the temperature of the superconductor is raised, some heat must be provided to increase the energy of the n-electrons which it already possesses, but extra heat will also be required in order to convert a certain number of super-electrons to n-electrons so that the correct value of the concentration x is maintained as the specimen is warmed. At very low temperatures we see from (6.14) that the change in x is small, but as the temperature is increased the decrease in x is sufficiently rapid that the extra heat required to produce the n-electrons enables c_s to become considerably greater than c_n.

6.8. *The specific heat at very low temperatures*

Until fairly recently it was generally accepted that c_s was proportional to T^3. Since at temperatures of T_c and below, the lattice specific heat by itself also has this temperature dependence (1.13), it follows that the electronic specific heat of a superconductor, c_{es}, should also vary as T^3. Equation (6.14) was indeed selected in order to yield this temperature dependence. If we assume that the lattice specific heat, which we write as βT^3, is the same in both states so that c_s is of the form

$$c_s = \alpha T^3 + \beta T^3 \qquad (6.15)$$

and c_n is given by

$$c_n = \beta T^3 + \gamma T \qquad (6.16)$$

as in (1.26), then the difference in the specific heats is

$$c_s - c_n = \alpha T^3 - \gamma T. \qquad (6.17)$$

If this expression is substituted in (6.12) and the equation is integrated twice to give H_c, then a parabolic law (6.1) is obtained. The calculation is given by Shoenberg (1952), p. 64. There is no doubt that the thermodynamical treatment of the transition is very well justified.

We have, however, already mentioned that the parabolic relation for H_c is only approximate (Fig. 6.5). Corroboration is given by recent experiments on the specific heat which show that, particularly at very low temperatures, c_{es} tends to deviate from the T^3 law and it exhibits an exponential dependence on the temperature. Some results on vanadium are shown plotted on a reduced scale in Fig. 6.10 in which an exponential relation is denoted by the straight dashed line. The T^3 law is given by the continuous curve and it is seen that, while there is little to choose between the two dependences at higher temperatures, for measurements taken below about $T_c/3$ the T^3 law is patently inaccurate. The results can be expressed in the form

$$c_{es}/(\gamma T_c) = a' \exp(-b'T_c/T),\qquad(6.18)$$

where γT is the usual electronic specific heat in the normal state

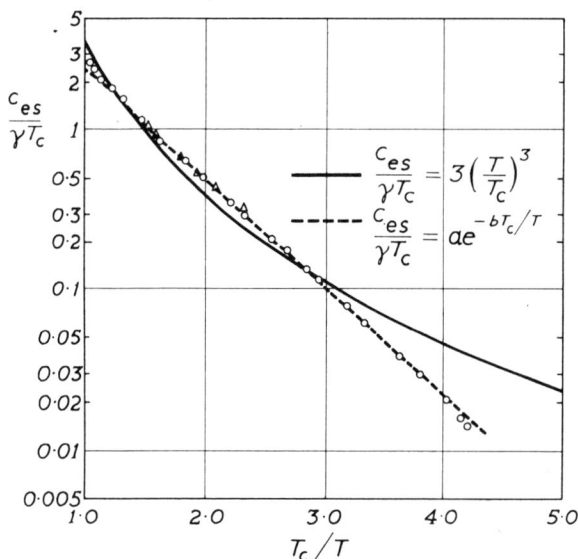

FIG. 6.10. The electronic specific heat c_{es} for vanadium in the superconducting state, showing the disagreement with a T^3 relationship (full line) at very low temperatures. The dotted line gives an exponential dependence and it is seen that this describes the experimental behaviour quite well (after Corak, Goodman, Satterthwaite, and Wexler, 1954).

(1.23). The constants a' and b' tend to differ from one metal to another but they usually have values of about 9 and 1·5 respectively.

A development of the two-fluid model yields this exponential dependence quite naturally. The concept of the super-electrons would imply that they exist in some special states which are separated by a finite energy gap from the n-electron states. If this is so, then the number of n-electrons which are thermally excited will depend exponentially on the temperature and hence this type of dependence would also be expected to be observed in the specific heat. We shall discuss the energy gap model in more detail in sections 6.28 to 6.35.

The thermal conductivity of a superconductor can also be explained in terms of a two-fluid model. As the temperature is reduced below T_c the heat transport falls below that in the normal state because there are fewer n-electrons available to carry the thermal energy. A fuller description of the effect is given in sections 5.13 to 5.18.

The thermopower of superconductors is discussed in section 8.10.

6.9. *Thermodynamics applied to the mechanical properties*

In addition to the correlation of specific-heat and transition-field data, thermodynamical relations can also be derived which show the changes in the mechanical properties and other related quantities during the superconducting-normal transition. The expressions may be derived from (6.8) if we remove the assumption which we have implicitly made, that the volume V was constant and, instead, treat it as a function of H. If one then uses the relation

$$V = (\partial G/\partial p)_T \tag{6.19}$$

it is quite straightforward to derive the expression for the change in volume at the transition as

$$V_n - V_s = (V_s H_c/4\pi)(\partial H_c/\partial p)_T \tag{6.20}$$

and the Clausius–Clapeyron equation

$$(\partial p/\partial T)_{H_c} = Q/(V_n - V_s). \tag{6.21}$$

The left-hand side of (6.21) is the change in pressure which is required to keep the critical field at the same value when the temperature is changed.

Differentiation of (6.20), with respect to T and p respectively, yields expressions for the change in the coefficient of volume thermal expansion $\beta_n - \beta_s$ and in bulk modulus $\eta_n - \eta_s$. For the transition from the normal to the superconducting state at $H_c = 0$, these are

$$\beta_n - \beta_s = (1/4\pi)(\partial H_c/\partial T)_p(\partial H_c/\partial p)_T, \tag{6.22}$$

$$\eta_n - \eta_s = (\eta^2/4\pi)(\partial H_c/\partial p)_T^2. \tag{6.23}$$

These expressions may be combined to give

$$\frac{dT_c}{dp} = VT_c\frac{(\beta_n - \beta_s)}{(c_n - c_s)} = \frac{\eta_n - \eta_s}{\eta^2(\beta_n - \beta_s)}. \tag{6.24}$$

The derivation and a discussion of these relations is given by Shoenberg (1952), p. 73. The changes are all very small but some of them have been measured. To give an idea of the magnitudes involved, $(\partial H_c/\partial p)_T$ is of the order of -10^{-8} oersted dyne^{-1} cm^2 and $\partial T_c/\partial p$ is about -5×10^{-11} deg dyne^{-1} cm^2. These values yield a volume change of about 1 part in 10^7 and a change in bulk modulus of about 1 part in 10^5 when a metal becomes superconducting. Due to the experimental difficulties involved in measuring such small changes, the accuracy is not usually very high, but the values obtained for these quantities do agree quite well with the relationships we have quoted. A review of the work is given by Serin (1956), p. 239.

6.10. Penetration effects

We have already mentioned that the expulsion of an applied magnetic field from a superconductor can be considered to be due to the opposing effect of the field produced by currents which flow on the surface of the specimen. It is clear, however, that when we stipulate some type of surface current, this current does not flow only on the surface—there must be some small but finite layer in which the current flows. Only within this layer will the full influence of the magnetic field due to the surface currents be felt and a proper Meissner effect be produced. In the layer itself the magnetic induction will not be zero. Thus on simple intuitive grounds we might expect that an applied magnetic field penetrates into the specimen a short distance. Such a penetration, which has been observed, was predicted by a theory of superconductivity which was proposed by F. and H. London (1935). This theory essentially suggests relations between the superconducting current (provided by the flow of super-electrons) and the electric and magnetic fields associated with that current. The expressions coupled with the ordinary Maxwell relations give a basis for a complete treatment of the electrodynamics of superconductors.

6.11. *The London theory*

We do not propose to give any justification for the basic equations of the theory since this cannot be done rigorously. In fact they are special cases of more generalized equations as is discussed (p. 53) in F. London's detailed monograph (1950). The relations are†

$$\operatorname{curl} \Lambda \mathbf{j}_s = -\mathbf{H}/c \qquad (6.25)$$

and
$$\frac{\partial}{\partial t}(\Lambda \mathbf{j}_s) = \mathbf{E} \qquad (6.26)$$

where $\mathbf{j}_s$ is the superconducting current density and Λ is a constant characteristic of the material which has the dimensions of (time)^2. It should be noted that if, as we have suggested, the super-currents $\mathbf{j}_s$ flow only in a thin surface layer, then over most of the body of the

---

† The equations in this section are in Gaussian units, to conform with most of the literature on the subject; i.e. **E** and **j** are in e.s.u. and **H** is in e.m.u.

metal the London equations (6.25) and (6.26) reduce to $\mathbf{H} = 0$ and $\mathbf{E} = 0$, which as we have seen typifies the bulk properties of a super-conductor. The total current density $\mathbf{j}$ will be given by the sum of $\mathbf{j}_s$ and the normal component $\mathbf{j}_n$

$$\mathbf{j} = \mathbf{j}_n + \mathbf{j}_s. \tag{6.27}$$

In order to derive the equations for the penetration of the magnetic field we assume for simplicity that static conditions exist so that $\mathbf{E} = 0$, i.e. $\mathbf{j}_n = 0$. We then use the Maxwell equation

$$\operatorname{curl} \mathbf{H} = 4\pi \mathbf{j}_s/c \tag{6.28}$$

to obtain

$$\operatorname{curl} \operatorname{curl} \mathbf{H} = (4\pi/c)\operatorname{curl} \mathbf{j}_s, \tag{6.29}$$

and substituting for $\operatorname{curl} \mathbf{j}_s$ from the first of the London equations (6.25) gives

$$\operatorname{curl} \operatorname{curl} \mathbf{H} = -(4\pi/\Lambda c^2)\mathbf{H}. \tag{6.30}$$

From the vector relation for $\operatorname{curl} \operatorname{curl}$, and recalling that $\operatorname{div} \mathbf{H} = 0$ we then obtain

$$\nabla^2 \mathbf{H} = \mathbf{H}/\lambda^2, \tag{6.31}$$

where we have written

$$\lambda = (\Lambda c^2/4\pi)^{\frac{1}{2}}. \tag{6.32}$$

In a similar manner, by taking the curl of (6.25) and using (6.28), we find that

$$\nabla^2 \mathbf{j}_s = \mathbf{j}_s/\lambda^2. \tag{6.33}$$

The solutions of (6.31) and (6.33) depend on the boundary conditions. The simplest case is that of a semi-infinite specimen which extends everywhere beyond $x = 0$. If the external field H_{ext} is parallel to the surface (i.e. in the y direction), (6.31) reduces to

$$d^2 H_y/dx^2 = H_y/\lambda^2. \tag{6.34}$$

Since we know that as $x \to \infty$, $H_y \to 0$ the solution for $x > 0$ is

$$H_y = H_{ext} \exp(-x/\lambda). \tag{6.35}$$

Thus the field falls off exponentially within the superconductor and this decrease is characterized by the quantity λ which is called the penetration depth. λ has been shown by experiment to be about 5×10^{-6} cm. From (6.25) the current $\mathbf{j}_s$ will be in the z direction and it is clear that this will also fall off within the specimen with the same type of exponential behaviour.

The equations for $\mathbf{H}$ and $\mathbf{j}_s$ can be solved for other boundary conditions, the most important being for a finite sheet or film of metal, a cylinder, and a sphere. Once the field has been calculated, m_v may be found by integrating (6.4) over the volume. The results are:

For a sheet of thickness $2a$ whose surfaces are at $x = \pm a$ with H_0 parallel to the surfaces

$$H = H_{\text{ext}}\frac{\cosh(x/\lambda)}{\cosh(a/\lambda)}, \qquad (6.36)$$

$$m_v = -H_{\text{ext}}/(4\pi)\left(1 - \frac{\lambda}{a}\tanh\frac{a}{\lambda}\right). \qquad (6.37)$$

For a cylinder of radius a with H_0 parallel to the axis where r is the radial distance from the axis

$$H = H_{\text{ext}}\frac{J_0(ir/\lambda)}{J_0(ia/\lambda)}, \qquad (6.38)$$

$$m_v = \frac{-H_{\text{ext}}}{4\pi}\left\{1 + 2i\frac{\lambda J_1(ia/\lambda)}{aJ_0(ia/\lambda)}\right\}, \qquad (6.39)$$

where $i = \sqrt{(-1)}$ and J_0 and J_1 are Bessel functions of orders zero and unity respectively.

For a sphere of radius a, if the field direction is used as the axis of reference for spherical polar coordinates:

for $r \leqslant a$,

$$H_r = 3H_{\text{ext}}\frac{\lambda a}{r^2}\left\{\frac{\sinh(r/\lambda)}{\sinh(a/\lambda)}\right\}\left\{\coth\left(\frac{r}{\lambda}\right) - \frac{\lambda}{r}\right\}\cos\theta, \qquad (6.40)$$

$$H_\theta = -\frac{3H_{\text{ext}}}{2}\frac{\lambda a}{r^2}\left\{\frac{\sinh(r/\lambda)}{\sinh(a/\lambda)}\right\}\left\{\coth\left(\frac{r}{\lambda}\right) - \frac{\lambda}{r}\left(1 + \frac{r^2}{\lambda^2}\right)\right\}\sin\theta, \qquad (6.41)$$

$$H_\phi = 0;$$

for $r \geqslant a$,

$$H_r = (H_{\text{ext}} + 2m_s/r^3)\cos\theta,$$

$$H_\theta = (-H_{\text{ext}} + m_s/r^3)\sin\theta, \qquad (6.42)$$

$$H_\phi = 0,$$

where m_s is the induced magnetic moment of the sphere and is given by

$$m_s = -\frac{H_{\text{ext}}a^3}{2}\left(1 - \frac{3\lambda}{a}\coth\frac{a}{\lambda} + \frac{3\lambda^2}{a^2}\right). \qquad (6.43)$$

For a very small sphere when $a \ll \lambda$, this reduces to

$$m_s = -H_{\text{ext}}a^5/(30\lambda^2). \qquad (6.44)$$

Thus for very small particles the magnetic moment per unit volume is much reduced. Some other solutions of the London equations are given by Shoenberg (1952), p. 233, Bardeen (1956), p. 290, and London (1950), p. 34.

6.12. *Measurements of the penetration depth*

It is clear from the solutions which we have quoted, that in order to observe the effects of penetration, it is advantageous to work with very small or very thin specimens which have a dimension of the order of λ. The first experiments were made by Shoenberg (1940) who measured the magnetic moment of a mercury colloid. He found, in qualitative agreement with (6.44), that m_s was very much smaller than that to be

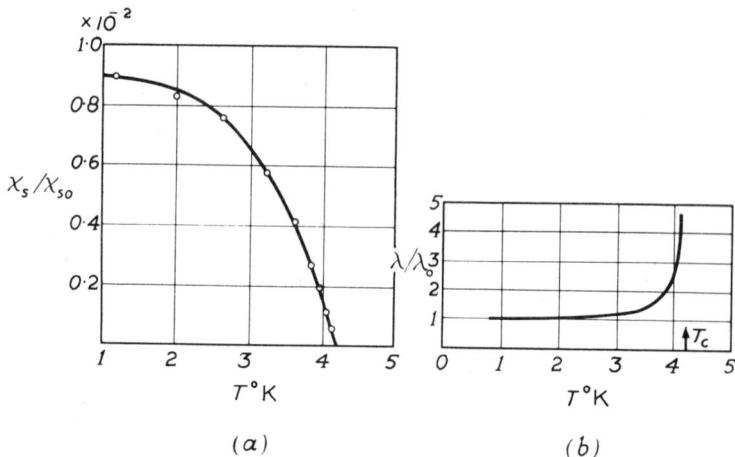

FIG. 6.11. (a) The ratio of the susceptibility, χ_s, of a mercury colloid to that, χ_{s0}, for a bulk specimen as a function of temperature. (b) The relative penetration depth, λ/λ_0, as deduced from the data of (a) (Shoenberg, 1940).

expected from a bulk specimen containing the same amount of mercury, and he also observed that m_s increased as the temperature was reduced below T_c. This was interpreted as being due to the fact that the penetration depth decreases as the temperature is reduced.

Unfortunately it was not possible to estimate the value of λ from these experiments, because the radii of the colloidal particles was not constant, nor was its statistical distribution known. The results are shown in Fig. 6.11 (a) in which χ_s/χ_{s0} is plotted against temperature; χ_s and χ_{s0} are the magnetic susceptibility (i.e. m_v/H) of the colloid and of the bulk spherical sample with the same mass respectively. It will be noted that the decrease in χ_s for the colloid is quite considerable, being of the order of 10^{-2} of that of the bulk metal. The flattening off of the curve in Fig. 6.11 (a) at the lowest temperatures suggests that as $T \to 0^\circ$ K the penetration depth tends to a constant value λ_0. If we assume on the basis of (6.44) that χ_s is proportional to $1/\lambda^2$, then the relative change in λ compared with the value λ_0 is as drawn in Fig. 6.11 (b). This curve is expressed quite accurately by the relation

$$\lambda = \lambda_0 \{1 - (T/T_c)^4\}^{-\frac{1}{2}}. \tag{6.45}$$

Experiments have also been made on the magnetic moment of bundles of thin wires (Désirant and Shoenberg, 1948) and also on thin films deposited on a mica substrate (Lock, 1951). These experiments together with those already quoted confirm the penetration effect and a temperature dependence of the form (6.45). Accurate determinations of λ_0 have been difficult, however, because of the difficulty of producing large numbers of very thin specimens of identical dimensions and also of packing a sufficient number within the cryostat, but there seems to be little doubt that it is of the order of 5×10^{-6} cm.

It is possible to detect the change in penetration depth with temperature in a macroscopic specimen if it is made part of a mutual inductance (by winding primary and secondary coils around it). Since λ decreases with reduction in temperature, the effective diamagnetism of the specimen increases and so the mutual inductance is reduced. This change can be detected by precision a.c. measurements. This method, which was suggested by Casimir (1940), was first performed successfully by Laurmann and Shoenberg (1949). However, much more accurate determinations of the change in λ may be derived from the high-frequency experiments which are described in the succeeding sections.

6.13. High-frequency effects

In our brief description of the London theory (section 6.11) we only discussed the behaviour of the super-current and neglected the effect of any possible flow of normal electrons. Provided that the electric field is zero this is quite justifiable, but in an alternating magnetic field the variations of $\mathbf{j}_s$ will give rise to a non-zero value of $\mathbf{E}$ as is shown by the second of the London equations, (6.26). The normal electrons will therefore be accelerated and their influence must be taken into account.

In order to give a qualitative description of the effects which are observed it is necessary to recall the behaviour of a normal metal in a high-frequency field. It is a standard result from a simple application of Maxwell's equations that the electric field falls off exponentially within the metal in the manner

$$E = E_0 \exp(-x/d) \tag{6.46}$$

and the current flows essentially only within a small layer of thickness d

from the surface. d is known as the skin depth and is given by

$$d = (2\pi\omega\sigma)^{-\frac{1}{2}}, \tag{6.47}$$

where ω is the angular frequency of the applied field and σ is the electrical conductivity in e.m.u.

In the normal state there will be both resistive and reactive components of the current and each will have its characteristic value of d.[†] In the superconducting state, the skin depth for the n-electrons is not simple to estimate because their effective conductivity σ_s will be proportional to their density and we have already assumed in (6.14) that this changes with temperature, tending to zero as $T \to 0°$ K. Thus d will increase as the temperature is reduced.

As we saw in section (6.11) the super-electrons have their own characteristic penetration depth λ, and their current j_s is reactive. This is implicit in the formulation of (6.26). The ratio of j_n to j_s may be calculated from the London theory (see, for example, F. London, 1950, p. 84). It may be expressed as

$$\frac{|j_n|}{|j_s|} = \sigma_s \Lambda \omega c^2 = 2\left(\frac{\lambda}{d}\right)^2. \tag{6.48}$$

The term $(\lambda/d)^2$ follows immediately from (6.47) and (6.32). Thus if $\sigma_s = 10^{-1}$ e.m.u. (which is typical of a high-purity metal at low temperatures) then, for a frequency of 10^{10} c/s, $d \approx 5 \times 10^{-6}$ cm; i.e. it is of the same order as λ and hence the effect of j_n cannot be neglected.

In the treatment of high-frequency phenomena it is convenient to consider the reactance and the resistance separately. This we shall do in the next sections.

6.14. *The surface reactance; penetration depth*

As we have just seen, reactive effects will arise from both the n- and the super-electrons and these must be combined to yield a rather complicated reactance. The situation can be simplified, however, if we cool the superconductor to a temperature well below T_c. There will then be very few n-electrons present and hence d will be large compared with λ. Thus (6.48) suggests that the reactive effect of j_n will be very small compared with that of j_s, and, in fact, the surface reactance will be determined almost entirely by λ. Since, as we have already described, λ is temperature dependent, we would expect to be able to detect its

---

† In the classical case the values of d for the resistive and reactive components are equal, but they differ in the region where the anomalous skin effect (section 4.25) is operative (see Pippard, 1954); however, they still remain of the same order of magnitude.

variation by measuring the reactance. This principle has been applied very successfully by Pippard (1947*b*) who was able to detect changes in λ with a precision much greater than that obtainable by the static experiments which we have already quoted (section 6.12). Pippard measured the resonant frequency of a cavity which contained a super-conducting element. He found that at 1200 Mc/s the resonant frequency changed by about 1 part in 10^4 when superconductivity was destroyed by a magnetic field. For temperatures well below T_c, the frequency change, $\Delta\nu$, is of the form

$$\Delta\nu = \text{constant} \times (d_n - \lambda), \tag{6.49}$$

where d_n is the skin depth in the normal state. At these low tempera-tures the normal resistance is constant (section 4.5) and hence d_n is constant. Thus the variations in $\Delta\nu$ give a direct measure of the changes in λ (Fig. 6.12 (*a*)). Since frequency changes can be measured with a high degree of accuracy it was possible to detect changes in λ of 0·1 per cent. At low temperatures the results are in good agreement with those which we have already described, but at temperatures approaching T_c, especially at higher frequencies (9400 Mc/s) there were deviations due to the reactive effect of the *n*-electrons.

6.15. *High-frequency resistivity*

The resistive component of j_n will, of course, give rise to energy dissipa-tion (which was indeed measured directly by H. London, 1940). Hence, provided that j_n is sufficiently large (i.e. if d is smaller than, or of the same order as, λ), a superconductor ought to have a measurable resis-tance. This can occur (*a*) close to T_c, where λ increases rapidly, or (*b*) at very high frequencies. From (6.48) two types of behaviour can be predicted. (*a*) At a given frequency the value of d increases as the temperature is reduced, due to the decrease in the number of *n*-electrons. Thus j_n will be smaller and hence the high-frequency resistance should also decrease at lower temperatures. (*b*) At a given temperature the resistance will be larger for higher frequencies since from (6.47) d will then be smaller.

The experimental results, some of which are shown in Fig. 6.12 (*b*), bear out these predictions in the micro-wave region, but not in the infra-red (14 μ). The curves show the ratio of the resistivity in the superconducting state to that in the normal state, just above T_c. Up to 10 Mc/s no resistance is detected and a discontinuous drop to zero is obtained at T_c as in a d.c. measurement. At higher frequencies the change is not discontinuous and it becomes more gradual as the

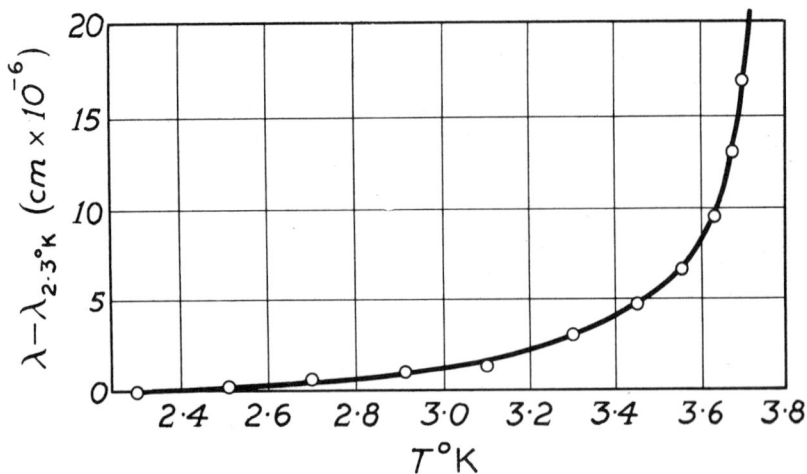

(a)

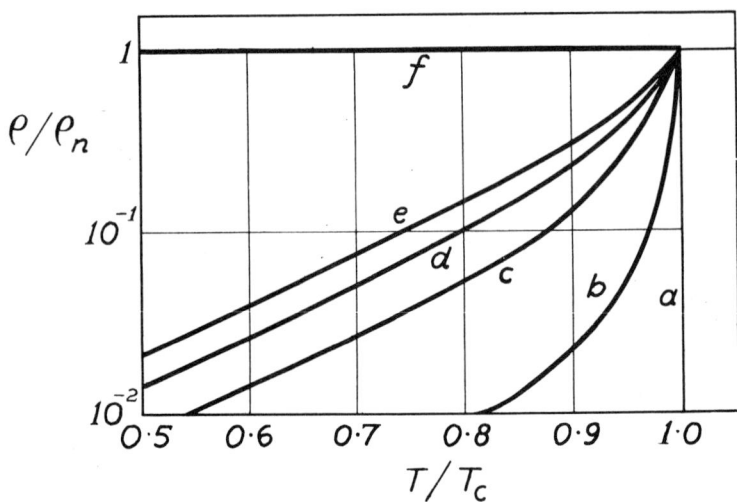

(b)

Fɪɢ. 6.12. (a) The penetration depth for tin deduced from high-frequency impedance measurements. Compare this curve with Fig. 6.11 (b) (Pippard, 1947b). (b) The ratio of the high-frequency resistance ρ to that in the normal state ρ_n for tin, at various frequencies; (a) low frequencies, (b) 1200 Mc/s (25 cm), (c) 10,000 Mc/s (3 cm), (d) 24,000 Mc/s (1·25 cm), (e) 36,000 Mc/s (8 mm), (f) 2·1 × 10⁷ Mc/s (14 μ) (after Pippard, 1954).

frequency is increased. Except for the infra-red results, which we shall discuss in sections 6.27 and 6.37, there is a definite trend to zero resistivity as $T \to 0°$ K in agreement with (a).

It is clear from (6.48) that since the high-frequency resistance is also a function of λ the change in this resistance with temperature may also be used to determine the temperature variation of the penetration depth. This was done by Pippard (1947a) who measured the quality factor of a cavity resonator as a function of temperature. This method, however, is not so accurate as that in which $\Delta\nu$ is measured (section 6.14).

In concluding this description of high-frequency effects, there are a few remarks which should be made. Firstly, on the theoretical side, the interpretation of the results is complicated by the fact that the experiments are usually made under conditions in which the anomalous skin effect (section 4.25) is operative. Thus (6.47) does not usually give the correct value of d and more complicated ones must be used (see Shoenberg, 1952, p. 197). Secondly, as with all experiments on surface phenomena, the results obtained are very dependent on the care with which the specimens are prepared. Heavily cold-worked surfaces must be avoided and if possible the specimens should be electro-polished.

Some more recent high-frequency experiments in the infra-red are described in a later section (6.37) where it will be more convenient to deal with their interpretation.

6.16. The critical field of small specimens

Another effect of the penetration of an external magnetic field within the superconductor is that the transition field of a thin or very small specimen is much higher than that for the bulk metal. From thermodynamic arguments this may be understood very simply. We have already seen from (6.5) that the free-energy difference between the normal and the superconducting states is given by

$$G_n(H_c) - G_s(0) = -V \int_0^{H_c} \mathbf{m}_v \, d\mathbf{H}. \tag{6.50}$$

Since, however, the field penetrates the specimen, m_v is less than that of a bulk sample. Hence if we assume that $(G_n - G_s)$ is independent of the specimen size, then the range of integration in (6.50) must be extended to higher fields in order to counteract the smaller values of m_v. Thus the effective value of H_c is increased.

Fig. 6.13 shows the magnetization curves of tin films in which the decrease in m_v for the thinner specimens is clearly demonstrated. The areas under the curves (i.e. the value of the integral in (6.50)) are, however, about the same for each specimen, indicating that our assumption regarding the constancy of $(G_n-G_s)/V$ is basically correct. Some small discrepancies in these and in other experiments are in fact observed and it has been suggested that this is because the difference in the surface energies of the two phases (section 6.19) has not been taken into account.

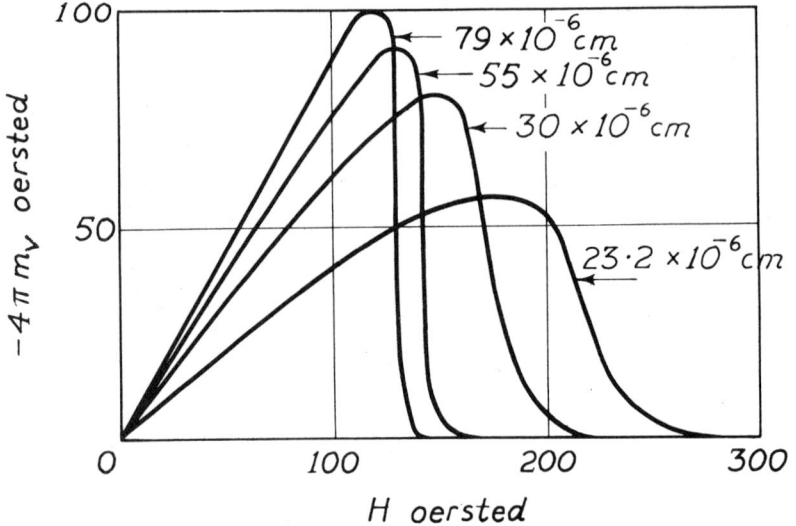

FIG. 6.13. Magnetization curves for thin tin films at 3° K showing the decrease in m_v and the increase in H_c for the thinner films (Lock, 1951).

This, however, is by no means fully established, particularly since the non-linearity of the magnetization curves has made it difficult to understand the details of the transition which appears to be much more smeared than in a large specimen. The reader is referred to Shoenberg (1952), p. 171, for further discussion.

The actual behaviour of the critical field H_s of a small specimen as a function of thickness and temperature is given in Fig. 6.14. At temperatures well below T_c, H_s/H_c for a particular specimen is constant, but as T_c is approached, it increases rapidly; this temperature variation is very reminiscent of that of the penetration depth (Fig. 6.11 (b)). Indeed the rapid rise of H_s/H_c near T_c is just due to a similar increase in λ in that temperature region. The large values of λ will reduce $|m_v|$ much more than at lower temperatures, and hence H_s/H_c rises near T_c.

At any one temperature it has been found that H_s/H_c depends on the specimen size in the following manner:

$$\text{for a film of thickness } 2a, \quad H_s/H_c = 1+b/(2a); \qquad (6.51)$$

$$\text{for a wire of radius } r, \qquad H_s/H_c = 1+b/r. \qquad (6.52)$$

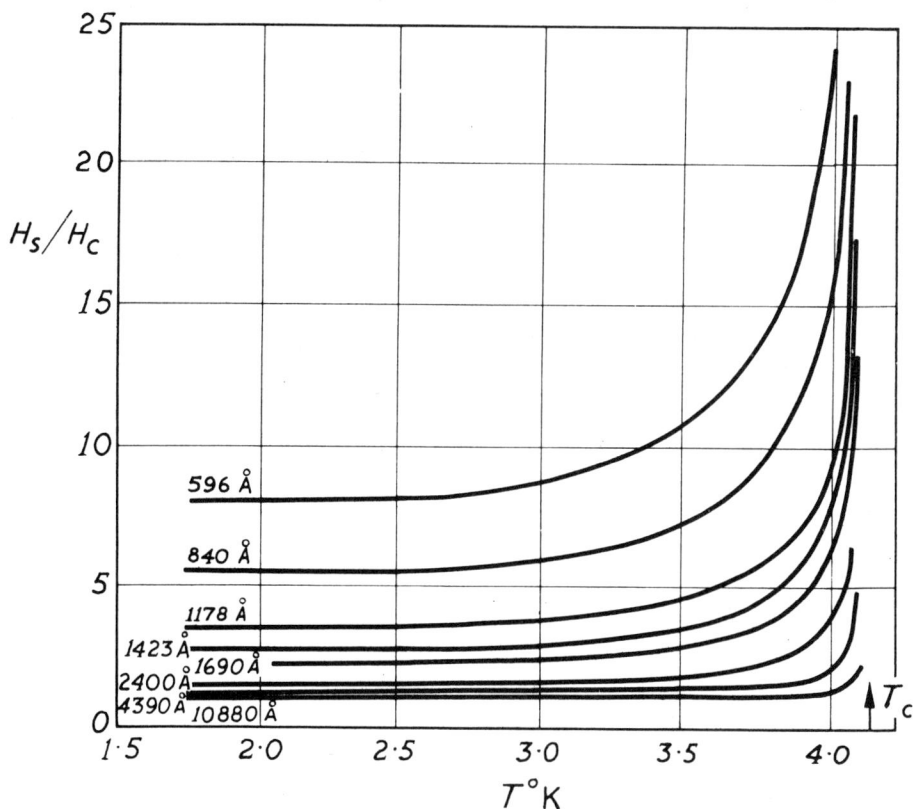

FIG. 6.14. The temperature variation of H_s/H_c for mercury films of various thicknesses (Appleyard, Bristow, London, and Misener, 1939).

At low temperatures b has a value of the order of 10^{-5} cm, i.e. it is of the same order of magnitude as λ. Its temperature dependence is also similar to that of λ.

This type of behaviour can be deduced very simply for a rather thick plane specimen in which $a \gg \lambda$. In this case we can assume to a first approximation that the field penetrates completely to a depth λ and thereafter it is zero. If m_v is proportional to H_s, then from (6.50), $G_n - G_s$

is proportional to $H_s^2 \times$ 'effective volume', i.e. to $H_s^2(2a-2\lambda)$. Thus the free energy difference per unit volume will be proportional to

$$H_s^2(2a-2\lambda)/\text{actual volume},$$

i.e. to

$$H_s^2(2a-2\lambda)/2a. \qquad (6.53)$$

For a very thick specimen H_s will be equal to H_c and the λ can be neglected in (6.53), which then becomes just H_c^2. Thus if we assume that the free energy difference per unit volume of large and small specimens is the same, then

$$H_c^2 = (2a-2\lambda)H_s^2/(2a) \approx H_s^2(1-\lambda/a); \qquad (6.54)$$

hence

$$(H_s/H_c)^2 \approx 1+\lambda/a, \qquad (6.55)$$

and

$$H_s/H_c \approx 1+\lambda/(2a), \qquad (6.56)$$

which is of the same form as (6.51) and shows why b is of the same order as λ.

6.17. The intermediate state

In the earlier sections of this chapter we described how the destruction of superconductivity by a magnetic field in a bulk specimen only occurred discontinuously at the field H_c when this field was parallel to the axis of a long cylindrical specimen. Otherwise, penetration of the field began for applied fields which were lower than H_c. In the special case of a cylinder in a transverse field the superconducting state began to be destroyed at $\sim \frac{1}{2}H_c$. This was shown both by the recovery of the electrical resistivity (Fig. 6.4) and the magnetic induction (Fig. 6.8 (b)), both of which then began to increase in an approximately linear fashion until at H_c the full normal state properties were attained. The corresponding behaviour of the heat conductivity is described in section 5.14. The region between that corresponding to the initial penetration of the field and H_c is known as the intermediate state and a large amount of both experimental and theoretical work has been concerned with a study of its nature.

The main conclusions of this work are as follows. When a superconductor is in the intermediate state it is split up into a number of laminae which are alternately normal and superconducting, the faces of the laminae being parallel to the applied field. As the field is increased the thickness of the normal laminae increases at the expense of that of the superconducting laminae, until at H_c the latter disappear altogether and the whole of the specimen is in the normal state. The steady increase in the resistivity which is observed for fields above $\sim \frac{1}{2}H_c$ is thus due to the increase in the thickness of the normal laminae. Within these laminae the magnetic induction is assumed to be equal to H_c whereas

in the superconducting ones it is zero. That the laminae must lie parallel
to the applied field can be seen from the general requirement that the
normal component of the magnetic induction must be continuous at
any boundary. Since in the normal state the induction will be parallel
to the field (except for end and edge effects) and in the superconducting
state it is zero, this condition can only be satisfied if the boundary
is parallel to the field. That the laminae must extend right across the

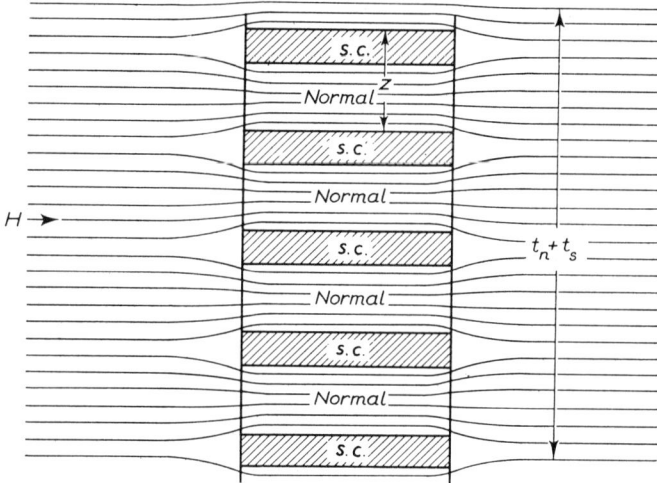

FIG. 6.15. The concentration of the magnetic lines of force through the normal regions
of a superconductor in the intermediate state when the field is transverse to the specimen.
The normal and superconducting laminae lie parallel to the field.

specimen is borne out by the non-zero resistivity of the intermediate
state which indicates that the current must traverse regions which are
in the normal state and hence there are no short-circuiting superconducting paths. In fact, Shubnikov and Nakchutin (1937) showed that
a sphere in the intermediate state did not have any resistivity in a
direction parallel to the field but only perpendicular to it.

The distribution of laminae and field is thought to occur in a manner
similar to that shown diagrammatically in Fig. 6.15, in which the lines
of force are distorted so that they all pass through the normal regions
(in which $B = H = H_c$). It is clear that if the specimen were plane and
t_n and t_s were the total thickness of the normal and superconducting
laminae respectively then, in a field which is equal to H in the region
distant from the specimen, we may write (excluding end effects)

$$H(t_n+t_s) = H_c t_n, \tag{6.57}$$

i.e.

$$t_n/(t_n+t_s) = H/H_c. \tag{6.58}$$

6.18. *The free energy of the intermediate state*

We now wish to show that on thermodynamic grounds it would seem reasonable that the intermediate state consists of regions of normal and superconducting metal.

If we consider a cylinder with the field H perpendicular to its axis, then the field at the surface of the specimen, H', is equal to $2H$ (section 6.4) and hence from (6.4)

$$m_v = -H'/(4\pi) = -H/(2\pi).\tag{6.59}$$

At $0°$ K, therefore, the free energy in a field H would be greater than that in zero field by an amount

$$V \int_0^H \mathbf{m}_v \, d\mathbf{H} = VH^2/(4\pi).\tag{6.60}$$

If, however, a fraction H/H_c of the specimen (6.58) becomes normal, then the free energy increase per unit volume of those regions will be $H_c^2/8\pi$ (6.8) and hence the total free energy of the normal regions will be

$$V(H/H_c)H_c^2/(8\pi) = VHH_c/(8\pi).\tag{6.61}$$

Under these circumstances the effective field on the superconducting regions will be greatly reduced, so that their contribution to the free energy can be neglected. We therefore see, on comparing (6.61) and (6.60), that provided $H > \frac{1}{2}H_c$, it is energetically more favourable for the specimen to split so that it contains a number of normal regions in which $B = H_c$ rather than remain entirely in the superconducting state.

6.19. *Surface energy*

The existence of a stable system of normal and superconducting laminae, however, necessitates the assumption that there is also a surface energy between the normal and superconducting regions. It is fairly clear that this energy must be positive; if it were not, there would always be a tendency for a superconductor in an external field to split up into a very small mosaic of normal and superconducting regions and the Meissner effect would then not be observed, but see end of section 6.40.

The existence of a surface energy between the normal and super-conducting regions may be taken to imply that the boundary layer between the laminae has a finite width δ (this is not to be confused with the penetration depth λ), which may be related to the surface energy in the following manner. Since the free-energy difference per unit volume at $0°$ K is $H_c^2/8\pi$, the boundary energy, α_{ns}, per unit area will be of the order of

$$\alpha_{ns} = \delta H_c^2/(8\pi).\tag{6.62}$$

This surface energy will tend to limit the number of laminae which are formed. Clearly, the less the distortion of the external field (see Fig. 6.15) the lower the energy of the system; this would favour a large number of very thin laminae. These considerations will, however, be offset by the large amount of surface energy which such a configuration would involve. A small number of laminae, on the other hand, would entail a large distortion of the field. Thus there will be one distribution of normal and superconducting regions which will have a minimum energy and which will therefore be stable.

The theories which have been developed on these lines (Landau, 1937; Kuper, 1951) are reviewed by Shoenberg (1952), p. 96. They are very complicated and since they only deal with idealized situations there would be no point in giving the results in detail. However, for a cylinder of diameter d in a transverse field they both yield, for the combined thickness, z, of one normal and one superconducting lamina

$$z = f(H/H_c)(\delta d)^{\frac{1}{2}}, \qquad (6.63)$$

where $f(H/H_c)$ is of the order of 10 from Landau's theory and is about half that value using Kuper's expression.

We note from (6.63) that the width of the laminae is proportional to $d^{\frac{1}{2}}$. Hence in a given field the volume of a lamina will be proportional to width $\times$ area, i.e. to $d^{\frac{5}{2}}$, whereas its surface area will only vary as d^2. Thus the larger the specimen the less important will be the effect of the surface area term. Hence in very large specimens α_{ns} may be neglected and then, as we saw in section 6.18, the transition will begin at $H = \frac{1}{2}H_c$. For smaller specimens, however, the effect of the surface energy will be to raise the value of the field which is able to penetrate initially, so that in tin, if for example d is 0·1 mm, the transition does not begin to occur until about $H = 0·6H_c$. This observation should not be confused with the fact that H_c itself is larger for the small specimens than for large ones (section 6.16).

6.20. *The experimental observation of the intermediate state*

The direct observation of the structure of the intermediate state has now been achieved in a number of very elegant experiments. Three different techniques have been exploited. The first, which was developed by Meshkovsky and Shalnikov (1947), utilized the very high magneto-resistance of bismuth (section 4.28) in order to detect the variations of field at the surface and within narrow channels between two super-conducting tin hemispheres. A very fine bismuth probe, which was only 0·05 mm in diameter and about 0·15 mm long, was used to detect the variations of field in a narrow gap (0·12 mm) between two tin hemispheres. A typical trace of the variations in the resistance of the probe as it was moved through the gap are shown in Fig. 6.16 (a) (Plate 1) where it will be seen that the discontinuous structure is clearly resolved. From these experiments the distance between two normal regions is about 0·2 mm.

The second technique is one in which powder patterns are produced on the surface of the superconductor in a manner similar to that used to delineate the domains in a ferromagnetic material. Balashova and Sharvin (1956) used a ferromagnetic powder which was attracted to the normal regions where the flux passed through the surface of the specimen. Schawlow (1956) used niobium powder which because of its high transition field remains superconducting and therefore diamagnetic when the specimen under observation is in the intermediate state. It is therefore pushed away from the normal regions where the flux has penetrated and it remains on the superconducting areas. Fig. 6.16 (b) shows a niobium powder pattern produced on a tin plate in the intermediate state in which the laminar structure is shown up in a most convincing manner.

The third method relies on the rotation of the plane of polarization when light passes through certain materials which are in a magnetic field (the Faraday effect). Such experiments were first carried out by Alers (1957) and others have since been made by DeSorbo (1960). DeSorbo used a thin (0·25 mm) piece of cerium phosphate glass on top of the superconductor which was illuminated with polarized light. The reflected beam passed through an analyser which was set for extinction. When the specimen was in the intermediate state the field from the normal regions acting on the glass rotated the plane of polarization and so these areas appear lighter. An example is shown in Fig. 6.16 (c). This method is particularly useful for studying time effects. In addition, the angle of rotation of the plane of polarization may be used to give a direct measure of the flux in various regions.

All these experiments show that whilst the general theoretical arguments regarding the existence of the intermediate state are correct, nevertheless the details are very complicated. The shapes of the normal and superconducting regions very often depend on the previous history of the sample, there are considerable hysteresis effects, and slight strains or impurities will also modify the patterns. With a change in the external field, a stable pattern is not always achieved until several seconds or even minutes afterwards.

PLATE 1

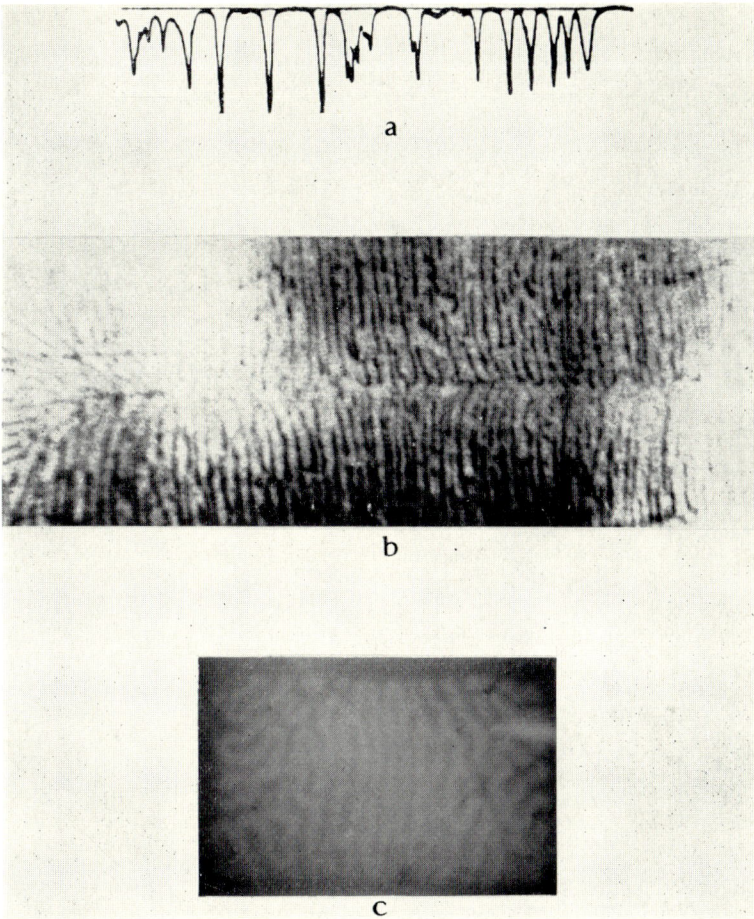

Fig. 6.16. Experimental observations of the structure of the intermediate state. (a) The field variation across the diameter in a gap 0·12 mm wide between two tin hemispheres using the bismuth probe technique. $T = 3·18°$ K, external field 72 oersted. The normal regions are those where the curve is high (Meshkovsky and Shalnikov, 1947). (b) Normal and superconducting laminae delineated by magnetic powder patterns on a tin plate; $T = 1·93°$ K, $H/H_c = 0·82$ (Schawlow, 1956). (c) The intermediate state structure as shown by the Faraday effect using cerium phosphate glass. High-purity tin single crystal, 1·4° K, transverse field reduced from H_c to $0·2H_c$ (photo by courtesy of Dr. DeSorbo).

6.21. *Measurements of* δ

In all these investigations the distance between two normal regions (i.e. z) is commonly a few tenths of a millimetre. If the specimen diameter is of the order of a few centimetres then substitution of these values in (6.63) or in a more accurate expression yields a value of δ of about 10^{-4} cm. The experiments show that, in fact, δ decreases at lower temperatures; thus in tin it changes from $\sim 5 \times 10^{-5}$ near T_c to $\sim 2 \times 10^{-5}$ at 2° K. It will be recalled that δ may be interpreted as being a measure of the width of the boundary between the normal and the superconducting states. The significance of such a dimension in the theory of superconductivity will be emphasized in a later section.

We have so far only described experiments and theory which attempt to explain and investigate the intermediate state when superconductivity is being destroyed by the application of an external magnetic field. Equally complicated situations are presented when the intermediate state is attained by passing an electric current through the specimen. The reader is referred to Shoenberg (1952), p. 130, for a discussion of this problem.

6.22. *Nucleation of superconducting regions*

We have already mentioned the hysteresis effects which are nearly always observed in experiments on superconductivity. In particular, when superconductivity has been destroyed by the application of a field $> H_c$ and the field is then reduced, the metal quite often remains in the normal state until the field has been reduced to a value H_{sup} which is considerably below H_c. H_{sup}/H_c is usually about 0·9 in tin but it can be as low as 0·37 in aluminium. This effect is commonly referred to as 'supercooling' by analogy with the effect observed in other phase transitions, although the temperature of the specimen is, in fact, kept constant.

It is assumed that the onset of superconductivity at H_{sup} occurs because a small stable superconducting nucleus is able to form. The most likely place for this to occur is at some flaw in the crystal. Experiments by Faber (1952) have demonstrated this in a most conclusive manner. In his work a tin rod had several short coils wound around it at various points along its length. It was then placed in a longitudinal field greater than H_c which was then reduced below H_c so that a slight amount of supercooling occurred. Current was then passed through one of the coils in such a direction that the local field in the region of the coil was further reduced. At some value of this field the region became superconducting and this spread throughout the specimen. It was found

that there were very wide variations of field for the initiation of superconductivity which were dependent on the position of the coil. This is
what one would expect if the flaw hypothesis was correct, because
clearly, some flaws will be more effective than others in acting as nuclea-
tion centres and hence, depending on the position of the coil, different
values of H_{sup} will be observed. By putting very short pulses of current
into one coil Faber was able to deduce that the superconducting nuclei
were formed very close to the surface of the specimen and they then
spread, first along the surface to form a superconducting sheath, and
finally inwards to the centre of the specimen. This type of behaviour
has been confirmed by visual observations using both the powder patterns and Faraday effect techniques which have been discussed in
section 6.20. Further details are given in the review article by Faber
and Pippard (1955).

6.23. Superconducting alloys and compounds

Whilst we have so far only described the behaviour of elements which
become superconducting, it has been found that many alloys and compounds show similar characteristics. One can divide these materials
into two groups; those whose properties approximate to an 'ideal'
superconductor and those which do not. The first group for the most
part consists of materials which only exhibit superconductivity in a
certain stoichiometric composition. It is not necessary for either element
to be a superconductor itself, although in this case one of the elements
must always be one which is next to a superconducting element in the
periodic table (Fig. 6.1). Many of these compounds, when carefully
prepared, have properties which are characteristic of a pure elemental
superconductor, i.e. a sharp transition to zero resistivity, a fairly low
value of H_c, a complete Meissner effect with little hysteresis when superconductivity is restored after the removal of a field greater than H_c.
This type of behaviour is shown, for example, in Au_2Bi (Shoenberg,
1938) although neither Au nor Bi is itself normally a superconductor.
Many of these compounds in the normal state are by no means very
metallic in nature, e.g. CuS and PbS together with many nitrides and
carbides of groups IVa, Va, and VIa of the periodic table. Serin (1956),
p. 268, lists four main types of compounds which become superconducting above $1°$ K: (a) Bi compounds, e.g. LiBi, (b) Metal+non-metallic
element, e.g. NbN, (c) Ni-As structure, e.g. PdSb, and (d) Mo and W
alloys, e.g. Mo_3Os.

The other group of superconductors consists of alloys which remain superconducting over a wide range of composition. In these one element is nearly always itself a superconductor. The properties of these alloys are very different from those of pure superconductors. The transition is broad, sometimes being spread over as much as a degree, and the field necessary to restore the electrical resistivity can be very high, over 10,000 oersted in some cases. There is a very large magnetic hysteresis and in many specimens no Meissner effect is observed; even when an applied field is reduced to zero the magnetic moment of the specimen is no different from that which it had when it was in the normal state. It seems evident that all this non-ideal behaviour is due to microscopic inhomogeneities in the specimens, although it is by no means clear in detail just how these give rise to the effects which are observed. The lack of a Meissner effect, for example, has been explained by Mendelssohn (1935) as being due to the fact that the specimen can be considered as containing a sponge-like structure in which the material filling up the 'holes' of the sponge has the normal transition field, whereas that making up the skeleton of the sponge has a very high value for H_c. When the field is reduced from this high value of H_c, a multiply connected superconducting network is formed. Thus when the 'holes' attempt to become superconducting at the lower field, the flux cannot be expelled from them because it is trapped by the sponge (as in the case of the flux through a superconducting torus, Fig. 6.7). Nevertheless, no satisfactory explanation has yet been given as to precisely why a sponge structure should exist which has such a large value of H_c.

The fact that large regions of an alloy specimen have quite a low value of H_c is in fact shown by experiments in which the initial penetration of the field is detected using a bismuth probe (as in section 6.20). This is found to occur at fields very much less than that required to restore the electrical resistance. It has also been shown (Mendelssohn and Moore, 1935; Rjabinin and Schubnikov, 1935) that B becomes equal to the external field for values of H which are much less than H_c. This indicates that most of the specimen is already in the normal state, but that there is a superconducting network present which is able to short-circuit the resistance of the normal regions. Fig. 6.17 shows the B–H curve for Pb+2 per cent In in which the field at which the resistivity is restored is also indicated. This figure also shows the complete absence of the Meissner effect when the external field is reduced to zero.

In recent years a large number of experiments have been made to try and clarify the behaviour of superconducting alloys (a review is given by Matthias, 1957). From this work some empirical rules can be formulated which give a guide as to the manner in which the transition temperature of an alloy will change with composition. For example, Matthias (1955) suggested that superconductivity was favoured, i.e. T_c would be high, for alloys which contained 5 and 7 valence electrons per atom. This he has demonstrated using several alloy series. An idealized curve from Matthias's paper is shown in Fig. 6.18 (a) in which the variation of T_c is plotted against the number of valence electrons. In

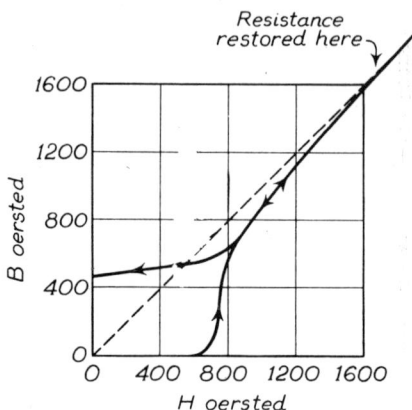

FIG. 6.17. The non-ideal behaviour of alloy superconductors. The B–H curve for a Pb-In 2 per cent alloy at 1·95° K showing that nearly all the specimen has attained the normal state at a field very much less than that for which the resistance is completely restored. The complete absence of a Meissner effect after the field has been reduced to zero should also be noted (Shubnikov, Chotkewitsch, Shepelev, and Rjabinin, 1936).

Fig. 6.18 (b) the experimental values of T_c for Zr-Rh alloys are shown and these clearly indicate a maximum in T_c for the samples which have just under 5 electrons per atom. Similar results have been obtained for compounds and alloys containing Mo or W which show a maximum for T_c at around 7 electrons per atom.

The effect of very small amounts of impurity on T_c have been determined by Chanin, Lynton, and Serin (1959). They have found that for Sn, In, and Al the addition of another element always initially decreases T_c, approximately linearly. For higher concentrations, however, two types of behaviour were found. For an alloying element which had a lower valency than the solvent, T_c tended to saturate. If, however, the valency was higher than the solvent, T_c started to increase rapidly. Typical curves for In alloys are shown in Fig. 6.19 in which the reciprocal of the electron mean free path is plotted (from residual resistance measurements) instead of the amount of impurity.

The most unusual superconducting alloys which have been investigated are those which contain rare earth elements. Some of these have
been found to be both superconducting and ferromagnetic at the same
time. For example, $ThRu_2$ is a superconductor whose transition tem-

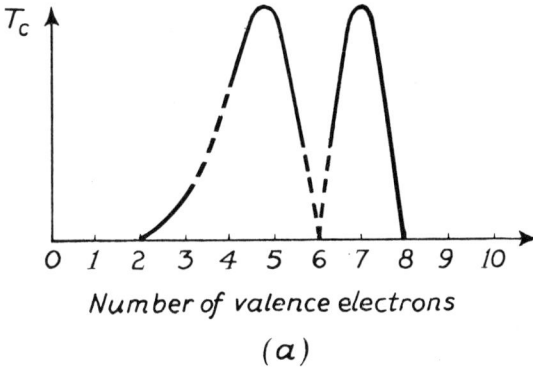

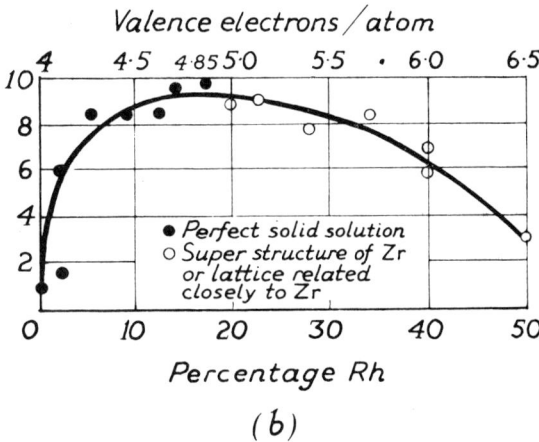

T_c

Number of valence electrons

(a)

Valence electrons / atom

● Perfect solid solution
○ Super structure of Zr
  or lattice related
  closely to Zr

Percentage Rh

(b)

FIG. 6.18. (a) A diagram showing the suggested variation of T_c with the number of
valence electrons per atom. According to this, T_c should be a maximum at about 5 and
7 valence electrons per atom. (b) Experimental values of T_c for rhodium-zirconium
alloys which show a maximum in T_c at approximately 5 valence electrons per atom
(Matthias, 1955).

perature is reduced by the addition of $GdRu_2$. With more than about
6 per cent $GdRu_2$, however, the system is ferromagnetic with rather a
low Curie temperature. As more $GdRu_2$ is added the Curie temperature
is raised but T_c is lowered. As Fig. 6.20 shows, there is a narrow range
of composition and temperature in which the alloy is both ferromagnetic
and superconducting. Theoretically this is a difficult situation to resolve;

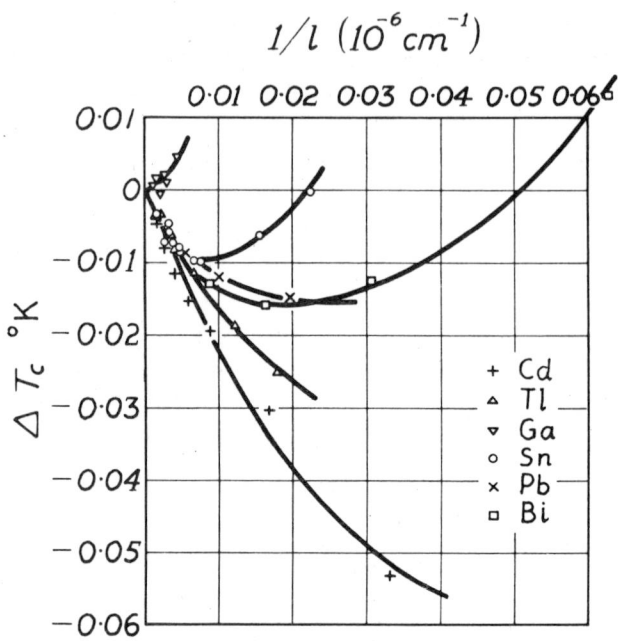

FIG. 6.19. The lowering of T_c for indium due to impurities plotted as a function of the electronic mean free path, l (Chanin, Lynton, and Serin, 1959).

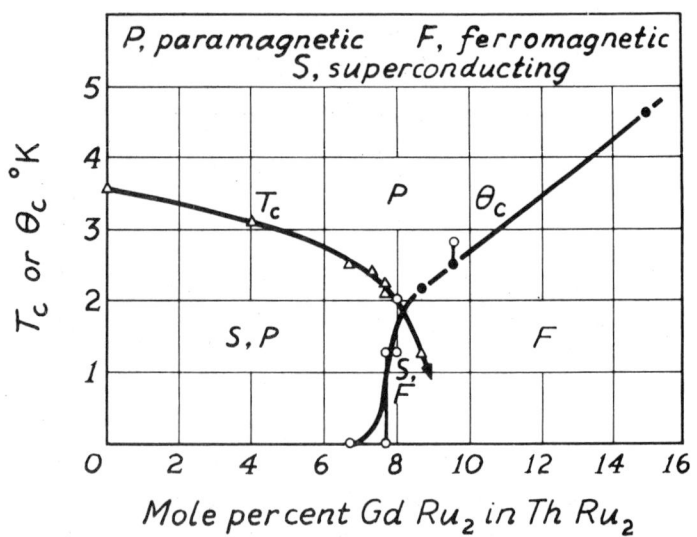

FIG. 6.20. Superconducting transition temperatures, T_c, and Curie temperatures (θ_c) for solid solutions of GdRu$_2$ in ThRu$_2$. In the central region between the two curves, the alloy is both superconducting and ferromagnetic (Bozorth, Matthias, and Davis, 1960).

some workers have suggested that the alloy is not completely homogeneous, so that the ferromagnetic and the superconducting regions are separate parts of the specimen, although the experimenters themselves do not agree with this.

The use of superconducting alloys in the construction of high field solenoids is discussed in section 6.40.

6.24. Experiments which assist towards our understanding of the nature of the superconducting state

The experiments and properties which we have discussed so far are, from the practical point of view, the most important. Nevertheless, by themselves they give very little clue as to why superconductivity occurs and why it does not appear to be present in all metals. Before giving a very simplified explanation of the modern microscopic theory of superconductivity, therefore, we shall first describe results of some of the experiments which have assisted in determining the important factors which are involved. After our discussion of the theory we shall then mention some more recent experiments which have been made to verify it.

The most important facts about the nature of superconductivity are the following.

(a) Superconductivity arises not by the interaction of the electrons among themselves, but by their interaction with the crystal lattice, i.e. the phonons. If this interaction is sufficiently strong, superconductivity can occur.

(b) An important feature of the superconducting state is that the electron's sphere of influence or its wave function extends over a considerable distance in the crystal so that the behaviour of one electron is rigidly correlated with that of many others.

(c) There exists some kind of an energy gap between the superconducting ground state and the excited states.

In the succeeding sections we shall discuss some of the experiments which have led to these conclusions.

6.25. *The influence of the lattice; the isotope effect*

The earliest observation which indicated the importance of the crystal lattice was that, whereas ordinary white tin was a superconductor, the allotropic form, grey tin, was not. This showed that superconductivity was not a property of the atom, but that the crystal structure was an important factor. Similar conclusions could be drawn from experiments on bismuth, which is normally not a superconductor. Bückel and Hilsch (1954) found that bismuth films were superconducting if they were prepared by condensing the vapour on to a substrate which was at liquid-helium temperature. Under these conditions a very disordered crystal lattice is formed. Chester and Jones (1953) showed that bismuth also became superconducting, with $T_c \sim 7°$ K, if the metal was subjected to a very high pressure (20,000 atmospheres).

The most important experiments, however, were those in which H_c and T_c were measured for different isotopes of the same element (Maxwell, 1950; Reynolds, Serin, Wright, and Nesbitt, 1950). These workers found that T_c depended on M, the isotopic mass, and that it varied as $M^{-\frac{1}{2}}$. These observations showed, for the first time, that it was not the crystal structure as such which determined whether a metal was a superconductor at any given temperature (since the crystal structure of the different isotopes is the same); instead the important fact was seen to be the interaction of the electrons with the lattice vibrations. The spectrum of these vibrations will change slightly for the different values of M and this gives rise to the different values of T_c. It is interesting to note that the dependence of T_c on $M^{-\frac{1}{2}}$ was predicted independently by Fröhlich (1950) on the basis of a theoretical treatment of the electron-phonon interaction.

6.26. *The electron coherence length*

Whilst the principle has been held for many years that some type of 'long-range order' must exist in the superconducting state (see, for example, F. London, 1950, p. 3) a crystallization of these ideas into a definite theory was not made until fairly recently by Pippard (1953), who introduced the concept of a coherence length. Since such an idea is necessary for the understanding of the nature of superconductivity, we briefly outline some of the arguments relating to it in this section.

In our discussion of the intermediate state we showed that in order to account for the laminar structure which was observed it was necessary to postulate the existence of a positive surface energy, α_{ns} per unit area, which was associated with the boundary between the normal and the superconducting regions (section 6.19). This energy can be considered to be a property of a surface layer of finite thickness δ and the results of experiments suggest that δ is of the order of 10^{-4} cm. Compared with atomic dimensions, or even with the penetration depth, λ ($\sim 10^{-6}$ cm), this length is very large. It is indicative of the fact that the boundary between the normal and the superconducting regions is not sharp, but rather, that there is a gradual change in properties between them over a distance of 10^{-4} cm. Using the two-fluid model (section 6.7) one could interpret this by saying that whereas the superconducting state is characterized at any given temperature by a certain ratio, x, of superconducting to normal electrons (in the normal regions $x = 0$) then because of the rather wide boundary region it appears that x is not able to vary very rapidly.

Another reason for suggesting that some kind of long-range collective action exists is given by the very sharp nature of the resistance change at T_c. Under favourable circumstances the transition can take place within a temperature interval of 0·002 degree (Doidge, 1956). For this to occur the collective action of large numbers of electrons must be involved; otherwise there would always be some local fluctuations which would tend to round off the transition curve.

It is also possible to deduce by the following simple argument that a long-range parameter must enter into the theory. Let us make the reasonable assumption that for superconductivity to occur, the electron states which are involved are those which lie within kT_c of the Fermi surface. Then those states must be defined to within a value $\Delta \mathscr{E} \approx kT_c$ and hence by the uncertainty principle their lifetime, τ, must be given by

$$\Delta \mathscr{E} \tau \approx \hbar, \tag{6.64}$$

and hence if v is the electron velocity, the wave function must extend over a distance $l = v\tau$. Thus, substituting for τ, we may write

$$\Delta \mathscr{E} \, l/v \approx \hbar, \tag{6.65}$$

and so

$$l \approx \hbar v/\Delta \mathscr{E} \approx \hbar v/(kT_c). \tag{6.66}$$

Since the Fermi velocity is of the order of 10^8 cm sec^{-1} we find that l is about 10^{-4} cm. This is of the same magnitude as the experimental values of δ which we have quoted.

These arguments and others of a more sophisticated and powerful nature (which have been deduced from observations of the change of penetration depth with magnetic field and with impurity), have been put forward by Pippard (1953) to show that there is good reason for believing that there is some type of long range interaction of the electrons which gives rise to what he calls 'coherence'. The effect of this, for example, is that the London equation (6.25) which defines the current density, j_s, must be replaced by an integral equation in which one can only determine the average value of j_s over a volume of the order of δ^3. We shall discuss the nature of this long-range order in a later section. The Pippard theory is reviewed very fully by Bardeen (1956), p. 299.

6.27. *The energy gap*

We have already discussed in section 6.8 that some type of energy gap between the superconducting and the normal electrons might be expected to exist, and some of the early evidence for such a gap was provided by measurements at very low temperatures $(T \ll T_c)$, which showed that in the superconducting state the specific heat had an exponential temperature dependence (Fig. 6.10). This suggested some type of excitation process across a finite energy gap. The other experiments which we have also described (section 6.15) were those in which the high-frequency resistance was measured; it will be recalled that whilst in the micro-wave region the high-frequency resistance tended to zero as $T \to 0^\circ$ K, in the infra-red there was no difference between the normal and the superconducting state resistivity (Fig. 6.12 (*b*)). This result can be interpreted by assuming that the infra-red quanta have sufficient energy to excite the superconducting electrons across the gap so that they behave as normal electrons. Thus the resistance is apparently unchanged even in the superconducting state. More recently there have been many more experiments which have demonstrated the existence of an energy gap in most convincing ways, but we shall delay a description of these (sections 6.37, 6.38) until we have outlined the theoretical results.

6.28. The modern theory of superconductivity

Current explanations for the existence of the superconducting state stem from the theory which was proposed by Bardeen, Cooper, and Schrieffer (1957) and which is now called the BCS theory. Since its first publication it has been developed so that it is now able to account quite adequately for many of the properties of superconductors. In addition, it has stimulated a large number of experiments, all of which have been most useful in corroborating many of the theoretical predictions. In the following sections we intend to introduce and describe the theory with only the simplest mathematics so as to emphasize the physics involved. For further enlightenment the reader is referred to the review article by Bardeen and Schrieffer (1961) and to an introduction to the theory by Cooper (1960).

In principle the BCS theory considers the various types of interaction which can occur between the electrons and it then singles out one particular mechanism which it shows can be attractive and will therefore result in a reduction in energy of the system. This attraction is due to a special type of electron-phonon interaction (section 6.25) and it can operate over relatively large distances—reminiscent of the coherence length (section 6.26). It yields a superconducting ground state which is separated from the excited states by an energy gap of the order of kT_c (section 6.27). The theory is able to account qualitatively if not always quantitatively for the specific heat, the thermal conductivity, the Meissner effect, zero resistance, penetration depth, and many other properties associated with the superconducting state.

6.29. *Electron-phonon interaction*

One of the important things to realize about the transition to the superconducting state is that the average decrease in energy per electron is very small. From (6.8) we saw that the free-energy difference per unit volume between the normal and the superconducting states is $H_c^2/8\pi$. Since H_c is usually a few hundred oersted this means that the energy difference is a few thousand ergs per cm³ which is equivalent to about 10^{-8} eV per electron. However if we assume that the interaction energy, $\mathscr{E}_x$, is confined to electrons which have an energy which lies within $\mathscr{E}_x$ of the Fermi energy $\mathscr{E}_F$, then the mean energy averaged over all the electrons is of the order of $(\mathscr{E}_x/\mathscr{E}_F)\mathscr{E}_x$ per electron. If this is equated to 10^{-8} eV then, since $\mathscr{E}_F$ is a few eV, $\mathscr{E}_x$ is about 10^{-4} eV. Thus $\mathscr{E}_x$ is equal to kT when T is a few degrees absolute, i.e. when $T \approx T_c$. Hence it is reasonable to assume that in the transition to superconductivity, it is only the electrons which lie within kT_c of the Fermi surface which are affected. In the BCS theory an interaction is selected which is of this order of magnitude, and it is assumed that all other interactions are the same in both the normal and the superconducting states and can therefore be neglected.

The process which is considered is one which can yield either an attraction or a repulsion between two electrons. This interaction is not of a direct type such as ordinary coulomb repulsion; in order for it to occur a 'virtual phonon' must be transferred from one electron to the other. Processes involving 'virtual' excitations are concepts of field theory which are quite easy to understand. We consider an electron with wave vector $\mathbf{k_1}$ which changes its state to $\mathbf{k_1'}$ with the emission of a phonon whose wave vector is $\mathbf{q}$. From the conservation law (4.18) we then have

$$\mathbf{k_1'} = \mathbf{k_1} - \mathbf{q}. \tag{6.67}$$

Normally, as we have seen in section 4.16, we must satisfy not only the wave-vector conservation equation, but also that for energy conservation (4.20). If, however, the phonon $\mathbf{q}$ is very rapidly reabsorbed by another electron, so that its lifetime is short, then, by the uncertainty relation, $\Delta\mathscr{E}\tau \approx \hbar$, the uncertainty in its energy will be sufficient to cover any deficiency in the energy relation. We therefore suppose that a second electron with wave vector $\mathbf{k_2}$ absorbs the phonon thereby changing its state to $\mathbf{k_2'}$ such that

$$\mathbf{k_2'} = \mathbf{k_2} + \mathbf{q}. \tag{6.68}$$

The result of this process is that both electrons have changed their states, but that, as is right and proper, the total wave vector is conserved, i.e.

$$\mathbf{k_1} + \mathbf{k_2} = \mathbf{k_1'} + \mathbf{k_2'}. \tag{6.69}$$

This process takes place via the very rapid emission and absorption of the phonon $\mathbf{q}$ so that energy conservation is not required. This short-lived phonon is called a virtual phonon. The mechanism is similar to the Raman process which will be described in section 9.27. If the energy of the electron pair is changed by this process, then we can talk of an interaction via a virtual phonon. A reduction in energy will lead to an attraction between the two electrons whilst an increase in energy would cause a repulsion. It is assumed that in the superconducting state there is an attraction between electron pairs.

6.30. *Electron pairs*

It is fairly straightforward to give a physical picture of the nature of this interaction. When an electron moves through the lattice it will distort the motion of the positive ions, (unless they are infinitely heavy or rigidly fixed). This extra distortion of the lattice can then affect the motion of a second electron so that its state is changed whilst the ions return to their original state. All this happens very rapidly so that, in effect, electron 1 interacts with electron 2 via a very short fluctuation of the ionic lattice. It is this fluctuation which we have called the virtual phonon. For a further elementary discussion on virtual excitations see Peierls (1955).

We have yet to explain why this might lead to an attractive interaction. We must realize that the electrons and the ions move very quickly. The electrons are always passing through the field of ions, which are themselves vibrating rapidly, and the effect of the virtual phonon on the ionic vibrations depends very much on their relative frequencies. If the lattice frequency, ν_g (usually taken to be about one-half of the maximum Debye frequency), is much higher than that of the virtual phonon, ν_v, then as in the case of ordinary forced vibrations, the system will tend to oscillate so that it is in phase with the driving phonon. If, however, $\nu_g \ll \nu_v$ then the ionic fluctuation will be out of phase with the original virtual phonon. Thus the original charge fluctuation which occurred when electron 1 changed its state, and which under almost static conditions would have tended to be compensated by the ionic motion, will now be under-compensated if the ions move out of phase, so that the Coulomb repulsion on electron 2 is magnified, or it can be overcompensated if the ions move in phase so that the Coulomb repulsion is diminished. Under this last condition it is possible for the ionic influence to overcome that of the original electron charge fluctuation so that electron 2 is actually attracted to electron 1 rather than being repelled from it. If this occurs the energy of the system will be reduced.

We should at this stage make it clear that this attraction is one which is only produced by rapid fluctuations of the charge density. The ordinary Coulomb field of the electrons moving through the lattice is still repulsive and is assumed to be the same in both the normal and the superconducting states. By interacting with the lattice in the way which we have described, however, an attractive force can be produced which results in a reduction in the energy of the system, and thus there will be a tendency for two electrons to remain together so that their motions are correlated. One can form a picture of a pair of electrons pulsating rapidly and so moving closer to one another. If the anti-phase condition held, then the electrons would tend to move away from one another. Processes involving the simultaneous interaction of three or more electrons might also be possible but they are not necessary in order to account for superconductivity.

Under what conditions will an attraction take place? Clearly if the interaction between electrons and phonons was very weak, then there would be little likelihood of a virtual phonon exchange. Thus one can account for the strange circumstance that substances such as the alkali and noble metals, which are normally very good electrical conductors, and which do not therefore interact strongly with the phonons, are not superconductors, whereas many of the bad conductors are.

Having established that some type of pair interaction or correlation exists in the superconducting state we must investigate the nature of the pairs. Do the electrons link up with any partner or is there some special relationship between the members of a pair? This problem is essentially a statistical one. Whilst many types of pair might interact there is only one relationship which is overwhelmingly probable. It is very simple to see what the rule is. We first assume that, whatever the nature of the interaction, it will only involve electrons which have energies close to the Fermi energy, of the order of kT_c. Thus the wave vector of the interacting electrons will lie within the narrow shell shown on the left-hand sphere of Fig. 6.21, which has limiting radii $k_F + dk$, $k_F - dk$, where k_F is the wave vector at the Fermi surface and dk is the change in k corresponding to an energy charge of kT_c. The wave vectors of the electrons before pairing are shown as $\mathbf{k}_1$ and $\mathbf{k}_2$ together with their resultant $\mathbf{K}$. Since after the interaction the total wave vector, $\mathbf{K}$, is conserved (6.69), the volume of phase space into which $\mathbf{k}_1$ and $\mathbf{k}_2$ may be scattered can be found by constructing a similar spherical shell centred on the right-hand end of $\mathbf{K}$. The only possible wave vectors which the two electrons can have after the interaction, and yet maintain $\mathbf{K}$ unchanged, will be those which lie in the shaded overlap region. Thus for the situation as drawn, the probability of interaction is not very large. It is clear, however, that if the two spheres overlap completely so that their centres coincide, the amount of phase space available for transitions will be a maximum. When this occurs the total wave vector $\mathbf{K} = 0$. Hence the most probable type of interaction will be one in

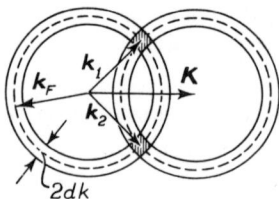

FIG. 6.21. Electron pairing within a thin shell at the Fermi surface. When two electrons with wave vectors $\mathbf{k}_1$ and $\mathbf{k}_2$ interact, their final wave vectors must lie within the shaded region if the total wave vector $\mathbf{K}$ is to remain constant. Thus the greatest probability for interaction occurs if $\mathbf{K} = 0$, i.e. if the spheres completely overlap (after Cooper, 1960).

which the individual electrons of a pair have equal but opposite $\mathbf{k}$ values. From general considerations of exchange interaction (section 9.16) it also seems probable that the two electrons should have opposite spin $(\downarrow\uparrow)$. Thus our picture of the new state is one in which pairs of electrons in states $+\mathbf{k}\uparrow$, $-\mathbf{k}\downarrow$, are attracted together by a pulsating dynamic interaction. Of course, a particular electron will not pair off with another special electron for any length of time: there will be a continual interchange of partners so that at any instant all the pairs within a coherence volume δ^3 ($\approx 10^{-12}$ cm³) will be interacting with each other. Since the fraction of the electrons involved in superconductivity is $kT_c/\mathscr{E}_F \approx 10^{-4}$, the number of interacting pairs in a volume δ^3 will be

$$10^{22} \times 10^{-12} \times 10^{-4} = 10^6.$$

We therefore see how the theory gives rise to a correlation mechanism suggestive of Pippard's coherence concept (section 6.26).

At absolute zero the occupation of pair states is rather different from that of the normal metal. At first sight one might think, since from the Fermi–Dirac statistics all states up to k_F are occupied whilst those above are vacant, that the occupied states could pair off amongst themselves very easily, with a consequent reduction in energy of the system. This cannot occur, however, because the virtual phonon process specifically requires a *change* of state of electron 1 from $\mathbf{k}_1$ to $\mathbf{k}_1'$, and similarly for electron 2. Thus the $\mathbf{k}_1'$, $\mathbf{k}_2'$ must be vacant before the process can take place and this will not be possible with the ordinary distribution. Thus for pairing to occur, some states below k_F must be empty and some above must be occupied—even at $0°$ K (Fig. 6.22). The increase in energy due to the occupation of these higher $\mathbf{k}$ values is more than compensated for by the virtual phonon interaction.

6.31. *The superconducting ground state*

In the BCS theory we consider the pairing of electrons which each have a normal energy ϵ_k relative to the Fermi surface (i.e. not taking

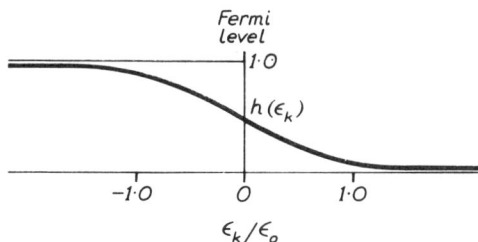

Fig. 6.22. The probability $h(\epsilon_k)$ that a state of energy ϵ_k is occupied in the superconducting ground state (6.70).

account of the correlation energy). Thus in the normal state the two electrons would have a total energy $2\epsilon_k$. The probability function $h(\epsilon_k)$ for the occupation of pairs is then calculated by determining the conditions under which the free energy of the system is a minimum when the correlation energy is taken into account. At $T = 0°$ K the results are

$$h(\epsilon_k) = \tfrac{1}{2}(1 - \epsilon_k/\mathscr{E}_k), \tag{6.70}$$

where

$$\mathscr{E}_k = (\epsilon_k^2 + \epsilon_0^2)^{\frac{1}{2}} \tag{6.71}$$

and

$$\epsilon_0 \approx 2h\nu_g \exp\{-1/F(0)V'\}. \tag{6.72}$$

$F(0)$ is the density of states of the normal metal for electrons of one spin at the Fermi surface and V' is the interaction energy associated with the virtual phonon process.

The interpretation of these expressions can be a little confusing. Let us emphasize that in Fig. 6.22, which shows the behaviour of $h(\epsilon_k)$, the energy ϵ_k is only the kinetic energy of the electron and not the total energy. The latter will be less because it is reduced by the pairing process. $\mathscr{E}_k$ can in fact be interpreted as the total energy *below* the Fermi surface when the pair correlation has been taken into account. Thus an electron which originally had kinetic energy of either $+\epsilon_k$ or $-\epsilon_k$ can now be considered to have an energy of $|\mathscr{E}_k|$ below the Fermi surface. One of the important consequences of the theory (Cooper, 1956) is that no matter how large the value of ϵ_k for a single electron, if pairing takes place the resultant energy will be *below* the Fermi level. Conversely we can consider that $+|\mathscr{E}_k|$ is the energy required to take one electron from a pair and raise it to the Fermi energy. It is important to realize, however, that all these statements are oversimplified. Because of the correlations which exist we cannot really speak about exciting one electron from a particular pair as in reality the whole configuration of pairs within the volume δ^3 is involved.

6.32. *Excited states and the energy gap*

It is now possible to see why this new state has rather special properties. If we try to excite an electron from the pair which has wave vectors $+\mathbf{k}$ and $-\mathbf{k}$, so that one electron remains, say, in state $+\mathbf{k}$ whilst the other goes to a new state $\mathbf{k}'$, then, although we have only altered the wave vector of one of the electrons, *both* of them have really been excited because we have destroyed the possibility of two sets of pairs, those for $\pm\mathbf{k}$ and for $\pm\mathbf{k}'$. Thus the energy required to do this will not be just $\mathscr{E}_k$, but it will be $\mathscr{E}_k+\mathscr{E}_{k'}$. From (6.71) we see that even if the excitation is as small as possible (ϵ_k and $\epsilon_{k'} = 0$), the excitation energy will still be $2\epsilon_0$. It will therefore be impossible to break up a pair unless energy of at least $2\epsilon_0$ is available, and hence the configuration of electron pairs is a very stable one. In this way an energy gap $2\epsilon_0$ between the ground and the excited states is introduced. This was another of the empirical results (section 6.27) which had to be explained.

6.33. *Density of states of a superconductor*

This concept of an energy gap can also be derived in a manner in which another aspect of the energy distribution may be described. Let us calculate the density of states per unit energy interval for the superconducting state. In the normal metal this can be written as $dN(\epsilon_k)/d\epsilon_k$, but as we have already described in the preceding section, in the superconducting state electrons which had an energy ϵ_k now have an energy $-|\mathscr{E}_k|$. Hence the density of states should be written as $dN(\mathscr{E}_k)/d\mathscr{E}_k$. Since, however,

$$\frac{dN(\mathscr{E}_k)}{d\mathscr{E}_k} = \frac{dN(\epsilon_k)}{d\epsilon_k}\frac{d\epsilon_k}{d\mathscr{E}_k} \tag{6.73}$$

we obtain using (6.71)

$$\frac{dN(\mathscr{E}_k)}{d\mathscr{E}_k} = F(0)\frac{\mathscr{E}}{(\mathscr{E}_k^2 - \epsilon_0^2)^{\frac{1}{2}}} \equiv F(\mathscr{E}_k), \qquad (6.74)$$

where $F(0)$, the density of states at the Fermi surface, has been written for $dN(\epsilon_k)/d\epsilon_k$.

From this we see that within a region ϵ_0 of the Fermi level there are no available states. It is also interesting to observe that the function

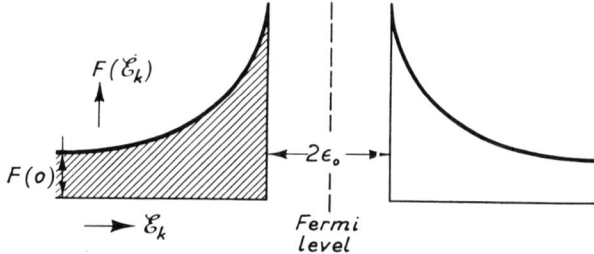

FIG. 6.23. The density of states, $F(\mathscr{E}_k)$, as a function of $\mathscr{E}_k$ for the superconducting state. At 0° K all the shaded states are occupied, but at any higher temperature there is some excitation across the energy gap to states on the right-hand side of the diagram.

(6.74) rises to a singularity at the edge of the gap. The density of the excited states, with energies $+|\mathscr{E}_k|$ above the Fermi level, will be exactly symmetrical to (6.74). Thus the density of states curve will be as shown in Fig. 6.23, in which the correlated pair states are separated from the excited states by an energy gap $2\epsilon_0$. The rapid rise in the density of states near the gap edge is due to the fact that the states which in the normal metal occupied the energy gap are now pushed into the regions on either side of it.

6.34. Zero resistance

The concept of pairing allows us to understand how zero resistance may be achieved, although no rigorous theoretical treatment has yet been made. When a current flows it is assumed that the electrons still pair off, but since there must be a finite resultant electron momentum, the total wave vector of all pairs cannot be zero. Under these circumstances, it will probably still be most favourable for all pairs to have the same total wave vector, although this will now be non-zero. One would also assume that energy $\sim 2\epsilon_0$ will still be necessary to break a pair. From this the explanation for the zero resistance of a superconductor follows quite naturally. For, unless in the collision process sufficient energy is available to overcome the pairing energy, the electrons will not be scattered and hence zero resistance will result.

6.35. *The energy gap above 0° K*

So far we have only discussed the situation at $T = 0°$ K. At higher temperatures there is always the possibility of thermal excitation from the virtual pair states. Because of these excitations the number of possible pairings will be reduced, since some states will now not be

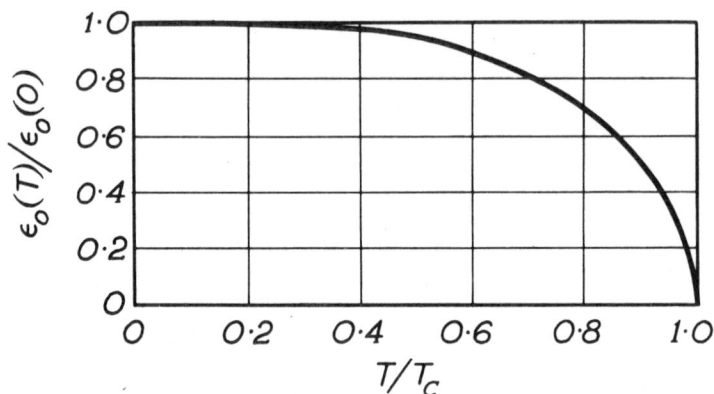

FIG. 6.24. The variation of the energy gap parameter $\epsilon_0(T)$ with the temperature as predicted by the BCS theory.

available for pair interaction. This reduces the value of the energy-gap parameter $\epsilon_0(T)$. Its temperature variation is as shown in Fig. 6.24. The decrease in $\epsilon_0(T)$ at higher temperatures occurs because there are fewer electrons which can be correlated and hence, when a pair is broken up, not so many other electrons are involved. At T_c, where we would expect that all pairs are excited so that the ordinary Fermi–Dirac distribution is restored, $\epsilon_0(T)$ becomes zero.

We should perhaps emphasize that $\epsilon_0(T)$ is not itself the interaction energy of a pair of electrons. This is assumed to be independent of temperature in the BCS theory. $\epsilon_0(T)$ arises because the pairs cannot be treated as isolated units but they must be considered with the other electrons as a complete ensemble. Since the pairing is a dynamic process, the strength of the interaction depends on the number of pair states which are involved, and since as the temperature is increased there are fewer pair interactions, the excitation energy of the system will be reduced.

The temperature at which $\epsilon_0(T)$ becomes zero is identified with T_c. From the theory this occurs when

$$kT_c = 1{\cdot}14h\nu_g \exp(-1/F(0)V').\tag{6.75}$$

This expression yields the isotope effect (section 6.25) since from the ordinary equation of motion of a harmonic oscillator the lattice vibrational frequency will be proportional to $M^{-\frac{1}{2}}$. If we compare (6.75) and (6.72) we find that the energy gap parameter at 0° K is given by $2kT_c/1{\cdot}14$, i.e.

$$\epsilon_0 = 1{\cdot}75kT_c\tag{6.76}$$

so that at $T = 0°$ K the energy gap itself will be

$$2\epsilon_0 = 3{\cdot}5kT_c.\tag{6.77}$$

We shall see that experiments in which the energy gap is measured directly give good agreement with (6.77).

From the expressions for the virtual pair distribution (6.70) and the excited-state distribution functions it is possible to calculate the free energy of the system as a function of temperature and from this, as we have seen in section (6.6), we can calculate the specific heat and the critical field relationships. As one would expect, the constant energy gap as $T \to 0°$ K yields an exponential specific heat at very low temperatures (Fig. 6.10). The experiments, however, give an exponent of $1{\cdot}5kT_c/kT$ which is what we should expect for an energy gap of $1{\cdot}5kT_c$. It should be recalled, however, that a single particle state requires the excitation of *two* electrons (section 6.32) and hence the energy gap from these experiments is really $\sim 3kT_c$ which is in reasonable accord with (6.77).

The BCS theory has so far been developed in detail only for the case of weak coupling between electrons and phonons, i.e. for $T_c \ll \theta$, where θ is the Debye temperature. Whilst this assumption is valid for most superconductors, it is not justified for two rather important ones, Pb ($T_c = 7{\cdot}2°$ K, $\theta \approx 90°$ K) and Hg ($T_c = 4{\cdot}2°$ K, $\theta \approx 70°$ K). It is interesting to note that some of the superconducting properties of these metals do differ from those of other superconductors; e.g. the deviations from the parabolic H_c relationship have opposite sign (Fig. 6.5), the heat conductivity at T_c shows a discontinuous instead of a gradual change of slope (Fig. 5.8) and, for Pb, the experimental value for the energy gap is considerably greater than $3{\cdot}5kT_c$ (sections 6.37, 6.38). Thus any detailed predictions from the theory should not be applied to these metals.

From this account of the BCS theory we can see its connexion with the two-fluid hypothesis which we discussed earlier (section 6.7). The paired electrons in states below the energy gap correspond to the super-electrons, whilst those in the excited states may be considered as the n-electrons. At $T = 0°$ K there is no excitation across the energy gap and thus there are no n-electrons. As the temperature is increased some excitation occurs and since $2\epsilon_0$ becomes smaller at higher T, the number of n-electrons rapidly increases, until at T_c, where the gap is zero, all the electrons are in normal states once again.

6.36. Experiments to measure the energy gap

The most outstanding success of the BCS theory has been in the prediction of a temperature-dependent energy-gap parameter. In addition to the specific-heat results which we have just mentioned there are two other types of investigation in which this parameter has been measured and the results have been found to agree very well with the theory. These are described in the following two sections.

6.37. *Infra-red transmission and reflection*

In the first set of experiments the infra-red transmission was measured through thin superconducting films. We have already described in section 6.27 the early work which showed that at 14 μ the surface resistance was unchanged when a metal became superconducting. This observation is easily explained, because the electromagnetic quanta will have an energy of 0·09 eV (equivalent to 1000° K) and this will be sufficient to excite many electron pairs across the energy gap, where they will behave like normal electrons. Since at 1° K, $kT \approx 10^{-4}$ eV which corresponds to 1 mm photons, the influence of the energy gap will only be detected if experiments are made with infra-red radiation of much lower energy (i.e. longer wavelength). Experiments to exploit this range were made by Glover and Tinkham (1957) and Ginsberg and Tinkham (1960) who measured the transmission of infra-red radiation of from 0·1 to 1 mm wavelength through films which were thin compared with the penetration depth. From measurements of the frequency dependence of the power transmitted in the normal and in the super-conducting states it was possible to deduce the energy-gap parameter.

Experiments which are capable of a more straightforward interpretation, however, are those in which the change in reflectivity of a bulk metal is measured when superconductivity is destroyed (Richards and Tinkham, 1960). The specimens are made in the form of a non-resonant cavity with a bolometer mounted inside. The radiation reaching the bolometer is a measure of the reflectivity of the cavity. Results of these experiments are shown in Fig. 6.25 in which $(P_s - P_n)/P_n$ is shown plotted against the frequency, where P_s, P_n are the powers absorbed by the bolometer when the cavity is superconducting and normal, respectively. Quite a sharp absorption edge is found. This occurs at a frequency

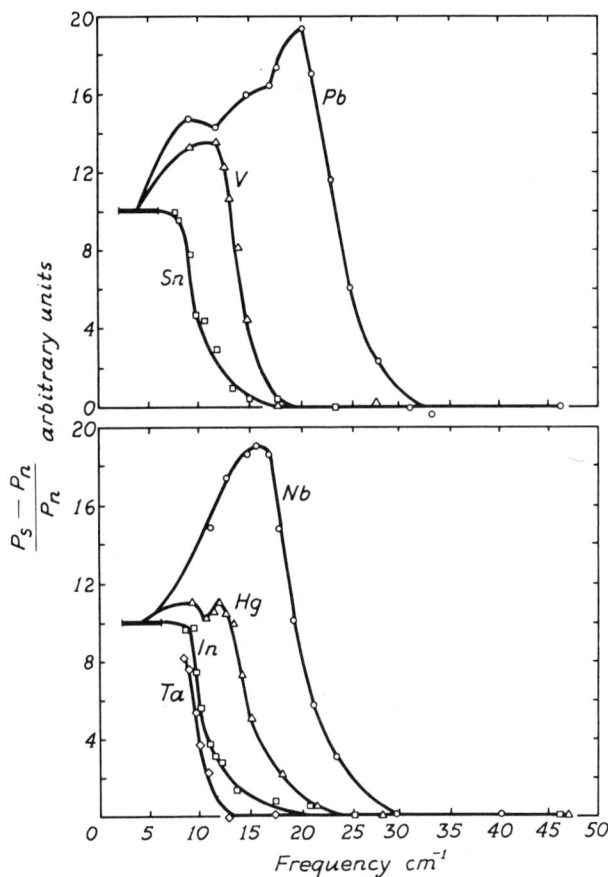

FIG. 6.25. Low-temperature bolometer absorption curves from reflectivity experiments for seven superconductors showing the sharp absorption edge which is identified with the parameter $2\epsilon_0$ (Richards and Tinkham, 1960).

which is identified with the energy gap parameter. The values of $2\epsilon_0$ which are obtained from these curves are given in Table 6.1.

6.38. *Electron tunnelling experiments*

In addition to the infra-red observations described in the preceding section there are some extremely elegant experiments in which a direct measure of the energy gap $2\epsilon_0(T)$ can be made. The principle of the experiments, which were first made by Giaever (1960) is as follows. Two thin superconducting films are deposited on top of one another, separated by a very thin (about 20 Å) oxide insulating film. If the metals were both in the normal state and a potential was applied between

TABLE 6.1

Values of the energy-gap parameter

(Richards and Tinkham, 1960)

Metal	Ta	Nb	V	Pb	Sn	Hg	In
$2\epsilon_0$ in units of kT_c	$< 3\cdot0$	$2\cdot8$ $\pm0\cdot3$	$3\cdot4$ $\pm0\cdot2$	$4\cdot1$ $\pm0\cdot2$	$3\cdot6$ $\pm0\cdot2$	$4\cdot6$ $\pm0\cdot2$	$4\cdot1$ $\pm0\cdot2$

the two films a current would flow between them, through the insulating layer, because of the quantum mechanical tunnel effect. The potential shifts the energy distributions of the two films relative to one another (Fig. 6.26). The current will depend on the product of the number of vacant states in one film multiplied by the number of occupied states of the same energy in the other. At zero potential equal numbers of electrons will flow in both directions and so the net current is zero. For small potential differences the current which flows is proportional to the potential, provided that the density of states of the two metals does not change in the potential range which is used. In a superconductor, however, we have seen (Fig. 6.23) that the density of states rises very rapidly at the edges of the energy gap. As the gap is approached, therefore, we should expect to see a marked variation in the current.

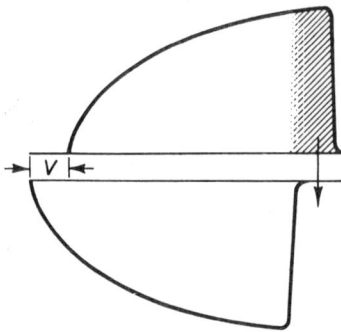

Fig. 6.26. The density of states function for two metals which have a potential V between them. The distributions are shifted so that a net tunnelling current flows from the upper to the lower specimen as shown.

The most informative experiments have been those in which both films are in the superconducting state. If they are both of the same metal the sequence of Fig. 6.27 occurs. Since the temperature is above $0°$ K there will be some excitation from the ground to the excited states across the gap. With zero potential (a) the Fermi levels will be coincident and zero current flows. If film 1 is made positive with respect to 2 its distribution moves to the left. At (b) only a small electron current can flow from 2 to 1 because the carriers in the left-hand band of 2 face the band gap of 1 and only the few excited electrons in the right-hand band of 2 can tunnel to vacant states in 1. When, however, the right-hand band of 1 becomes level with the left-hand band of 2, (c) then the large number of empty states in the right-hand band of 1 are available for the electrons from the left-hand band of 2. Hence there will be an increase in the tunnelling current. The potential for which this occurs will be

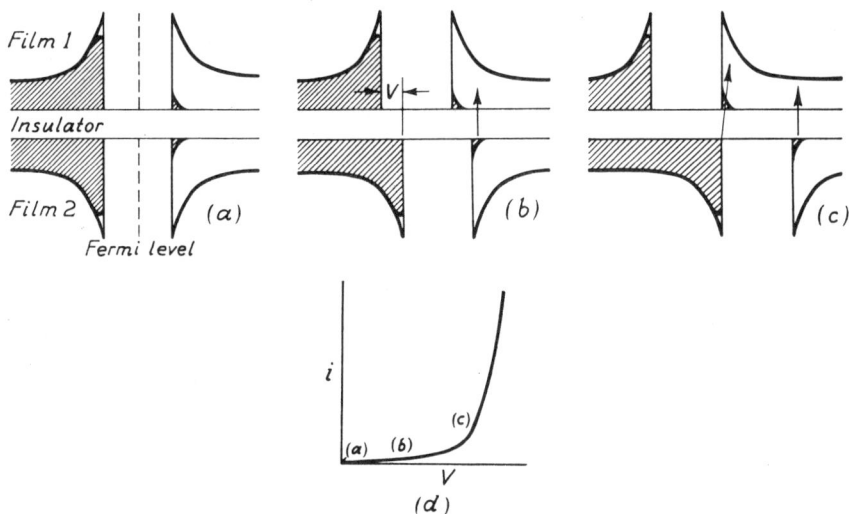

FIG. 6.27. Tunnelling between two superconducting films of the same metal. (a) Zero potential, no current flow; (b) potential less than $2\epsilon_0(T)$, very small current flow; (c) potential equal to or greater than $2\epsilon_0(T)$, large current due to tunnelling from ground states of lower film to excited states of upper film. (d) Diagram showing the relation between current and voltage in which the situations at (a), (b), and (c) are indicated.

equal to $2\epsilon_0(T)$, the energy gap. In Fig. 6.27 (d) a diagram of the current–voltage relation is given and Fig. 6.28 shows some of the earlier experimental curves from Giaever's work on an Al—Al_2O_3—Al sandwich. The rapid rise in current as condition (c) is reached is quite obvious. It will be noted that the potential for which it occurs increases with decreasing temperature showing that the band gap increases as the temperature is reduced, as in Fig. 6.24.

Unfortunately with Al ($T_c \approx 1.2°$ K) it has not yet been possible to make experiments at a temperature which is sufficiently low for the limiting value of $2\epsilon_0$ to be reached (6.77). This has now been achieved with metals which have a high value of T_c (e.g. Pb). The most interesting experiments have been those in which the two films were of different metals. A typical set of investigations was that made on an aluminium-lead sandwich (Giaever, 1960b) in which both metals were super-conducting. The situation is now slightly different from that shown in Fig. 6.27, since the Al gap is much narrower than that for Pb and there is more thermal excitation to the states above the gap. Initially with zero potential between the films the Fermi levels are coincident and the

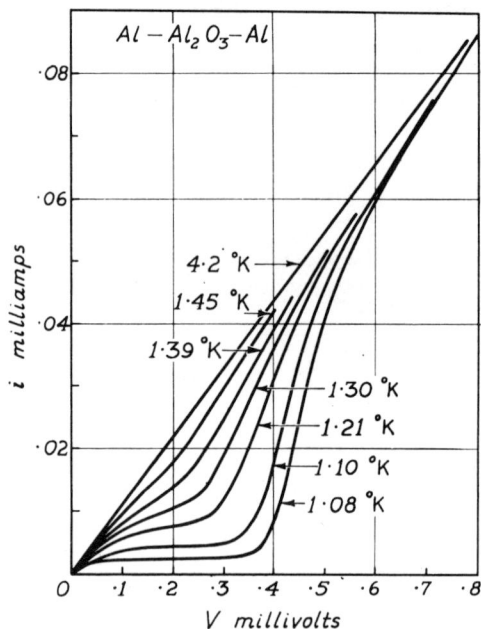

FIG. 6.28. The tunnel current between two aluminium films separated by an Al_2O_3 film, as a function of voltage. This gives an experimental verification of Fig. 6.27 (d) (Giaever, 1960a).

situation is as in Fig. 6.29 (a). If a positive potential is applied to film 1 there will be a net flow of electrons from 2 to 1 and as the left-hand edges of the two gaps come into coincidence, (b), at $V_{max} = \epsilon_0(2) - \epsilon_0(1)$ there will be an increase in current. This arises because at the temperature of the experiment ($\sim 1.1^\circ$ K) there are still a considerable number of vacant states in the lower band of metal 1 (Al) since its band gap is not very large. At 0° K this would not occur. As soon as the potential is increased farther so that the band gap of film 1 starts to pass over the left-hand band edge of film 2, the current drops rapidly because there are now few states available, (c). This continues until we reach stage (d) when the right-hand band edge of film 1 comes into coincidence with the left-hand edge of that of film 2 at $V_{min} = \epsilon_0(2) + \epsilon_0(1)$. Beyond this, the current will start to increase once again. Thus the current will increase from (a) to (b), decrease from (b) to (d) and it will increase again beyond (d). This is shown very clearly in Fig. 6.30. The potential difference between the minimum and maximum of the curve is $V_{min}-V_{max}$ and this gives a direct measure of $2\epsilon_0(1)$ the energy gap of film 1 (i.e. that with the smaller energy gap). Once $\epsilon_0(1)$ is known, $\epsilon_0(2)$ can be found from either V_{max} or V_{min}. In this way $2\epsilon_0$ was found to be $4{\cdot}35\pm0{\cdot}10$

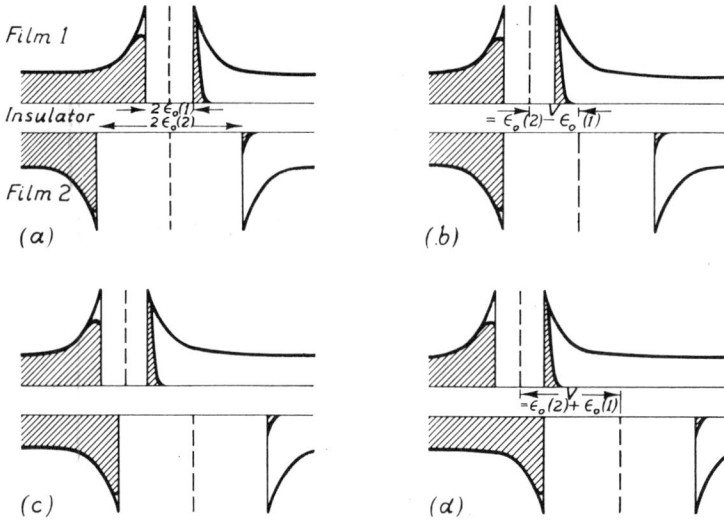

FIG. 6.29. Tunnelling between two superconducting films of different metals (a) $V = 0$; (b) $V = \epsilon_0(2)-\epsilon_0(1)$; (c) V slightly greater than in (b); (d) $V = \epsilon_0(2)+\epsilon_0(1)$.

kT_c for Pb and $3{\cdot}63\pm0{\cdot}1$ kT_c for In. These values are in reasonable agreement with those obtained by Richards and Tinkham (1960) from infra-red reflection experiments (Table 6.1, p. 200).

One interesting point about the curve of Fig. 6.30 should be noted. The superconducting tunnel sandwich has a negative resistance over part of its range and it therefore lends itself to being developed as an electronic device which would be capable of being used as an oscillator or an amplifier.

In this chapter we have only mentioned a few of the aspects of super-conductivity which can be accounted for by the BCS theory. It has, however, been developed to give a good quantitative description of most of the important properties of superconductors—penetration depth, ultrasonic attenuation, nuclear spin relaxation, thermal conductivity, etc. For further details of this work and for a review of the developments by other theoretical groups the reader is referred to the article by Bardeen and Schrieffer (1961).

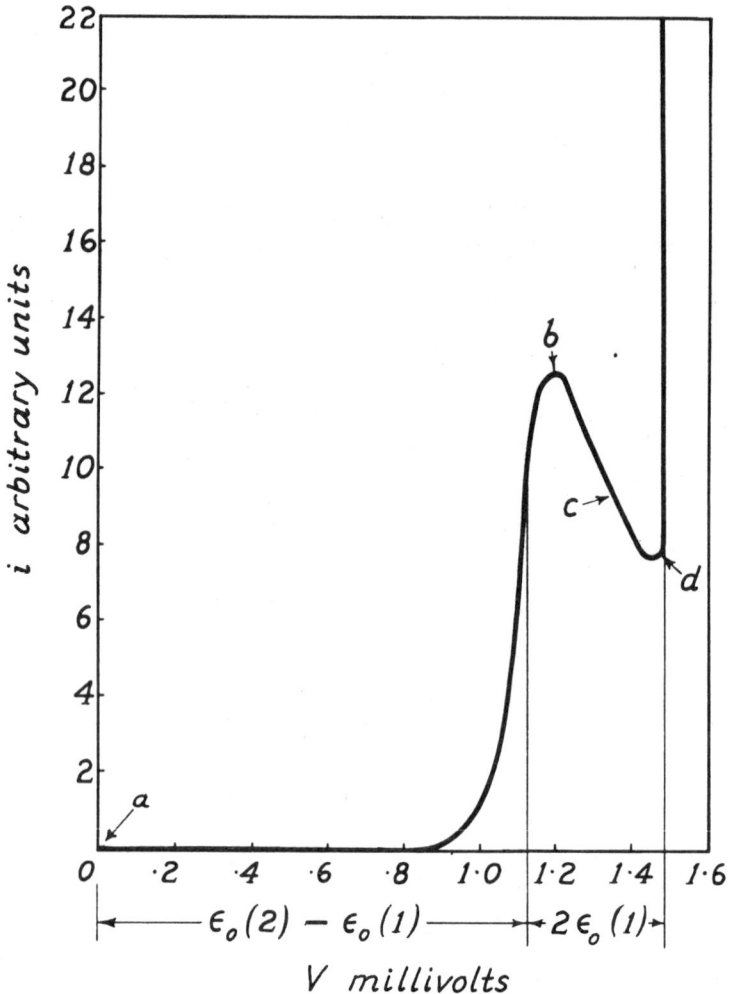

FIG. 6.30. Current–voltage characteristics for the tunnelling current between aluminium and lead. The positions for the situations (a), (b), (c), and (d) of Fig. 6.29 are indicated (Giaever, 1960b).

6.39. Superconducting devices

It is not surprising that advantage has been taken of the unique properties of superconductors in order to produce devices which can operate with no energy dissipation. Some of the more important and interesting of these are described in the following sections.

6.40. *Solenoids*

The existence of alloys with very high critical fields (section 6.23) soon suggested the possibility that some of them might be able to be used to make solenoids in which no power was dissipated. It was found, however, that the current required to destroy superconductivity was very much less than that which would produce a field H_c at the surface of the specimen; it was, in fact, much closer to that at which the field first penetrated the specimen. Very recently, however, alloys have been discovered which are capable of carrying extremely high current densities in high fields without losing their superconducting properties. Outstanding amongst these is Nb_3Sn, which has the highest transition temperature (18° K) yet found (Matthias, Geballe, Geller, and Corenzwit, 1954) and which can remain superconducting with a current density of 10^5 amp cm^{-2} in a static field of 88,000 oersted. Some typical curves are shown in Fig. 6.31. In pulsed fields (the only method, as yet, of making magnetic measurements in fields of over 10^5 oersted) Nb_3Sn has been found to remain superconducting up to 185,000 oersted (Arp, Kropschott, Wilson, Love, and Phelan, 1961). Unfortunately it is rather a difficult material to fabricate and it has not yet been possible to make it in the form of thin wires except by making it as a core within a niobium tube. For smaller fields $Nb+25$ or 33 at. per cent Zr has been used with success (Berlincourt, Hake, and Leslie, 1961). The alloys can be produced as wires (down to 0·005 inch diameter) and solenoids are commercially available which will produce a field of 60,000 oersted at 4·2° K.

The basic simplicity of these superconducting solenoids compared with the complicated engineering and cost of a conventional high power, water-cooled electromagnet is such that a very considerable effort is now being put into their development. It now seems clear that these alloys fall into a special category called type 2 superconductors in which the coherence length δ is *smaller* than the penetration depth λ. It can then be shown that the surface energy between the normal and superconducting phases will be negative (cf. section 6.19) and so there will be a tendency for the intermediate state to split up with a large number of fine superconducting threads which have a high transition field.

6.41. *The cryotron*

Several devices have been designed which exploit the unusual properties of a superconductor but none has been more assiduously developed than the high-speed circuit switching and memory elements which have been called cryotrons. The original device which was invented by Buck

(1956) is shown diagrammatically in Fig. 6.32. It consists of a tantalum wire, known as the gate, which is surrounded by a niobium coil called the control winding. Both circuits are superconducting at 4·2° K but

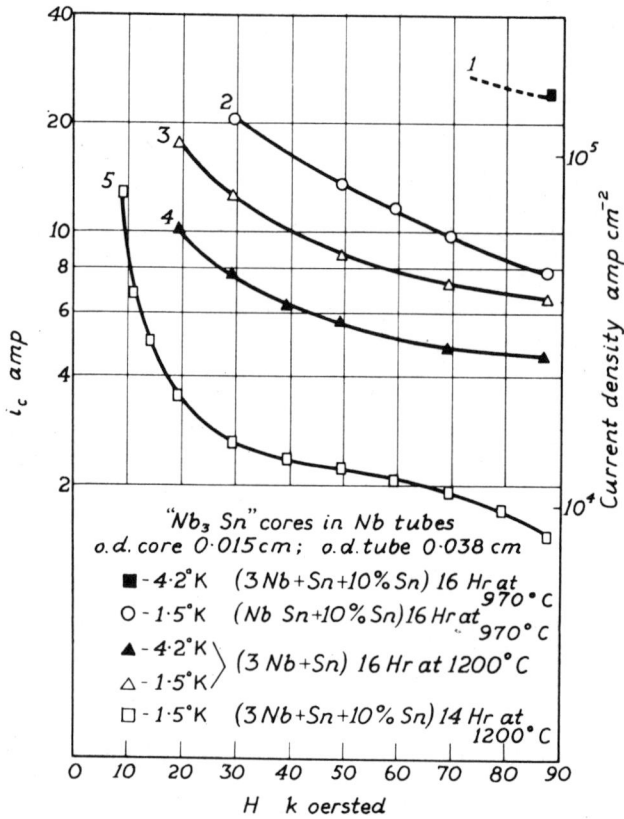

FIG. 6.31. The critical current as a function of the transverse applied magnetic field for Nb-clad cores of Nb₃Sn. The specimens are made by heating a Nb tube filled with the appropriate amounts of Nb and Sn for the time and at the temperature indicated on the figure. '+10%' means that 10 per cent by weight more Sn was added than is required to form Nb₃Sn, assuming that there is no reaction with the tube itself. Each experimental point represents the maximum current which could be passed through the specimen, for which no voltage drop along the sample was observed (Kunzler, Buehler, Hsu, and Wernick, 1961).

because Nb has a much higher transition field than Ta, a small current through the Nb control winding can produce sufficient field to make the Ta gate wire become normal without destroying the superconductivity of the Nb. The change in the gate resistance from zero to a finite value is equivalent to breaking the gate circuit. Thus a large current through the gate may be controlled by a small one through the control winding, i.e. the device acts like a relay.

To demonstrate how cryotrons may be used, consider the simple two-

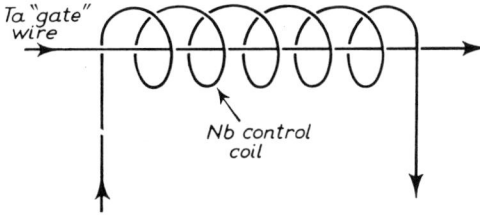

FIG. 6.32. A diagram showing the principle of the original Nb-Ta cryotron.

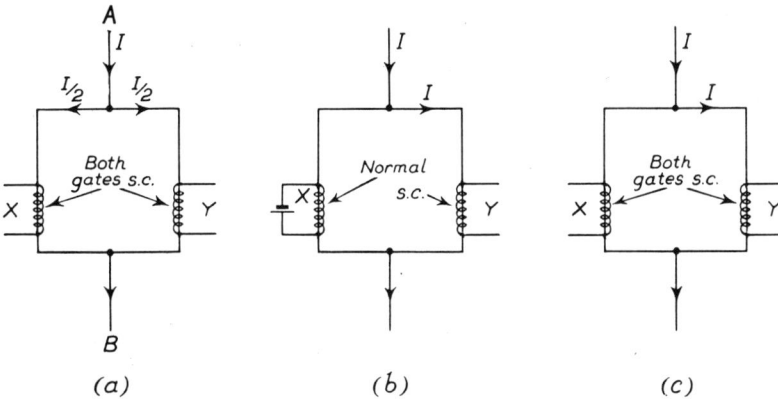

FIG. 6.33. A two cryotron memory circuit. (a) Both gates superconducting. On connecting AB to a battery equal currents flow through each arm X and Y. (b) X gate made normal, all the current flows through Y. (c) X gate again made superconducting, but all the current continues to flow through Y.

cryotron circuit shown in Fig. 6.33 in which the cryotrons are connected by superconducting wires. If a potential is applied between A and B, then current will flow. Since the circuit is superconducting, the ratio of the current in the arms X and Y will be inversely proportional to the ratio of the self inductance of the arms, so that, if they are similar to one another, equal currents will flow in each arm. Let the control winding of cryotron X be energized. The gate will become normal and hence all the current will now flow through arm Y (Fig. 6.33 (b)). At this stage an interesting situation can develop; because if cryotron X is de-energized so that its gate becomes superconducting once more, the current *will still continue to flow through Y only*, Fig. 6.33 (c). The current is not re-established through X because a potential will be necessary to enable it to overcome the back e.m.f. from the self inductance of X. Thus a short pulse through the control winding of X will divert the main current through Y and it will continue to flow through Y after the pulse has finished. The circuit can therefore act as a memory element. It is still necessary, however, to be able to 'read out' this information. This can be done with two more cryotrons which are

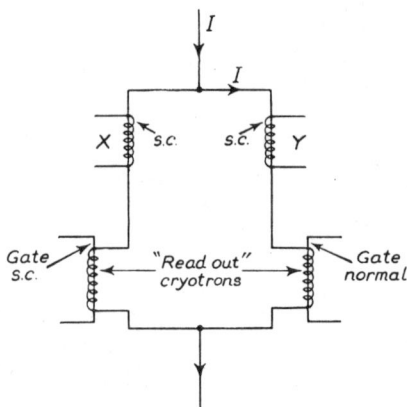

FIG. 6.34. The circuit of Fig. 6.33 with read-out cryotrons added, with the current situation as in Fig. 6.33 (c).

connected, one in each arm, as in Fig. 6.34, so that their control windings are in series with the gate circuits of the two original cryotrons. It is clear that the arm through which the current is flowing will energize the read-out cryotron in series with it and hence this read-out gate will become normal. This can be detected by another circuit in series with the gate. These examples are, of course, only very simple circuits compared with those which can be devised. Some others are described in the review article by Young (1959).

Nb-Ta cryotrons have been the forerunners of much cheaper and more compact devices which are made in the form of evaporated thin films. One such arrangement is shown in Fig. 6.35 in which the control and gate circuits are thick and thin strips of superconducting metal (e.g. Pb) respectively, which are insulated from one another. The active part of the circuit is just in the region where the strips cross over one another. Other types have been devised in which the gate and control films are parallel. These cryotrons have a much lower self inductance than the Nb-Ta type and so have a very fast response time. They can be produced cheaply in large numbers by evaporation complete with their connecting circuits, and they can be accommodated in a small volume. All of these factors are important in modern computer design and a very considerable research effort is being expended in their development to this end.

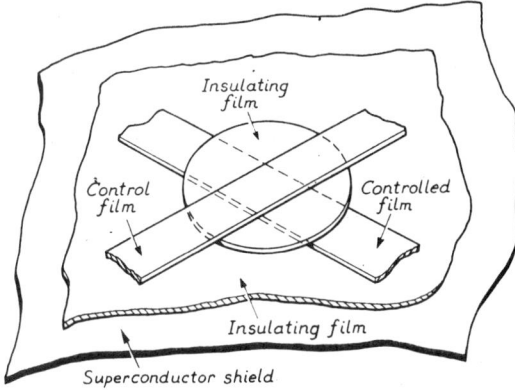

Fig. 6.35. An example of a thin film cryotron in which the control and the gate circuits are crossed films which are insulated from one another (Young, 1959).

6.42. *Superconducting bolometers*

The very rapid change in the electrical resistivity of a metal at its transition temperature has been used in order to make very sensitive infra-red detectors. A thin wire (usually tin) is attached to an absorber —aluminium foil is very suitable—which is maintained very accurately at a constant temperature within the transition range of tin. Very small heat inputs then cause relatively large changes in the electrical resistance of the tin and this can be amplified and recorded. One such bolometer is described by Hulbert and Jones (1955) which is able to detect 3×10^{-5} erg sec^{-1}.

Other superconducting devices such as modulators and transformers have been designed for the measurement of very small potentials, particularly those encountered in low-temperature thermo-electric experiments. These are discussed in the review article by Young (1959).

7

SEMICONDUCTORS

In no branch of solid-state physics has so much effort been expended during the last decade as in the investigation of the properties of semiconductors. Many hundreds of papers on the subject are published each year and whilst most of these do not have any direct low-temperature interest, even so, we can do little more in this book than survey in a general fashion the main types of experiment which have been made at low temperatures and the reasons why these are necessary. For the most part we shall specifically consider the elemental semiconductors, germanium and silicon, although much of what we say can be applied with certain modifications to the group III–V compound semiconductors such as indium antimonide, gallium arsenide, etc.

General principles

7.1. Pure semiconductors

A semiconductor is a material which differs from a metal in that, at the absolute zero, one of its Brillouin zones is completely filled by electrons, whilst the next higher zone is completely empty. There is no overlap of the energies associated with the two zones. In the case of metals (see section 4.2) it will be recalled that the zones are never completely filled and so it is always possible for electrons to be excited to higher states by the application of an electric field. In this way an unbalance in the momentum distribution may be achieved and an electric current flows. In a semiconductor at the absolute zero no current can flow because all the possible electron states in the valence band are filled, Fig. 7.1 (*a*).

A semiconductor differs from an insulator, however, in that the energy gap, $\mathscr{E}_g$, between the filled (or valence) band and the empty (or conduction) band is sufficiently small (of the order of an electron-volt) that at ordinary temperatures some electrons can be excited from the valence to the conduction band (Fig. 7.1 (*b*)). Electrical conduction can then take place when an electric field is applied to these electrons and it will be enhanced by the motion of the corresponding holes which have been created in the valence band. As the temperature is increased, more electrons will be excited into the conduction band and so the electrical conductivity will increase. The behaviour of semiconductors is therefore in direct contrast to that of metals for which (see section 4.5) the electrical conductivity *decreases* as the temperature is raised. The value of the energy gap for some substances is shown in Table 7.1. The difference between that for diamond (which is usually a good insulator)† and the semiconductors should be noted. It is clear from these figures that studies involving the change in population of the levels must be taken to low temperatures before a complete picture of semiconducting behaviour can be obtained.

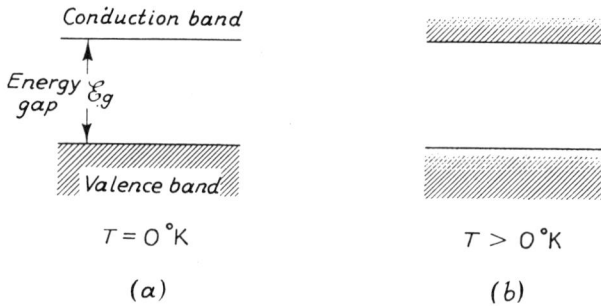

FIG. 7.1. Simplified energy-level diagrams of an ideally pure semiconductor (a) at 0° K, so that the valence band is full and the conduction band empty and there is no conduction; (b) for $T > 0°$ K, there is some excitation of electrons across the band gap which enables conduction to take place within the bands.

TABLE 7.1

	Diamond	Si	Ge	PbTe	InSb
Energy gap, $\mathscr{E}_g$, in eV	6 to 7	1·1	0·75	0·63	0·18

7.2. Impurity levels

If the energy levels of semiconductors were indeed just as is shown in Fig. 7.1 then their behaviour would be quite simple and, in fact, rather uninteresting. In practice the tremendous interest in this subject has been stimulated by the fact that the properties of semiconductors can be strongly influenced by the presence of quite small quantities of impurity atoms. The atoms which are usually studied are those which are in the groups of the periodic table which lie on either side of the semiconductor, e.g. for germanium, In and Ga (group IIIa) and Sb and As (group Va), and for silicon B (group IIIa) and P (group Va), although some other impurities, e.g. Cu in Ge and Au in Si, have also been studied extensively. The effect of these impurity atoms is to introduce extra

† Some samples of diamond do show semiconducting behaviour and have a relatively low electrical resistivity.

energy levels into what was the forbidden energy gap. For the group III and group V atoms there is only one level and this is very close (of the order of 10^{-2} eV) to the top of the valence band or to the bottom of the conduction band respectively. Such levels are called 'shallow' states. Other impurities often have levels which can be anywhere within the forbidden band. Idealized schemes are shown in Fig. 7.2 (a), (b).

The reason for the existence of the shallow states is quite easy to see. A 5-valent atom which is in a substitutional position in the diamond

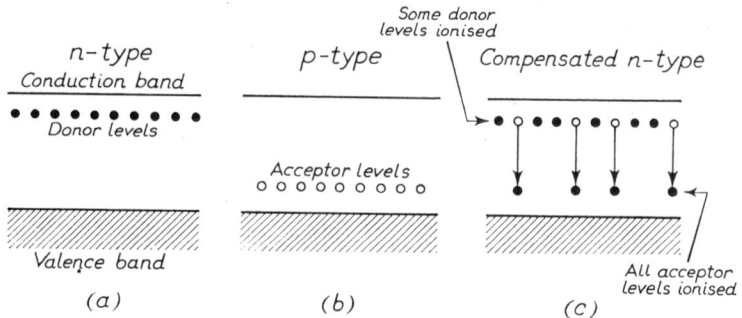

FIG. 7.2. Simplified energy-level diagrams showing the position at 0° K for a semiconductor containing (a) donor impurities (e.g. group V elements), (b) acceptor impurities (e.g. group III elements), and (c) both donor and acceptor impurities, with an excess of donors, producing a compensated n-type sample in which, even at 0° K all the acceptors and some of the donor levels are ionized.

type lattice of a quadrivalent element such as germanium has one electron more than the germanium atoms and so when it joins up with all four covalent bonds of its surrounding germanium neighbours it then has one electron over which is not strongly bound. The energy required to remove this electron from the influence of its atom will be relatively small, particularly since it is moving in a medium of high dielectric constant, ϵ, which will reduce its potential energy. An estimate of the energy can be made by assuming that the ionization energy of the impurity atom is the same as that for a hydrogen atom, reduced by a factor of $1/\epsilon^2$. Since ϵ is about 16 for Ge and 12 for Si this ionization energy is very small—a few hundredths of an electron-volt. In terms of the band theory one can say that this very small amount of energy has enabled an electron to be excited from the impurity atom to a slightly higher state in the conduction band where it is free to travel through the crystal. Thus electrical conduction can occur. An impurity atom, such as the one from group V which we have considered, which donates an electron to the conduction band is called a *donor atom* and the state associated with it before it is ionized is called a *donor* level. Since the donor level has an energy relative to the bottom of the conduction band which is very much less than that of the energy gap, it can be seen that the presence of such levels can alter the electrical properties very profoundly; no longer does thermal activation of electrons right across the gap have to occur before conduction can take place.

An entirely analogous situation arises with a group III impurity. This has one less electron than the germanium, and its bonding would be more satisfactory if it could obtain an extra electron from the valence band, thereby leaving a mobile hole there. We can treat this situation in an analogous manner to that which we used for the 5-valent impurity if we assume that before the excitation the atom had already acquired the extra electron plus a bound hole (which will cancel out the extra electron). On excitation, the bound hole must be freed from the impurity atom so that it is mobile. The ionization energy of the hole to the valence band is of a similar order of magnitude to that for the extra electron in the 5-valent impurity. In terms of the energy-band model we can say that a hole has been excited by a very small amount of energy to the valence band. Thus once again electrical conduction can occur by a mechanism involving a very small activation energy. An impurity atom such as the one from group III, which *accepts* an electron from the valence band, leaving a hole there, is called an *acceptor atom* and the state associated with the atom before it is ionized is called an *acceptor* level. The conductivity associated with acceptor atoms can be most easily considered by the motion of the excited holes in the valence band. Material which has a predominance of donor atoms in which the conduction is by the negative charges in the conduction band is termed n-type, whereas samples which have a predominance of acceptor atoms so that conduction is by the positive holes in the valence band are termed p-type.† Properties of the semiconductors which are characteristic of the pure material, i.e. in which excitation occurs from the valence to the conduction band across the main gap $\mathscr{E}_g$ are called 'intrinsic', whilst those which are concerned with the impurity levels are called 'extrinsic'.

7.3. *Multiple levels and compensation*

In practice the behaviour of impure semiconductors is complicated by the following facts. (*a*) A given impurity atom can have more than one donor or acceptor level associated with it. This occurs if the valency

---

† Mnemonic—do*n*or *n*egative, acce*p*tor *p*ositive.

difference between the semiconductor and the impurity is greater than unity, because then double or triple ionization becomes possible. (b) However carefully a sample is prepared it always contains *both* acceptor and donor atoms. The electrons in the donor levels will then prefer to fall into the lower acceptor levels rather than be excited into the conduction band. This then neutralizes the effect of some of the acceptor atoms since there will now be no inducement for electrons to be excited from the valence band to those levels (Fig. 7.2 (c)). If there were exactly the same number of acceptor as donor levels, the material would behave electrically as the simple semiconductor depicted in Fig. 7.1. This neutralization of the effect of one set of levels by another is called *compensation*.

7.4. *Data which determine the properties of a semiconductor*

The most important data which we require for a given sample of a semiconductor are the effective number of charge carriers which are excited at any temperature, the numbers of the various donor and acceptor atoms, and the actual position of the impurity energy levels. In addition to this we must also know the mobility of the electrons or the holes in the bands. Enthusiasts also require the lifetimes of these charge carriers, i.e. the time before they recombine, either with an ion, or with a carrier of opposite charge (electron-hole recombination) after which they are unable to take part in the current flow. We shall see that low-temperature measurements are needed in order to obtain all this information.

The two standard measurements which are taken on semiconductors are the electrical resistivity and the Hall coefficient. The latter quantity is important because, as we shall see, it enables us to calculate both the effective density n_c of charge carriers and their sign. Once we know n_c and the resistivity, ρ, the mobility, μ, may be calculated from the relation which is essentially the definition of the conductivity,

$$1/\rho = e\mu n_c \tag{7.1}$$

where e is the electronic charge.

7.5. *The Hall coefficient*

If a current flows along a conductor which is in a magnetic field whose direction is perpendicular to the direction of current flow, then an e.m.f., the Hall e.m.f., is generated across the specimen in a direction perpendicular to the magnetic field, i.e. the current, the magnetic field, and the Hall e.m.f. are mutually perpendicular (Fig. 7.3). This phenomenon is called the transverse Hall effect. In general terms it is quite simple to see how it arises. The application of the magnetic field will cause the moving charges to deviate from their motion along the specimen and their paths will curve towards the sides, where a space charge will accumulate. This will produce an electric field which will eventually be sufficient to counteract the effect of the magnetic field on the other moving charges and they will then traverse the length of the specimen without deviation. It is this electric field due to the space charge at the sides of the specimen which gives rise to the Hall e.m.f.

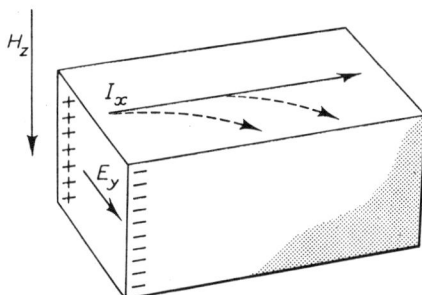

FIG. 7.3. The principle of the Hall effect. When an electron current I_x flows perpendicularly to the magnetic field, H_z, a space charge is set up as shown and this gives rise to the Hall field E_y. If the current was produced by a flow of positive charges, E_y would be in the opposite direction.

If we use the axes shown, then for electrons with a velocity v_x in a magnetic field H_z, the Hall field will be E_y in the direction indicated. If the charges are to be undeviated, the force due to E_y must be equal and opposite to that produced by H_z,

i.e. $$eE_y = ev_x H_z/c = j_x H_z/(n_c c), \tag{7.2}$$

where $j_x = n_c e v_x$, is the current density. Now the Hall coefficient R_H is defined as

$$R_H = \frac{V_H z}{I_x H_z}, \tag{7.3}$$

where I_x is the current through the specimen, V_H is the measured Hall e.m.f., and z is the thickness of the specimen in the z direction. Since $V_H = E_y y$ and $I_x = j_x yz$, we obtain

$$R_H = E_y/(j_x H_z) = 1/(n_c ce). \tag{7.4}$$

Thus, using (7.1), the mobility is

$$\mu = R_H c/\rho. \tag{7.5}$$

It will be noted that R_H depends on the sign of e and the reader should verify that if E_y is in a certain direction for a flow of negative charges, then it will.be in the opposite sense for a flow of positive charges in the reverse direction.

We should remark here that (7.5) is not always quite accurate. It must be modified slightly depending on the precise model of the system. Further details are given in section 7.13.

In the monovalent metals, measurements of the Hall effect give a negative value of R_H consistent with our belief in a flow of negatively charged particles; the magnitude of R_H is such that there is of the order of one moving charge per atom. In the more complicated metals, particularly those in which there is band overlap, R_H can be positive (e.g. in zinc and cadmium) and here it is assumed that most of the conduction occurs by the motion of positive holes. There has been very little interest in the Hall effect of metals at low temperatures, however, for the simple reason that the number of carriers in a metal is independent of temperature, and hence we shall not pursue this part of the subject any further in this book. For semiconductors in which carriers are excited into their conduction bands by thermal activation, low-temperature measurements are of vital importance and in the next sections we discuss how the carrier concentration should vary as the temperature is changed.

7.6. *The Fermi energy*

The treatment of carrier excitation.is very often obscured by a discussion and calculation of the Fermi energy $\mathscr{E}_F$ and we therefore intend to dispose of this topic in this section. The probability $g(\mathscr{E})$ that an electron will occupy a state of energy $\mathscr{E}$ is governed by the Fermi–Dirac statistics, i.e.

$$g(\mathscr{E}) = 1/\{\exp(\mathscr{E}-\mathscr{E}_F)/kT+1\}. \tag{7.6}$$

$\mathscr{E}_F$ is in reality a normalizing parameter. It is determined by ensuring that $\int_{-\infty}^{\infty} F(\mathscr{E})g(\mathscr{E})\,d\mathscr{E}$ is equal to the total number of electrons in the bands, where $F(\mathscr{E})$ is the density of states function; i.e. $F(\mathscr{E})\,d\mathscr{E}$ is the number of states between $\mathscr{E}$ and $\mathscr{E}+d\mathscr{E}$.

In a metal $\mathscr{E}_F$ has a straightforward physical significance. It is the energy of the highest occupied state at the absolute zero. In a semiconductor such an interpretation of $\mathscr{E}_F$ is not possible. Since, however, the number of electrons which are excited to higher states must be equal to the number of lower states which have been vacated, the normalizing condition for $\mathscr{E}_F$ leads to another kind of interpretation. If we assume that $F(\mathscr{E})$ has a symmetrical distribution about $\mathscr{E}_F$, then $\mathscr{E}_F$ lies in a position such that the number of electrons excited to an energy $\Delta\mathscr{E}$ above $\mathscr{E}_F$ is equal to the number of vacant states (or holes) at the same energy $\Delta\mathscr{E}$ *below* $\mathscr{E}_F$. This can be seen from the following argument. The number of electrons with energies lying between $\mathscr{E}'$ and $\mathscr{E}'+d\mathscr{E}$, $(\mathscr{E}' > \mathscr{E}_F)$ will be

$$F(\mathscr{E}')\,g(\mathscr{E}')\,d\mathscr{E} \;=\; F(\mathscr{E}')\,d\mathscr{E}/\{\exp(\mathscr{E}'-\mathscr{E}_F)/kT+1\}, \tag{7.7}$$

and since as we shall see $(\mathscr{E}'-\mathscr{E}_F) \gg kT$, this is approximately equal to

$$F(\mathscr{E}')\exp\{(\mathscr{E}_F-\mathscr{E}')/kT\}\,d\mathscr{E}. \tag{7.8}$$

The number of vacant states with energies lying between $\mathscr{E}''$ and $\mathscr{E}''+d\mathscr{E}$ $(\mathscr{E}'' < \mathscr{E}_F)$ will be

$$F(\mathscr{E}'')\{1-g(\mathscr{E}'')\}\,d\mathscr{E} \;\approx\; F(\mathscr{E}'')\exp\{(\mathscr{E}''-\mathscr{E}_F)/kT\}\,d\mathscr{E} \tag{7.9}$$

for $|\mathscr{E}''-\mathscr{E}_F| \gg kT$. Comparing (7.8) and (7.9) we see that *if*

$$F(\mathscr{E}') = F(\mathscr{E}'')$$

the condition for the electron occupation at $\mathscr{E}'$ to be equal to the hole occupation at $\mathscr{E}''$ is

$$\mathscr{E}_F-\mathscr{E}' = \mathscr{E}''-\mathscr{E}_F, \tag{7.10}$$

i.e.

$$\tfrac{1}{2}(\mathscr{E}'+\mathscr{E}'') = \mathscr{E}_F. \tag{7.11}$$

Thus $\mathscr{E}_F$ lies midway between the pairs of energies which have an equal probability of being occupied or vacant, respectively. It is fairly clear that at $0°$ K (7.11) holds independent of the condition $F(\mathscr{E}') = F(\mathscr{E}'')$, and hence $\mathscr{E}_F$ will lie midway between the highest occupied and the lowest unoccupied energy states.

If therefore we first consider a high-purity semiconductor in which there are no donor or acceptor levels, then at $0°$ K, the valence band is full and the conduction band is empty. Thus $\mathscr{E}_F$ will lie midway between the two, i.e. at $\tfrac{1}{2}\mathscr{E}_g$ (Fig. 7.4 (a)). As the temperature is raised and electrons are excited into the conduction band, then if the density of states at the bottom of the conduction band is equal to that at the top of the valence band, $\mathscr{E}_F$ will remain at the midpoint of the gap (Fig. 7.4 (b)). If, however, $F(\mathscr{E})$ is not the same for the two bands then $\mathscr{E}_F$ will move to higher or lower energies as T is increased so that the normalization condition still holds.

The behaviour of $\mathscr{E}_F$ in an n-type semiconductor can be deduced in a similar manner. Let the donor states be at an energy $\mathscr{E}_d$ above the valence band. Then at $0°$ K all states up to $\mathscr{E}_d$ are filled (assuming that there is no compensation) and all states above $\mathscr{E}_g$ are vacant. Thus $\mathscr{E}_F$ will lie midway between the donor levels and the edge of the conduction band, i.e. at $\tfrac{1}{2}(\mathscr{E}_d+\mathscr{E}_g)$ (Fig. 7.5 (a)). In this case, however, since $F(\mathscr{E})$

is *not* the same for the donor levels as for the conduction band, $\mathscr{E}_F$ will *not* remain constant when the temperature is increased. If $F(\mathscr{E})$ at the bottom of the conduction band is assumed to be the same as that for a free electron system then, for N_d donor levels,

$$\mathscr{E}_F = \tfrac{1}{2}(\mathscr{E}_d + \mathscr{E}_g) + \tfrac{1}{2}kT\log_e\{(4N_d h^3)/(8\pi m_e kT)^{\frac{3}{2}}\}. \qquad (7.12)$$

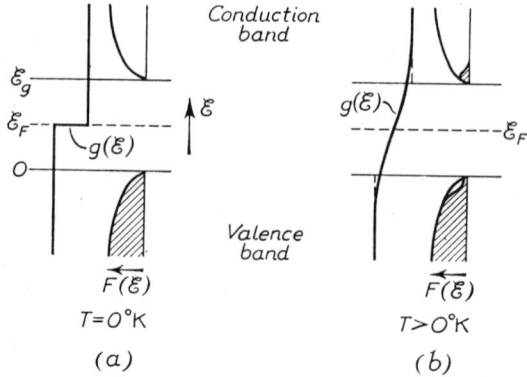

Fig. 7.4. The Fermi level $\mathscr{E}_F$ for an intrinsic semiconductor. (a) At $0°$ K $g(\mathscr{E})$ is a step function and the electrons only occupy the shaded states in the valence band. (b) Above $0°$ K $g(\mathscr{E})$ is still symmetrical about the mid-point of the band gap, provided that $F(\mathscr{E})$ has the same form in both bands. The excitation to the conduction band is indicated by the shading.

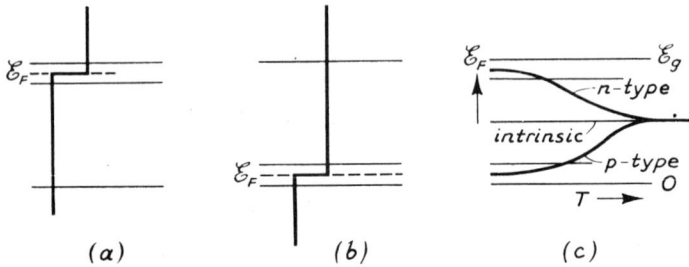

Fig. 7.5. The Fermi function at $0°$ K for (a) n-type, (b) p-type material, (c) the variation of $\mathscr{E}_F$ with temperature. At high temperatures when most of the excitation is intrinsic $\mathscr{E}_F$ will be the same for all samples of a given material.

A derivation is given by Dekker (1957), p. 311. The logarithmic term in (7.12) is usually negative (except for very high values of N_d) and so, as the temperature is increased, $\mathscr{E}_F$ decreases below the mid-point of the donor energy gap (Fig. 7.5 (c)). At sufficiently high temperatures all the donors will be ionized and most of the electrons in the conduction band will be excited from the valence band, just as in a pure sample. Thus eventually we return to the position where $\mathscr{E}_F$ is at $\tfrac{1}{2}\mathscr{E}_g$.

For a p-type specimen an analogous situation occurs. At $0°$ K, $\mathscr{E}_F$ will be midway between the acceptor levels and the top of the valence band (Fig. 7.5 (b)) and it will rise as the temperature is increased until at sufficiently high temperatures it will be at $\frac{1}{2}\mathscr{E}_g$.

We have given this discussion of the behaviour of the Fermi energy because $\mathscr{E}_F$ characterizes the type of material with which we are dealing. The most important use of the concept of the Fermi energy, however, lies in a topic which is outside the scope of this book—the properties of a junction between two different semiconductors.

7.7. Carrier excitation

We are now in a position to calculate the variation of the carrier concentration as a function of temperature. The number of electrons, n_e, which are excited into the conduction band will be

$$n_e = \int_{\mathscr{E}_g}^{\infty} F_c(\mathscr{E}) g(\mathscr{E}) \, d\mathscr{E}, \tag{7.13}$$

where $F_c(\mathscr{E})$ is the density of states in the conduction band. For $(\mathscr{E}-\mathscr{E}_F) \gg kT$ and using (7.6) we obtain

$$n_e = \int_{\mathscr{E}_g}^{\infty} F_c(\mathscr{E}) \exp\{(\mathscr{E}_F-\mathscr{E})/kT\} \, d\mathscr{E}. \tag{7.14}$$

This can be simplified by substituting $x = (\mathscr{E}-\mathscr{E}_g)/kT$ and hence

$$n_e = \exp\{(\mathscr{E}_F-\mathscr{E}_g)/kT\} \int_{0}^{\infty} \mathscr{F}_c(x, T) \exp(-x) \, dx, \tag{7.15}$$

or

$$n_e = N_c \exp\{(\mathscr{E}_F-\mathscr{E}_g)/kT\}, \tag{7.16}$$

where

$$N_c = \int_{0}^{\infty} \mathscr{F}_c(x, T) \exp(-x) \, dx. \tag{7.17}$$

The number of holes n_h which are excited in the valence band will be given by

$$n_h = \int_{-\infty}^{0} F_v(\mathscr{E})\{1-g(\mathscr{E})\} \, d\mathscr{E}, \tag{7.18}$$

where $F_v(\mathscr{E})$ is the density of states in the valence band. In the same way as we derived (7.16), we may find an analogous expression for n_h after making the substitution $y = -\mathscr{E}/kT$ (if $(\mathscr{E}_F-\mathscr{E}) \gg kT$), i.e.

$$n_h = N_v \exp(-\mathscr{E}_F/kT), \tag{7.19}$$

where

$$N_v = \int_{-\infty}^{0} \mathscr{F}_v(y, T) \exp(-y) \, dy. \tag{7.20}$$

On multiplying (7.16) and (7.19) we obtain the important result

$$n_e n_h = N_c N_v \exp(-\mathscr{E}_g/kT). \tag{7.21}$$

It should be noted that this expression is independent of the value of $\mathscr{E}_F$ and, at a given temperature, it is a constant for any sample of a semiconductor, whether it is pure or has been doped n or p-type.

For a pure semiconductor with no donor or acceptor levels the number of electrons in the conduction band is equal to the number of holes in the valence band, i.e. $n_e = n_h$. From (7.21) we can therefore write

$$n_e = n_h = (N_c N_v)^{\frac{1}{2}} \exp(-\mathscr{E}_g/2kT). \tag{7.22}$$

It is interesting to note that the index of the exponential in (7.22) is $-\mathscr{E}_g/2kT$ and *not* $-\mathscr{E}_g/kT$ as one might at first think for a process of thermal excitation across a gap $\mathscr{E}_g$.

If in the derivation of (7.22) we now assume that $F_v(\mathscr{E})$ and $F_c(\mathscr{E})$ are the same as the density of states function for a free-electron system (1.17), then we may obtain

$$n_e = 2(2\pi m_e kT/h^2)^{\frac{3}{2}} \exp(-\mathscr{E}_g/2kT). \tag{7.23}$$

The full calculation for (7.23) is given by Dekker (1957), p. 308.

From (7.23) we see that for an ideal semiconductor in which the activation of carriers only occurs from the valence to the conduction bands, a graph of $\log n_e$ against $1/T$ should be a straight line with a slope of $-\mathscr{E}_g/2k$. This, of course, neglects the variation of $T^{\frac{3}{2}}$ which will be small compared with that of the exponential, although for greater precision $\log(n_e T^{-\frac{3}{2}})$ can be plotted. Instead of n_e, we could use the Hall coefficient R_H, since this is proportional to $1/n_e$ (7.4). Thus a plot of $\log R_H$ (or $\log(R_H T^{\frac{3}{2}})$) against $1/T$ should be linear with a slope of $\mathscr{E}_g/2k$. Such behaviour is indeed shown by all semiconductors at high temperatures.

We must now consider how the presence of donor and acceptor levels affects these conclusions. Without working through the algebra it is fairly easy to see the type of behaviour we should expect. If we are at a sufficiently low temperature, so that the intrinsic excitation can be neglected compared with the excitation from, say, the donor levels to the conduction band in n-type material, then we can follow through the same argument that led to (7.16) for n_e and to (7.19) for n_h and obtain a similar result to (7.21) for $n_e n_h$, except that n_h will now refer to the number of ionized donor levels, which to avoid confusion we shall call n_d; we shall then have

$$n_d n_e = N_d N_c \exp(-\Delta\mathscr{E}_d/kT), \tag{7.24}$$

where $\Delta\mathscr{E}_d = \mathscr{E}_g - \mathscr{E}_d$ is the energy gap between the donor levels and the conduction band, and N_d is the total number of donor levels. If, then, our specimen is a compensated n-type sample (section 7.3) so that the number of donor states, N_d, is greater than the number of acceptor states, N_a, all the acceptors will be ionized by electrons from the donor impurities. The number of excited donors will be equal to N_a plus the number of electrons which are excited into the conduction band, i.e. $n_d = N_a + n_e$. Substituting for n_d in (7.24) we obtain

$$(N_a + n_e)n_e = N_d N_c \exp(-\Delta\mathscr{E}_d/kT). \tag{7.25}$$

Since at low temperatures $N_a > n_e$, we may neglect the n_e^2 on the L.H.S. of (7.25) and hence

$$n_e \approx (N_d/N_a)N_c \exp(-\Delta\mathscr{E}_d/kT). \tag{7.26}$$

From (7.26) we see how important is the influence of the minority carriers in determining the low temperature properties.

If the effect of the compensation may be neglected (i.e. if $n_e > N_a$) then we may put $N_a = 0$ in (7.25) and then we obtain

$$n_e = (N_d N_c)^{\frac{1}{2}} \exp(-\Delta\mathscr{E}_d/2kT). \tag{7.27}$$

Analogous expressions to the three preceding ones may, of course, be derived for p-type specimens. In a detailed treatment (7.26) and (7.27) are multiplied by factors of $\frac{1}{2}$ and $\sqrt{\frac{1}{2}}$ respectively. This arises because the Fermi–Dirac function (7.6) must be modified in order to take account of the fact that once an electron has attached itself to a donor atom in a certain state, no other electron can occupy the remaining states. This restriction does not apply in a conduction band. If one considers that there are just two possible states (with opposite spin) then the factors are as given above. A discussion is given by Wilson (1953), p. 327 and by Putley (1960), p. 122.

From (7.26) and (7.27) we see that at low temperatures where intrinsic excitation may be neglected, a plot of $\log R_H$ against $1/T$ should be linear with a slope of either $\Delta\mathscr{E}_d/k$ or $\Delta\mathscr{E}_d/(2k)$, depending on the temperature region and the degree of compensation. It should be emphasized that all these expressions are only valid when $\mathscr{E}_F - \mathscr{E}_d$ is greater than kT and since $\mathscr{E}_F$ is somewhere near the midpoint of the impurity gap, this is not a very high temperature. Beyond this temperature the unity in the denominator of (7.6) cannot be discarded and the full expressions must be evaluated.

A simple calculation based on (7.23) shows that in a pure semi-conductor the number of electrons excited into the conduction band at room temperature is given by $n_e = 2 \cdot 5 \times 10^{19} \exp(-20\mathscr{E}_g)$ when $\mathscr{E}_g$ is in eV. Thus for germanium ($\mathscr{E}_g = 0 \cdot 75$ eV), n_e is about $7 \cdot 5 \times 10^{12}$ cm^{-3} and hence if we wish to observe effects due to these carriers we must ensure that there are no impurities which can give impurity levels with a concentration greater than about 10^{12} cm^{-3}. This requires great care in the preparation of specimens and in recent years a very considerable effort has been expended on the development of methods for the production of high-purity semiconductors. It is clear that at low temperatures the intrinsic excitation will become even less and so the properties of the material will become even more dominated by the impurity levels. The greater the energy gap the more will this be accentuated. In silicon, for example, with $\mathscr{E}_g = 1 \cdot 1$ eV, the number of intrinsic electrons which are excited should be about 1000 times less than that for germanium and hence the experiments on silicon require samples in which the purity is even more carefully controlled. In particular, even if one is looking for specific impurity effects (as indeed is the case for most investigations) then one must make sure that the effect is due to a known added impurity rather than to the presence of unknown residual impurities.

7.8. The mobility

The mobility of a charge will be determined by the various scattering processes which can affect its motion. We have already discussed a similar problem when we considered the electrical resistivity of a metal (section 4.5) where it will be recalled that the important scattering processes were those due to the thermal vibrations of the crystal lattice and to the effect of ionized impurity atoms. In semiconductors, the impurity atoms are not always ionized and so we must also consider the scattering by *neutral* impurity atoms.

Whilst the general principles for calculating the scattering cross-section and relaxation time are the same as those used for metals, the details and the results are quite different. We have already assumed (section 7.7) that for a semiconductor at low to moderate temperatures, we can use Maxwell–Boltzmann statistics, i.e. we can neglect the unity in the denominator of (7.6). The electrons in the conduction band will not be very numerous and for any particular energy there will always be some empty states available into which an electron from a neighbouring state can be scattered. In a metal all states except those near the maximum electron energy are fully occupied. Thus whereas in a metal it is only the electrons which are close to the Fermi surface, which can be scattered into nearby vacant states (section 4.2), in a semiconductor we must consider the effect of interactions with electrons of all energies. As in the case of metals a full calculation of the scattering cross-section and the relaxation time is very complicated and we shall only present qualitative arguments. Reference should be made to Ziman (1960), p. 421 for a more complete mathematical treatment.

We have already quoted the kinetic theory expression for the electrical resistivity

$$\rho = m_e^* v/(ne^2 l), \tag{7.28}$$

or

$$\rho = m_e^*/(ne^2 \tau). \tag{7.29}$$

If we combine this with (7.1), assuming that only one type of carrier is present, we obtain

$$\mu = e\tau/m_e^*. \tag{7.30}$$

For semiconductors, as we have just stated, we have to take account of interactions with all the carriers in the band and since, in general, the relaxation time, τ, will be a function of the energy, $\mathscr{E}$, of a carrier, we must in (7.30) use a value of the relaxation time $\langle \tau(\mathscr{E}) \rangle$ which is an average over all the occupied states. Thus

$$\mu = \frac{e}{m_e^*} \langle \tau(\mathscr{E}) \rangle. \tag{7.31}$$

Now the value of a function of the form $\mathscr{E}^j$ when averaged over a Maxwellian energy distribution is of the order $(kT)^j$. Thus if τ is a function of the form constant $\times \mathscr{E}^j$ we can write

$$\langle \tau(\mathscr{E}) \rangle = \langle \text{constant} \times \mathscr{E}^j \rangle \approx \text{constant} \times (kT)^j. \tag{7.32}$$

The problem is therefore to find the form of $\tau(\mathscr{E})$ for the three scattering mechanisms we have mentioned in the first paragraph of the section.

7.9. *Scattering by thermal vibrations*

Since the phonon energy will usually be considerably less than that of the electron, scattering will only take place to nearby vacant states. Thus the scattering probability for an electron with an energy $\mathscr{E}$ will be dependent on the number of nearby vacant states which are available for it to be scattered into. The number of vacant sites will presumably be proportional to the total density of states at the energy $\mathscr{E}$. For free electrons this is proportional to $\mathscr{E}^{\frac{1}{2}}$ (1.17). The scattering will also depend on the square of the amplitude (i.e. the energy) of the thermal vibrations of the lattice. Thus the total scattering probability will be proportional to $kT\mathscr{E}^{\frac{1}{2}}$ and hence the relaxation time will be of the form

$$\tau_g(\mathscr{E}) = \text{constant} \times (kT\mathscr{E}^{\frac{1}{2}})^{-1}. \tag{7.33}$$

Using (7.32) to obtain the average value of τ_g gives

$$\langle \tau_g(\mathscr{E}) \rangle = \text{constant} \times (kT)^{-\frac{3}{2}} \tag{7.34}$$

and hence from (7.31) μ_g will also be proportional to $T^{-\frac{3}{2}}$. Full calculations have been made by Seitz (1948) and by Bardeen and Shockley (1950). Seitz gives

$$\mu_g = \frac{2^{\frac{1}{2}} \cdot 6^{\frac{1}{3}}}{4\pi^{\frac{5}{6}}} \frac{N^{\frac{1}{3}} e\hbar^2 k^2 \theta^2 M}{m_e^{*\frac{5}{2}} C^2 (kT)^{\frac{3}{2}}}, \tag{7.35}$$

where N is the number of atoms per cm^3, M the atomic mass, and θ is the Debye characteristic temperature. C is a constant with the dimensions of energy which is a measure of the electron-phonon interaction. It is usually found empirically from experiment; for germanium it has a value of about 5 eV. The dependence of μ_g on $m_e^{*-\frac{5}{2}}$ should be noted. Very high mobilities (of the order of 10^5 cm^2 volt^{-1} sec^{-1}) can be obtained with InSb, which has a very small effective mass. It is also of interest to consider the behaviour of the mean free path, l, of the electrons. This is equal to τv. The velocity, v, is equal to $(2\mathscr{E}/m_e^*)^{\frac{1}{2}}$ for electrons near the bottom of the band and we have just seen (7.33) that τ is proportional to $\mathscr{E}^{-\frac{1}{2}}$. Thus the mean free path of the electrons, when they are being scattered by phonons, is independent of their energy, which is rather a surprising result.

7.10. Scattering by ionized impurities

As in the case of a metal (section 4.7) we can treat this problem by using the Rutherford scattering formula. It will be recalled that this gives a scattering cross-section which is proportional to $e^4(\Delta Z)^2/\mathscr{E}^2$, where ΔZ is the effective charge of an ion (for singly ionized impurities this will, of course, be unity) and $\mathscr{E}$ is the energy of the incident electron. In a metal we did not consider the dependence on $\mathscr{E}$ because, as we have already mentioned, only the electrons with an energy close to the Fermi energy could be scattered. In a semiconductor we cannot disregard this dependence. We must also take account of the high dielectric constant, ϵ, of the material (section 7.2), since this will reduce the interaction by ϵ^2. The mean free path, l, will therefore be of the form

$$l = \text{constant} \times \epsilon^2 \mathscr{E}^2 / \{(\Delta Z)^2 e^4\} \tag{7.36}$$

and since, once again, the velocity is equal to $(2\mathscr{E}/m_e^*)^{\frac{1}{2}}$ the relaxation time can be written

$$\tau_i(\mathscr{E}) = l/v = \text{constant} \times m_e^{*\frac{1}{2}} \epsilon^2 \mathscr{E}^{\frac{3}{2}} / \{(\Delta Z)^2 e^4\}. \tag{7.37}$$

When we take the average value using (7.32) and insert it into (7.31) we obtain

$$\mu_i = \text{constant} \times \frac{\epsilon^2(kT)^{\frac{3}{2}}}{m_e^{*\frac{1}{2}}(\Delta Z)^2 e^3}. \tag{7.38}$$

A full calculation on these lines by Conwell and Weisskopf (1950) yields

$$\mu_i = \frac{2^{\frac{7}{2}}\epsilon^2(kT)^{\frac{3}{2}}}{N_i \pi^{\frac{3}{2}}(\Delta Z)^2 e^3 m_e^{*\frac{1}{2}}} \{\log(1+36\epsilon^2 d^2 k^2 T^2/e^4)\}^{-1}, \tag{7.39}$$

where N_i is the concentration of ionized impurities and $2d$ is the average distance between them. A more rigorous treatment which gives the same order of magnitude for μ has been made by Brooks and Herring (see Brooks, 1955). This yields an expression which is very similar to (7.39) except that the logarithmic term is changed to

$$[\log(24\pi m_e^* \epsilon N_d k^2 T^2)/\{e^2 h^2 n'(2-n')\}]^{-1}, \tag{7.40}$$

where

$$n' = n_e + \left(1 - \frac{n_e + N_a}{N_d}\right)(n_e + N_a). \tag{7.41}$$

It is important to note that, neglecting any change in the logarithmic term, μ_i has the inverse temperature dependence to μ_g, and so we should therefore expect μ_i to become the dominant mechanism as the temperature is reduced.

7.11. Scattering by neutral impurities

At low temperatures many of the impurities in a semiconductor may not be ionized and these will have a different scattering effect on an electron. The situation has been treated by assuming that it is similar to the scattering of an electron by a hydrogen atom. Allowance has to be made for the dielectric constant of the semiconductor which will increase the spread of the hydrogen wave function, and the correct effective mass of the carriers (section 7.14) must be used instead of that of the free electron. Qualitatively it seems reasonable to suppose that the higher the velocity of an electron, the less will it be scattered; Erginsoy (1950) has shown that for electron energies up to 25 per cent of the ionization energy of the impurity atom, the mean free path should be directly proportional to the velocity. Since $\tau_0 = l/v$ this means that the relaxation time is independent of the velocity (and hence also of the energy). Thus from (7.31) μ_0 should be independent of the temperature. Erginsoy's calculation gives

$$\mu_0 = \frac{m_e^* e^3}{20 N_0 \hbar^3}, \tag{7.42}$$

where N_0 is the density of neutral impurity atoms.

The expressions for μ_i and μ_0 (7.39), (7.40), and (7.42) are only valid for low densities of impurity ions or atoms. Sclar (1956) has extended both theories to take account of a high density of scattering centres. In addition we should not expect these expressions to give exact agreement with experiment because, apart from the simplifications which have had to be introduced in order to make the calculations tractable, the values of N_i and N_0 will not remain constant. In general the amount of ionization will change with temperature and this will further influence the temperature dependence of μ_i and μ_0 except at very low or at high temperatures.

7.12. *The combination of mobilities*

The scattering due to all three types of process (Fig. 7.6) which we have considered will give rise to an effective mobility which must be calculated from (7.31) using the resultant relaxation time, τ, from

$$1/\tau = 1/\tau_g + 1/\tau_i + 1/\tau_0. \qquad (7.43)$$

This will be a complicated function of the energy and hence so will μ. Provided that one of the mechanisms is dominant, we can use the analogue of Matthiessen's rule and add the inverse mobilities directly,

$$1/\mu = 1/\mu_g + 1/\mu_i + 1/\mu_0, \qquad (7.44)$$

but this will be inaccurate if the two smallest mobilities are of about the same magnitude.

In addition to the effects we have discussed, scattering will also be produced by dislocations and by point defects, such as interstitials and vacancies. Although these will probably not be present in very large numbers in carefully prepared specimens, they will be produced after deformation or irradiation. Depending on whether they are charged or not, the point defects will presumably give a scattering similar to that which we have described by (7.39) or (7.42). The scattering by dislocations has been treated by Dexter and Seitz (1952).

Some complications

7.13. *The Hall mobility*

The simple formula $\mu = R_H c/\rho$ (7.5) enables us to determine the mobility once we know the Hall coefficient and the resistivity. In a rigorous calculation of R_H, however, we must take an average over the full range of electron energies in a manner similar to that which we have used for the calculation of mobilities in the preceding sections. It is then found that the expression for R_H differs from (7.5) by a numerical factor which depends on the way in which the relaxation time, τ, varies with energy (see Ziman, 1960, p. 490). If τ is independent of $\mathscr{E}$ (as in neutral impurity scattering) the factor is unity and so (7.5) is correct. For

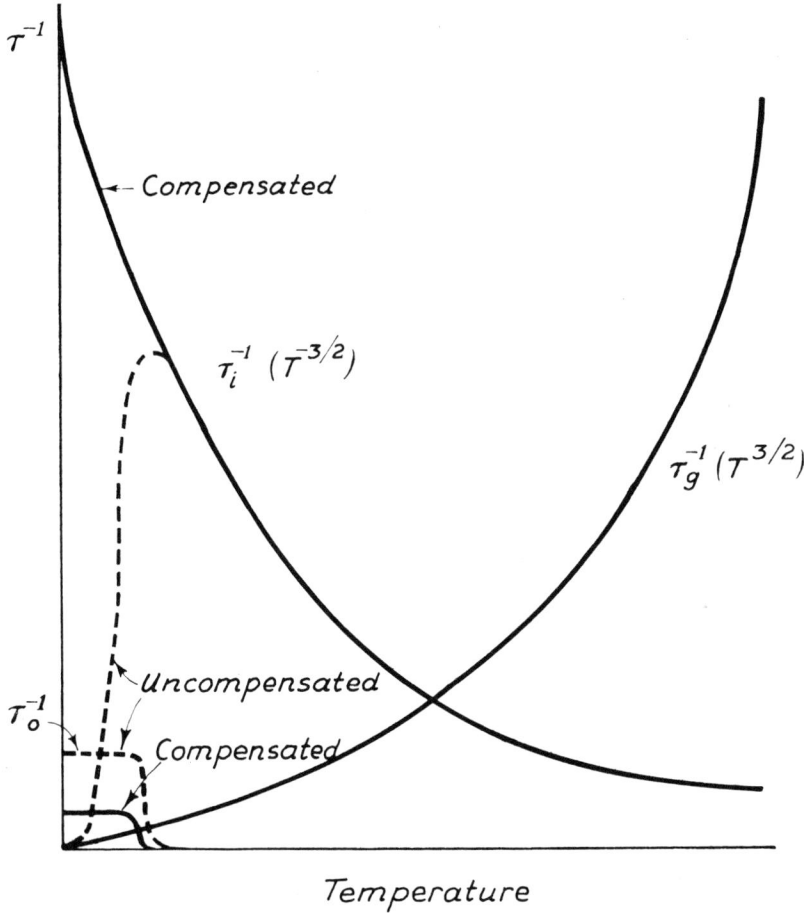

Temperature

FIG. 7.6. A diagrammatic representation of the variation of the inverse relaxation time, τ^{-1}, as a function of temperature for the three main scattering processes which determine the mobility in a semiconductor. The scattering by neutral atoms is shown as dropping to zero as the temperature is increased since these atoms will become ionized. The amount of ionized scattering which occurs at low temperatures will depend on the extent of the compensation in the sample. If the sample is uncompensated then τ_i^{-1} will drop to zero at $0°$ K, with a corresponding increase in τ_0^{-1} as is shown by the dashed curves.

phonon scattering (τ proportional to $\mathscr{E}^{-\frac{1}{2}}$) the factor is $\frac{3}{8}\pi$. For ionic scattering (τ proportional to $\mathscr{E}^{\frac{3}{2}}$) the factor is 1·93. Whatever the energy dependence, however, the factor always lies between 1 and 2.

If, however, we persist in using the simple formula (7.5) even when it does not really apply, then the mobility, $R_H c/\rho$, which we calculate from the experimental values of ρ and R_H is called the *Hall mobility*. The errors involved in assuming that this is equal to the actual drift mobility are usually much smaller than other uncertainties in the interpretation of the results. It is the Hall mobility which is usually quoted in the literature.

7.14. *The band structure of semiconductors*

In our treatment of mobility we have assumed that the energy levels and density of states at the top and bottom of the valency and the

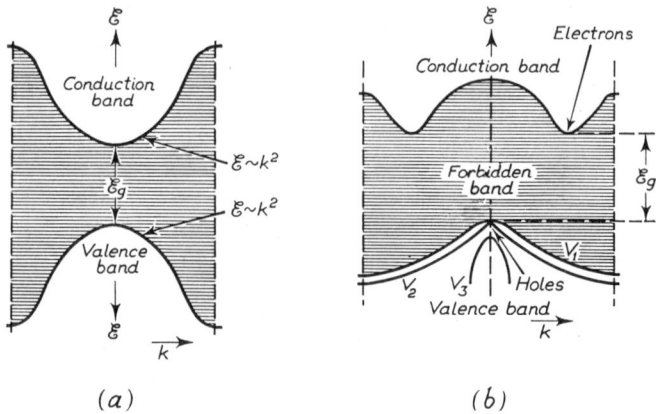

(a)   (b)

FIG. 7.7. (a) An idealized semiconductor in which the conduction and valence bands are the same as for a free electron metal. (b) The energy band structure as a function of **k** along a [100] axis for Si. Note that the valence band really consists of three bands, two of which coincide at **k** = 0. Whilst the maximum energy of the valence band occurs at **k** = 0, the minimum energy for the conduction band occurs for a higher value of **k**, as is also shown by the constant energy surfaces of Fig. 7.8.

conduction bands respectively, were similar to those for a free electron model (i.e. $\mathscr{E}$ proportional to k^2), Fig. 7.7 (a). Calculations by Herman (1954) and others have shown that this assumption is very much an oversimplification. The actual band structure of semiconductors is quite complicated. The relationship between the wave vector, **k**, and the energy is not spherically symmetrical but has cubic symmetry and, in addition, the bands themselves have a complex structure. In the conduction band the minimum energy does not occur at **k** = 0 but at values of **k** which lie on the cube diagonals for germanium, and on the main cube axes for silicon. The surfaces of constant energy around these minima are ellipsoids of revolution. These are shown diagrammatically in Fig. 7.8. These are the states which are initially occupied by the excited electrons. Whilst the valence band does have its maximum energy at $\mathbf{k} = 0$, it is really a combination of three bands, two of which touch at $\mathbf{k} = 0$ but which otherwise have different curvatures. This is shown in Fig. 7.7 (b). The energy surfaces are more nearly spherical than in the conduction band and it is not so necessary to consider their detailed shape in this treatment.

These results must modify the results and conclusions which we have drawn in several ways. First, any calculations which have involved the

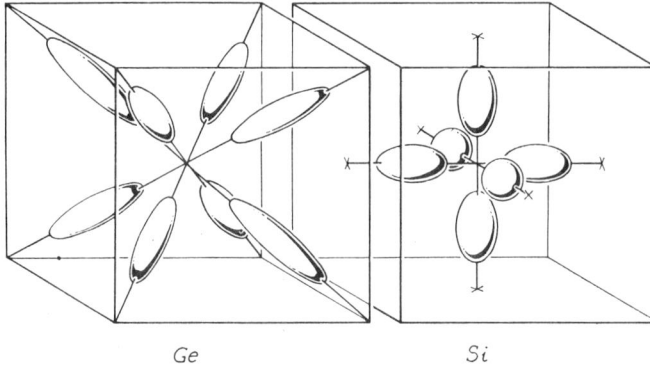

Ge   *Si*

FIG. 7.8. The surfaces of constant energy in k-space for the conduction bands of Ge and · Si. The states of minimum energy, i.e. those most likely to be occupied, will be those with $\mathbf{k}$-vectors which join the centre of the cube to the centres of the ellipsoids.

integration over the available energy states should really use a special density of states function rather than the free electron value of $F(\mathscr{E})$. Thus all our expressions for the Hall coefficients and the mobilities cannot be very accurate. Secondly we run into considerable complication when we consider the effective masses of the carriers. The effective mass is defined by

$$m_e^* = \hbar^2/(d^2\mathscr{E}/dk^2). \tag{7.45}$$

It therefore depends on the band structure of the material and in particular on the curvature of the $\mathscr{E} \sim \mathbf{k}$ surface. For a free electron system where $\mathscr{E} = \hbar^2 k^2/2m_e$, m_e^* from (7.45) is equal to the mass m_e of the electron, but otherwise m_e^* will in general be different from m_e. Since the valence and conduction bands have different shapes, not only will m_e^* be different from m_e, but m_e^* will not have the same value for holes as for electrons. To add the final blow we are bound to mention (a) that since there are two valence bands which coincide at $\mathbf{k} = 0$ they will give two kinds of holes—usually called light and heavy holes—and (b) due to the ellipsoidal energy surfaces in the conduction band (Fig. 7.8), the electron mass will vary, being dependent on the direction of motion, and it is characterized by two values, called the transverse and longitudinal mass, which are associated with the curvature of the energy surfaces perpendicular and parallel to the axis of revolution of the ellipsoids. We shall see in section 7.21 that these rather complicated results of the detailed band theory have been borne out remarkably well by cyclotron resonance experiments.

7.15. *The Hall effect for two types of carrier*

A further consequence of the fact that the valence and conduction bands have different shapes is that the mobilities of electrons and holes will be different from one another. If for simplicity we consider just one type of electron and one type of hole, then our formula for the conductivity (7.1) must now be modified so as to read

$$1/\rho = e(\mu_e n_e + \mu_h n_h), \tag{7.46}$$

where μ_e and μ_h are the mobilities of electrons and holes respectively.

The expression for the Hall coefficient will also be affected. The condition for a stationary state will no longer be that the Hall field and the effect of the magnetic field should be equal and opposite so that the carriers are not deviated. In general this cannot occur for two kinds of charges with different mobilities. A steady state will now be set up when the electrons and holes are so deflected that they arrive at the boundaries of the specimen at the same rate. Since they have equal but opposite charges they will not affect the space charge field and so a stationary state can be established. By equating the numbers of electrons and holes which arrive at the boundaries of the specimen under the combined influence of the Hall and the magnetic field it is simple to derive the following modified expression for the Hall coefficient R_H:

$$R_H = -(n_e \mu_e^2 - n_h \mu_h^2)/\{ec(n_e \mu_e + n_h \mu_h)^2\}, \tag{7.47}$$

where e is the magnitude of the electronic charge. This expression should be multiplied by the appropriate factor, depending on the relationship between the relaxation time and the energy as has been described in section 7.13.

7.16. Experiments on the Hall effect and the resistivity

At temperatures above the liquid-helium region Hall and conductivity measurements give results which are in fairly good agreement with the theory and we shall discuss these in this section. We shall defer a description of the unusual effects which occur at liquid-helium temperatures until later (section 7.20).

A typical series of measurements of the Hall coefficient of n-type germanium is shown in Fig. 7.9. At low temperatures intrinsic excitation from the valence to the conduction band is practically non-existent and

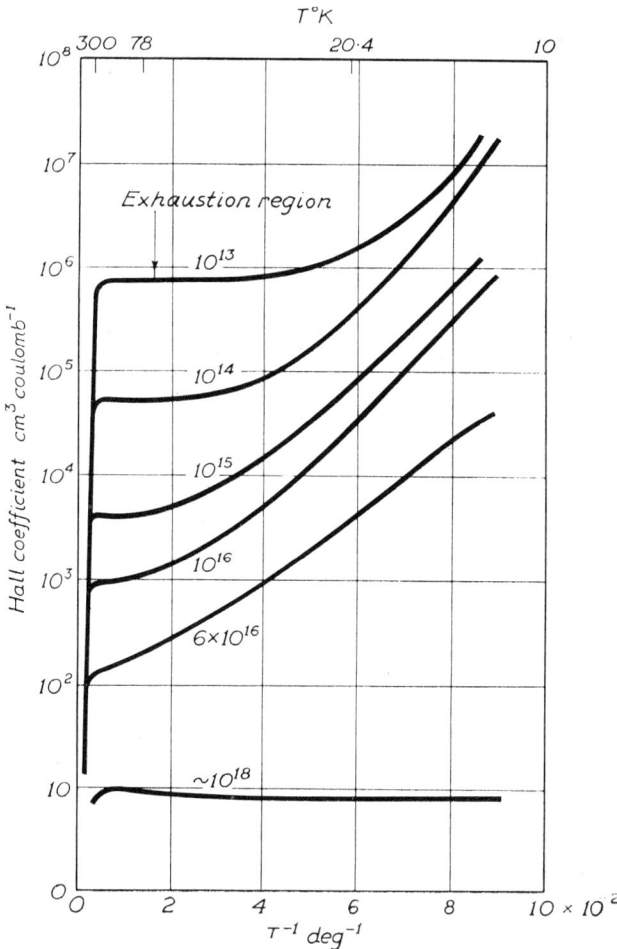

FIG. 7.9. The Hall coefficient of a set of n-type germanium crystals (arsenic doped) as a function of the inverse temperature down to about 11° K. The number on each curve is the approximate carrier density at room temperature (Debye and Conwell, 1954).

the only carriers which can be detected will be those electrons which are excited from the impurity donor states. The high value of $|R_H|$ at low temperatures shows that the number of charge carriers is indeed very small (7.4). As the temperature is increased the amount of donor excitation is increased as is shown by the decrease in $|R_H|$. Eventually the R_H curve flattens off because all the impurity states have been ionized. This exhaustion region, as it is called, continues until a temperature is reached at which thermal excitation can occur from the valence to the conduction band. This only takes place at temperatures above 300° K for germanium and above 500° K for silicon. There is then a sudden increase in the number of carriers, as can be seen from the rapid decrease in the value of $|R_H|$ at high temperatures. Specimens containing more impurity show similar behaviour to high-purity specimens, except that since there are extra donor states available, more carriers will be excited at any given temperature and so $|R_H|$ will be smaller. For all specimens the value of R_H approaches the common intrinsic curve at high temperatures. From the slope of this curve the intrinsic activation energy (see Table 7.1, p. 211) can be calculated by using (7.22), and from the low-temperature slope one can find the donor activation energy (see Table 7.4, p. 251). p-type material is essentially similar until the intrinsic region is approached. R_H then shows rather wayward behaviour because it changes sign over a very small temperature interval from positive at the lower temperatures to negative at the higher. This occurs because the electrons which are excited in the intrinsic range have a higher mobility than the holes which were the only type of carriers in the impurity region. Thus as soon as sufficient electrons have been excited, their effect will overshadow that due to the holes and R_H will become negative (μ_e/μ_h is about 3 for Si and about 1·5 for Ge).

Since both germanium and silicon are in the exhaustion region at room temperature so that all the impurities are ionized, a measurement of the room-temperature Hall coefficient (or the resistivity) gives a measure of the effective number of extrinsic carriers, i.e. in n-type material, $(N_d - N_a)$. Thus specimens are very often designated by their room-temperature resistivity. It should be noted, however, that this gives no indication of the degree of compensation (i.e. the absolute values of N_d and N_a) in the sample.

7.17. Degeneracy

The behaviour of specimens which have a large number of impurities is somewhat different from purer specimens because the assumption that we can use Maxwell–Boltzmann statistics (i.e. by neglecting the unity in the denominator of 7.6), is not valid. This is due to the fact that the number of electrons or holes which are excited is no longer small compared with the number of available states in the conduction or valence band which they may occupy. This will be particularly true at lower temperatures when the excitation will only be to levels near the bottom of the band, where the density of states is small. The exclusion principle must then be complied with, i.e. we must use the complete Fermi–Dirac function (7.6). Because the number of carriers depends on the density of impurity atoms, it is clear that the larger the number of impurities, the higher the temperature at which this effect will still occur. In the degeneracy region the Fermi energy rises into the conduction band (for n-type material) and, if $\mathscr{E}_F \gg kT$, all the low-lying states $\leqslant \mathscr{E}_F$ will be filled just as in a metal (sections 1.9 and 4.2). We can define a degeneracy temperature, T_d, when conditions are such that $\mathscr{E}_F = kT_d$. Using the expression for $\mathscr{E}_F$ from the theory of metals (e.g. from 1.18) we obtain

$$T_d = \frac{h^2}{2m_e^* k} \left(\frac{3}{8\pi}\right)^{\frac{2}{3}} n_e^{\frac{2}{3}} \,^\circ\text{K}. \tag{7.48}$$

If m_e^* is equal to the free electron mass, m_e, then

$$T_d \approx 4 \cdot 2 \times 10^{-11} n_e^{\frac{2}{3}} \,^\circ\text{K}. \tag{7.49}$$

Thus if $n_e = 10^{17}$ cm^{-3}, T_d will be 10° K and for $n_e = 10^{18}$ cm^{-3}, $T_d = 40^\circ$ K. If m_e^* is less than m_e, as is often the case (see Table 7.3, p. 242) then T_d will be correspondingly increased.

At very low temperatures, when excitation to the conduction band is very small, we should expect the specimen to become non-degenerate again. In many cases, however, degeneracy arises for specimens in which the density of impurity states is so high that they overlap into the conduction band. When this occurs no thermal excitation is necessary to produce carriers. The density of carriers is independent of the temperature and this is indeed shown by the constant value of R_H for very impure specimens, e.g. the lowest curve in Fig. 7.9. Since in such specimens the mobility will be determined by impurity ion scattering, the situation is the same as that which gives rise to the residual resistance of metals. μ will be independent of temperature and hence ρ will also be temperature independent. This is shown by the corresponding curves of Figs. 7.10 and 7.11.

7.18. *Resistivity*

The resistivity of semiconductors will, of course, reflect the behaviour of the Hall coefficient (in so far as it gives a measure of the carrier concentration) but this will be modified by the temperature dependence of the mobility (7.1). Resistivity curves for specimens of n-type germanium of various purities are shown in Fig. 7.10. At high temperatures most of the carriers are excited intrinsically but as the temperature is

reduced this source of carriers is rapidly removed. This accounts for the
initial steep rise in ρ for small values of $1/T$. The rise does not continue
indefinitely, however, because impurity excitation is still present and

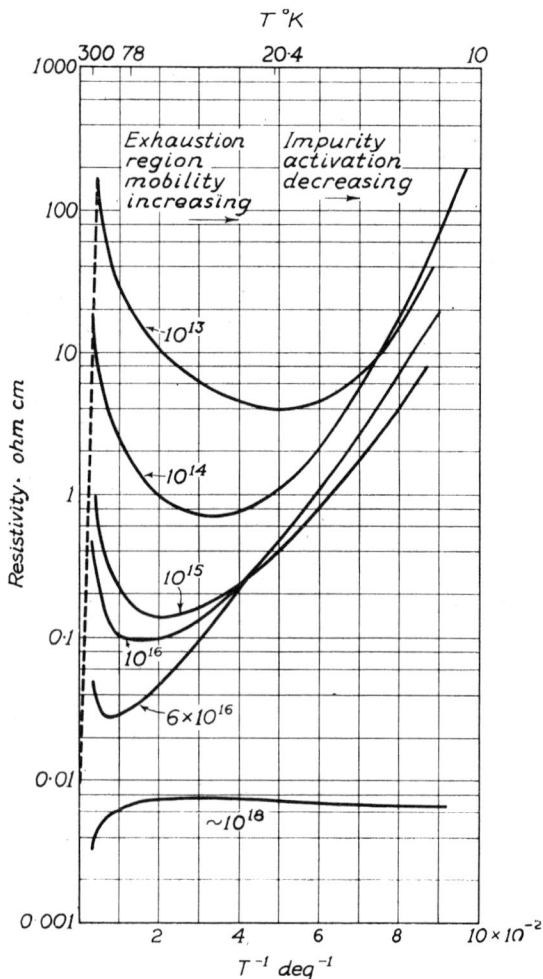

FIG. 7.10. The resistivity of a set of n-type germanium crystals (the same set as is shown
in Fig. 7.9) as a function of the inverse temperature (Debye and Conwell, 1954).

this will maintain a certain number of carriers in the conduction band.
This is the exhaustion region which we discussed in section 7.16. The
more impure the specimen, the greater the number of impurity carriers,
and hence the resistivity in this range is smaller. The resistivity in the
exhaustion range is not constant, however, because as the temperature
is reduced, the mobility μ_g increases (7.35). Thus the resistivity in this

region actually decreases as the temperature is reduced. Eventually the excitation of the impurity carriers stops and so at the lowest temperatures there is another rapid rise in the resistivity. For high impurity concentrations (i.e. the lowest curve in Fig. 7.10) when degeneracy and band overlap sets in (section 7.17) the resistivity becomes temperature independent.

7.19. Mobility measurements

From the experimental values of R_H and ρ we can calculate the Hall mobility using (7.5). Depending on the temperature and the impurity

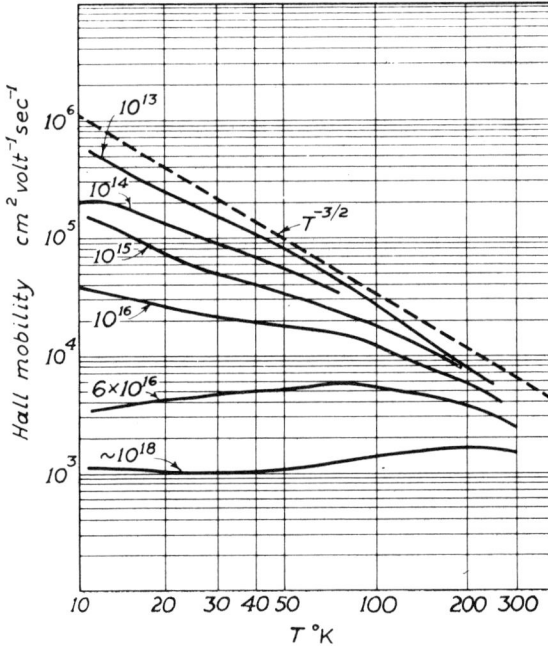

FIG. 7.11. The Hall mobility of a set of n-type germanium crystals calculated from the data of Figs. 7.9 and 7.10. The dotted line indicates a $T^{-3/2}$ behaviour (Debye and Conwell, 1954).

concentration, we might expect to find regions where μ is dominated by thermal vibrational scattering, μ_g (proportional to $T^{-1.5}$); ionic scattering, μ_i (proportional to $T^{1.5}$); or neutral impurity scattering, μ_0 (independent of T). Except at high temperatures, however, where μ_g is dominant, there is considerable overlap of these mechanisms. Fig. 7.11 shows the Hall mobility of n-type Ge calculated from the same data as Figs. 7.9 and 7.10. At the higher temperatures μ_g is proportional to a negative power of T which is somewhat larger than $1 \cdot 5$. Careful experiments by Debye and Conwell (1954) and by Morin (1954) have shown that μ_g does in fact vary as $T^{-1 \cdot 66}$. Whilst it might not be thought that this discrepancy with the theory was very serious, the values of μ_g, given in Table 7.2 for p-type specimens of Ge and both n and p-type Si, leave no doubt that the theory is inadequate. Only in the case of indium antimonide, for which values of μ_g are also given, can one say that reasonable agreement is achieved between theory and experiment.

TABLE 7.2

$$\mu_g \ (\mathrm{cm^2\,sec^{-1}\,volt^{-1}})$$

Ge electrons	$4 \cdot 9 \times 10^7 T^{-1 \cdot 66}$	Morin and Maita (1954)
Ge holes	$1 \cdot 05 \times 10^9 T^{-2 \cdot 33}$	
Si electrons	$4 \cdot 0 \times 10^9 T^{-2 \cdot 6}$	Morin and Maita (1954a)
Si holes	$1 \cdot 5 \times 10^8 T^{-2 \cdot 3}$	
InSb electrons	$1 \cdot 04 \times 10^9 T^{-1 \cdot 68}$	Putley (1959)
InSb holes	$5 \cdot 4 \times 10^6 T^{-1 \cdot 45}$	Putley (1959a)

There are two main directions in which the theory could be improved. The first is to take a realistic account of the complicated structure of the valence and conduction bands. The second is to use the proper lattice vibrational spectrum of the solid and, in this regard, scattering of the carriers by the optical vibrational modes of the crystal should be considered. These are high-frequency vibrations which arise in a crystal possessing two atoms per unit cell (as in Ge and Si). In the optical modes these two atoms, instead of vibrating more or less in phase with one another, tend to move in opposite directions at any particular time. This gives a mobility of the form (Morin and Maita, 1954)

$$\mu_{\mathrm{opt}} = BT^{-0 \cdot 5}(e^{\theta/T} - 1). \tag{7.50}$$

There is no doubt that both these types of correction would improve the theory, although detailed calculations have not been carried out as yet.

Ionized impurity mobility is difficult to measure directly because thermal scattering is an important mechanism to quite low temperatures and hence the effect of the lattice scattering mobility cannot be ignored. Its value must be estimated (by extrapolation from high temperatures where it is dominant) and then, in conjunction with the measured Hall mobility, the ionic scattering mobility may be calculated. The details of this calculation are quite complicated, because as we have mentioned in section 7.12, it is the reciprocals of the mobilities which are additive and not the mobilities themselves, and account must be also taken of the correct averaging procedure (section 7.13). A reading of the papers

by Debye and Conwell (1954) or by Long and Myers (1959) will show
how much work is required for this type of analysis. From their experimental results on n-type Si, Long and Myers (1959) have shown that,
provided the impurity concentration is not too high, the Brooks–Herring
formula (7.40) for μ_i holds quite well. Due to the logarithmic term in
the expression for μ_i a strict $T^{1\cdot5}$ dependence is not usually found. At
low temperatures, in specimens which have little compensation, very
few of the impurity atoms will be ionized and the mobility will be limited
by neutral impurity scattering. Debye and Conwell (1954) have shown
that their results on n-type Ge between 11 and $20\cdot4°$ K do yield a
temperature-independent mobility which is in reasonable agreement
with (7.42).

The general position with regard to the carrier mobility is that good
quantitative agreement between experiment and theory has only been
achieved for μ_i. For μ_0 insufficient experiments have been made and for
μ_g it is clear that further theoretical improvements are required. We
should also refer at this stage, to the method of determining the mobility
from cyclotron resonance experiments (section 7.23).

It might be thought that at liquid-helium temperatures the complication of having to deal with more than one type of mobility might be
avoided. Lattice scattering can then be neglected and in the case of
germanium and silicon at least, there will be virtually no donor or
acceptor excitation so that N_i and N_0 will remain constant, with $N_i \ll N_0$
except in heavily compensated samples. Experiments have shown that
this idyllic situation does not usually arise. What happens is described
in the next section.

7.20. *Impurity conduction*

Conductivity and Hall effect measurements on germanium of moderate
purity ($\sim 10^{15}$ to $\sim 10^{17}$ carriers cm^{-3} at room temperature) show rather
surprising behaviour below about $20°$ K. The steady rise of R_H with
decreasing temperature which was shown in Fig. 7.9 is halted. R_H passes
through a maximum and at lower temperatures decreases rapidly, very
often by two orders of magnitude or more (Fig. 7.12). At the temperature
corresponding to the maximum in R_H, there is a very sharp decrease
in the slope of the resistivity curve as if some new conductivity mechanism had come into play. These effects have been observed in InSb, Si,
Ge, SiC, and CdS. The temperature at which the maximum in R_H occurs
depends both on the concentration and type of impurity used for doping,
e.g. for In-doped Ge with about 10^{15} carriers cm^{-3} the maximum in R_H

occurs at about 6·5° K, but with increasing impurity concentration the maximum is shifted to higher temperatures. In the other materials the maximum tends to be at higher temperatures than in Ge.

The suggestion was first made by Hung (1950) that the extra conduction might arise from a transport of charge among the impurity levels themselves. For a fairly high density of impurities it seems quite reasonable to suppose that the electronic wave functions might overlap sufficiently, so that there was a good probability that an electron could travel

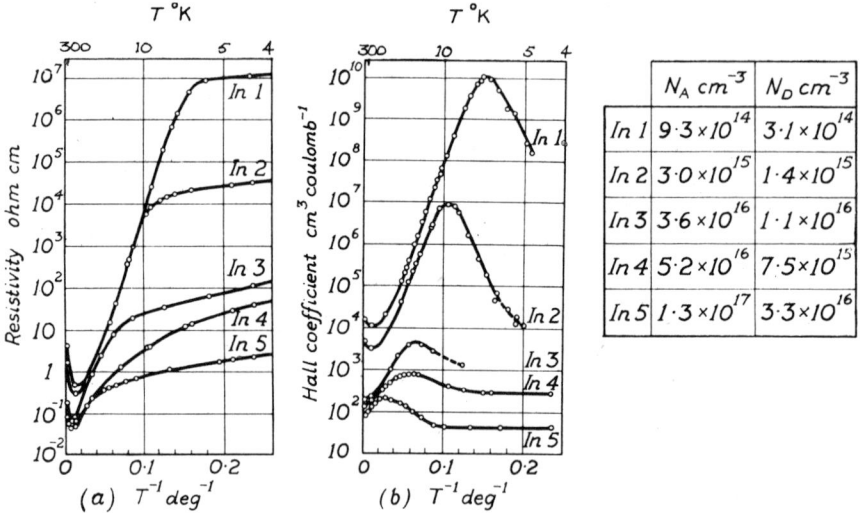

	N_A cm^{-3}	N_D cm^{-3}
$In\ 1$	$9·3 \times 10^{14}$	$3·1 \times 10^{14}$
$In\ 2$	$3·0 \times 10^{15}$	$1·4 \times 10^{15}$
$In\ 3$	$3·6 \times 10^{16}$	$1·1 \times 10^{16}$
$In\ 4$	$5·2 \times 10^{16}$	$7·5 \times 10^{15}$
$In\ 5$	$1·3 \times 10^{17}$	$3·3 \times 10^{16}$

Fig. 7.12. Impurity conduction. (a) The resistivity and (b) the Hall coefficient of a set of p-type (indium doped) germanium crystals at temperatures down to 4·2° K. Note the decrease in slope of the resistivity curves, which marks the onset of impurity conduction. This occurs at the same temperature as the maximum in the Hall coefficient.
The donor and acceptor concentrations are shown in the table (Fritzsche, 1955).

through the lattice. This overlap of wave functions would cause the discrete impurity levels to spread into a narrow band of states in which conduction would occur as in a metal. The difficulty in accepting this type of explanation arises because this phenomenon is observed for impurity concentrations which are so low ($\sim 10^{15}$ cm^{-3} in Ge and $\sim 10^{17}$ cm^{-3} in Si), that hardly any overlap seems possible. Calculations by Conwell (1956) indicate that the density of carriers which is necessary for band formation is greater than 10^{16} cm^{-3} for Ge and greater than 10^{17} cm^{-3} for Si.

It is now apparent that there are two kinds of impurity conduction. The first is indeed due to the formation of impurity bands and it occurs in rather impure specimens, as has been discussed above. The second arises in purer specimens and it is suggested that this is due to the

presence of compensating minority impurities (section 7.3). If we have
some n-type material containing N_d donor atoms which also has a small
number, N_a, of p-type acceptor levels, then at low temperatures N_a of
the donor atoms will lose their electron to ionize the N_a acceptors, since
this will be energetically favourable. Thus N_a donors will be lacking an
electron. This will give an incentive to an electron which is residing on
a nearby neutral donor atom to tunnel across to the ionized donor. In

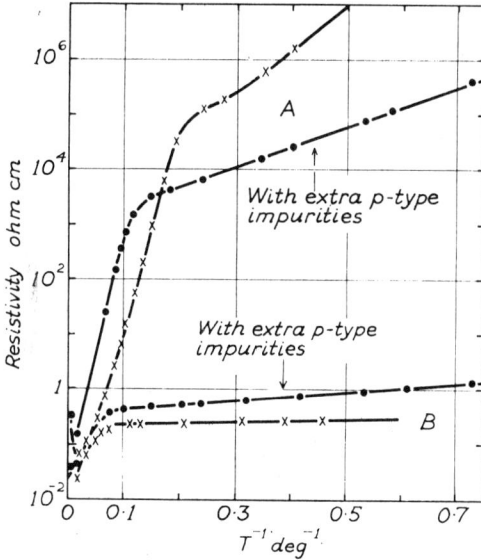

FIG. 7.13. Impurity conduction in n-type germanium crystals (Sb doped) showing the
effect of adding compensating impurities (copper). For the high-purity specimen, A,
extra acceptor levels *decrease* the impurity resistivity. For the less-pure specimen, B,
the compensating impurities *increase* the impurity resistivity (Fritzsche, 1958).

this way electrons can move from a neutral to an ionized donor and
hence a current will flow. A similar argument can be used for compensated p-type samples. If this mechanism is correct then one would
expect to produce changes in the resistivity when compensating impurities are added. Such experiments have been made most convincingly
by Fritzsche (1958) using n-type Ge (Sb doped), in which compensating
Cu impurities can be introduced very easily by diffusion at 700° C.
One would expect that for rather pure samples, in which the mechanism
which has just been described should apply, an increase in the number
of compensators would increase the number of ionized donors and hence
the impurity conduction should *increase*. For a less pure sample in
which there is wave function overlap and proper impurity band con-

duction is occurring, the extra compensators will still ionize the donors, but this results in removing carriers from the impurity band and also in increasing the number of scattering centres. Thus the conduction should *decrease*. These arguments are borne out by the experiments, some of which are illustrated in Fig. 7.13. The upper curves, A, relate to a sample containing $8 \cdot 5 \times 10^{15}$ donors cm^{-3}, whereas the lower curves, B, are for a sample containing $1 \cdot 2 \times 10^{17}$ donors cm^{-3}. On introducing Cu acceptors by diffusion at $700°$ C and quenching, the resistivity of the impure sample increased, whereas that of the purer sample decreased. In more recent experiments Fritzsche and Cuevas (1960) have introduced compensators by neutron irradiation.

We should note that impurities are distributed throughout the semiconductor in a random manner. They are not situated on a regular sub-lattice of their own and so while the band concept is a useful one, its shape cannot be calculated with any certainty. It has been pointed out (James, 1956) that there might be certain paths through the sample along which impurities are more concentrated and that conduction could take place along these paths. The kinds of neighbours which a given impurity atom has should also be taken into account. If, for example, in p-type material an electron is residing on an ionized acceptor atom and this happens to be near one of the ionized donors then it will be difficult to move the electron from that region to a neutral acceptor. It will be 'trapped' and might only move with the assistance of thermal activation (Mott, 1956). Thus the numbers of carriers which might be available for impurity conduction will be dependent on the temperature. Such considerations have been taken into account by Blakemore (1959) in the analysis of his experiments on impurity conduction in p-type Ge (In doped) and by Ray and Fan (1961) in their experiments on silicon.

7.21. Cyclotron resonance

We have already mentioned (section 7.14) that the complicated structure of the valence and conduction bands gives rise to multiple values for the effective masses of electrons and holes. One very direct method of studying m_e^* is by using the technique known as cyclotron resonance. The principle is very simple. If a magnetic field, H, is applied to a system of charge carriers then these will have, superimposed on their normal velocity, a spiralling motion about the axis of the field (Fig. 4.21); if there was no component of the velocity parallel to the field then the charges would travel in circles in planes at right angles to the field. The radius of curvature r, of the orbit will be determined by the fact that the centrifugal force on the charge must equal the force produced by the field, i.e.

$$m_e^* v^2/r = \pm Hev/c. \tag{7.51}$$

The angular frequency of rotation, $\omega = v/r$, is then

$$\omega = \pm He/m_e^* c. \tag{7.52}$$

The two possible signs of the expression denote the two senses of rotation which would occur for charges of opposite sign. It will be noted from (7.52) that ω does not depend on the energy of a particular carrier; all carriers will rotate at the same frequency, although those with higher energy will have orbits of a larger radius. Thus if one could measure H and ω, m_e^* could be calculated from (7.52). This can be determined at microwave frequencies by putting the sample in a resonant cavity into which microwave power is being fed. If the microwave frequency is equal to ω and the d.c. field H has the right value, the electrons will interact with the electromagnetic field in the same phase on each rotation; hence they will be able to gain energy from it and an absorption of radio frequency power will be detected. The analogy with the cyclotron accelerator is obvious. In practice it is much more convenient to keep the microwave frequency constant and to vary H until an absorption of power occurs.

In order that cyclotron resonance may be detected, it is clear that the charges must be able to complete at least one orbit before being scattered by either a lattice imperfection or by a phonon. If τ is the relaxation time which is characteristic of the scattering of the carrier then the condition for a resonance to be observed is that

$$\omega\tau \approx 1 \quad \text{or} \quad \omega\tau > 1. \qquad \tag{7.53}$$

Experimentally it is found that resonance can be detected for $\omega\tau \geqslant 0\cdot8$. At room temperature τ is 10^{-13} to 10^{-14} sec which would mean that experiments would have to be made using infra-red radiation, but at $4\cdot2°$ K τ is usually of the order of 5×10^{-11} sec which brings ω into the conventional microwave range.

7.22. *Experimental results*

The experiments were made first on Ge by Dresselhaus, Kip, and Kittel (1953) and a typical curve is shown in Fig. 7.14. Remarkable confirmation has been given to the theoretical calculations of the band structure (section 7.14) (Herman, 1954). Two types of holes have been found, termed light and heavy holes, which are associated with the two degenerate valence bands of different curvature which coalesce at $\mathbf{k} = 0$ (Fig. 7.7 (b)). The electrons have an effective mass which depends on the direction in which they are moving relative to the crystal lattice, which is what one would expect if the diagrams of the energy surfaces shown in Fig. 7.8 are correct. This mass is therefore a tensor and it can be designated by two principal masses which are for directions parallel and transverse to the axis of revolution of the energy surfaces

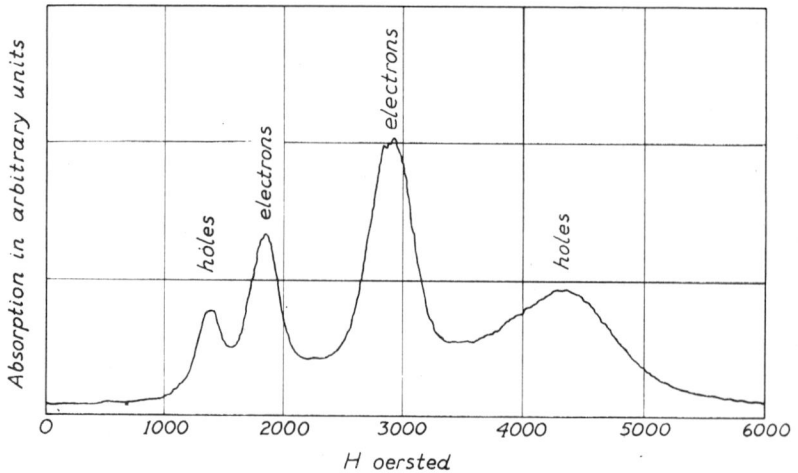

Fig. 7.14. Cyclotron resonance in Si at 24 kMc/s and 4° K, showing the absorption peaks which are ascribed to electrons and to holes. The static magnetic field is in a (110) plane at 30° from a [100] axis (Dresselhaus, Kip, and Kittel, 1955).

of Fig. 7.8. It should be emphasized that the two values of m_e^* for holes relate to two *different* types of hole, whereas the two values of m_e^* for electrons are two parameters referring to *one* electron. Experiments have been made on Ge, Si, and InSb and typical values of m_e^* are shown in Table 7.3.

TABLE 7.3

Substance	m_e^*/m_e electrons		m_e^*/m_e holes	
	transverse	parallel	heavy	light
Ge	0·08	1·64	0·28	0·044
Si	0·19	0·97	0·49	0·16
InSb	0·02			

The energy surfaces in the valence bands do not have exact spherical symmetry and so, even for the holes, the value of m_e^* will show a slight dependence on direction.

From the values of m_e^* it is possible to determine the shapes of the energy surfaces, such as those in Fig. 7.8. The experiments are reviewed by Bagguley and Owen (1957) and by Lax and Mavroides (1960).

This type of investigation is a very suitable one to be made on semiconductors, particularly at low temperatures, because their electrical conductivity is so low that the radio-frequency field can penetrate the specimen easily (it is this lack of penetration which makes cyclotron resonance experiments on metals so difficult). Nevertheless, the number of carriers present at low temperatures is normally too small to give a detectable resonance absorption. The number of carriers must therefore be artificially increased. This can be done in three ways; by optical excitation, by passing a current through the specimen, or by increasing the microwave power until ionization occurs. Of the three methods, optical excitation (from a tungsten filament lamp) seems to be the most satisfactory since it usually gives the narrowest resonance lines, probably because the other methods heat the specimen to some extent. Whilst in principle optical excitation should produce both electrons and holes, it usually produces electrons in n-type material and holes in p-type specimens. This may be due to some type of surface recombination or trapping effect, but the detailed cause is not known. With a small amount of Cu ($\sim 10^{14}$ atoms cm^{-3}) both electrons and holes are produced with white light.

7.23. *Width of the resonance line*

If cyclotron resonance is to be observed at higher temperatures where τ is much smaller, higher frequencies must be used and indeed some observations on InSb at room temperature have been taken using infrared radiation (Burstein, Picus, and Gebbie, 1956; Keyes, Zwerdling, Foner, Kolm, and Lax, 1956). Other experimenters have used microwaves in the 8 mm region (36,000 Mc/s) and have been able to detect resonance in Ge and Si up to about 100° K (Bagguley, Stradling, and Whiting, 1961). They have shown that m_e^* changes little if at all up to that temperature, which indicates that the shapes of the energy surfaces are not very temperature dependent. They do, however, find a marked temperature dependence in the width of the resonance line. At helium temperatures this extends over a few hundred oersted of the applied magnetic field, but as the temperature is raised it broadens very considerably. This occurs because the relaxation time is so short that the energy or the velocity of the carriers is not sufficiently defined (by the uncertainty principle) to give a very sharp resonance. Thus the width of the resonance lines gives a measure of the relaxation time or of the mobility. Experiments on the temperature dependence of the mobility from measurements of the width of the resonance lines have been made by Bagguley, Stradling, and Whiting (1961) from 2 to 100° K. For

p-type Ge and Si they find that the mobility can be analysed into a temperature-independent term due to neutral impurity scattering (section 7.11), and a term proportional to $T^{-1\cdot5}$ which corresponds well with the temperature dependence which is to be expected from phonon scattering (section 7.9). At high temperatures n-type Ge and Si also gave a $T^{-1\cdot5}$ mobility. There was no conclusive indication in these experiments of any ionized impurity scattering (section 7.10). It should, however, be realized that for cyclotron resonance to be observed it is necessary that any scattering which occurs should be sufficiently small so that the coherence of the spiralling of the carrier is not spoiled. Thus exact agreement between mobility behaviour from these experiments and from Hall and conductivity measurements would be rather surprising.

7.24. Optical and infra-red absorption

The presence in semiconductors of various energy bands and levels which are separated from one another by energies ranging from 10^{-2} eV to a volt or more, immediately suggests that the energy from electromagnetic radiation might be absorbed by inducing excitations from one level to a higher one, provided that the frequency, v, of the radiation was such that hv was equal to or greater than the energy difference between the two levels concerned. Since a photon with an energy of 1 eV has a wavelength of $1\cdot24\times10^{-4}$ cm ($1\cdot24$ micron) it will be seen that, for the energy levels of Ge and Si which we have discussed, the absorption spectra should be observed in a broad range covering the near and far infra-red.

The fact that many semiconductors, unlike metals, are quite transparent to large regions of the infra-red spectrum enables absorption experiments to be made on bulk samples rather than on thin films. The dependence of W, the intensity of radiation transmitted through a specimen, on the specimen thickness, x, is of the form

$$W = W_0 \exp(-\alpha x) \qquad (7.54)$$

where α is called the absorption coefficient and is the quantity which is usually quoted in the literature. If α is very small it is equal to the energy absorbed per unit thickness of material divided by the energy incident. It can be determined from a measurement of the ratio of the incident to the transmitted intensity of the radiation (see, for example, Fan, 1956). There are four main types of transition which can be induced by radiation.

(a) *Intrinsic absorption.* This occurs when an electron is excited across the main energy gap from the valence to the conduction band. For such excitation the shorter wavelengths are necessary, and the wavelength at which absorption starts can be used to estimate the energy gap. Radiation in the visible region of the spectrum is nearly always sufficient to produce intrinsic absorption and this causes most semiconductors to appear opaque. Only those with a large energy.gap such as CdS (2·4 V) are to some extent transparent to the optical spectrum.

(b) *Carrier absorption*. Radiation of much lower energy can be used to excite free carriers in the valence or conduction bands into vacant states of higher energy within those same bands.

(c) *Localized states*. The presence of impurity levels within the energy gap enables two kinds of excitation to occur. Electrons and holes can be excited to the conduction and valence bands respectively, giving rise to free charge carriers in those bands. This will increase the electrical conductivity of the specimen and it enables semiconductors to be used as infra-red detectors. Secondly the electrons and holes bound to the impurity levels can be excited to higher *bound* states by radiation of smaller energy than that needed to create free carriers.

(d) *Lattice absorption*. The interaction of electromagnetic radiation with lattice vibrations depends very much on the nature of the crystal. In compounds such as InSb, which has some polar character, the coupling between the field and the lattice will be strong, whereas it is much weaker in homopolar crystals such as Ge and Si. Nevertheless there is still sufficient interaction for some absorption to be observed which can be ascribed to phonon interaction.

7.25. *Intrinsic absorption*

We must now consider under what circumstances low-temperature measurements can be used with advantage in these absorption investigations. At first sight it might be thought that the intrinsic absorption across the main energy gap should be temperature independent, but it has been found that the width of the energy gap changes with temperature. This can be detected very easily by the shift in the absorption edge (Fig. 7.15). The change in the energy gap is due to two causes. (a) The dilatation of the crystal lattice with temperature; this changes the Brillouin zone shapes and their energy separation. (b) The energy levels of electrons in a fixed lattice differ slightly from those which occur if the lattice is vibrating, the shift becoming greater as the amplitude of oscillation is increased, i.e. at higher temperatures. Since at low

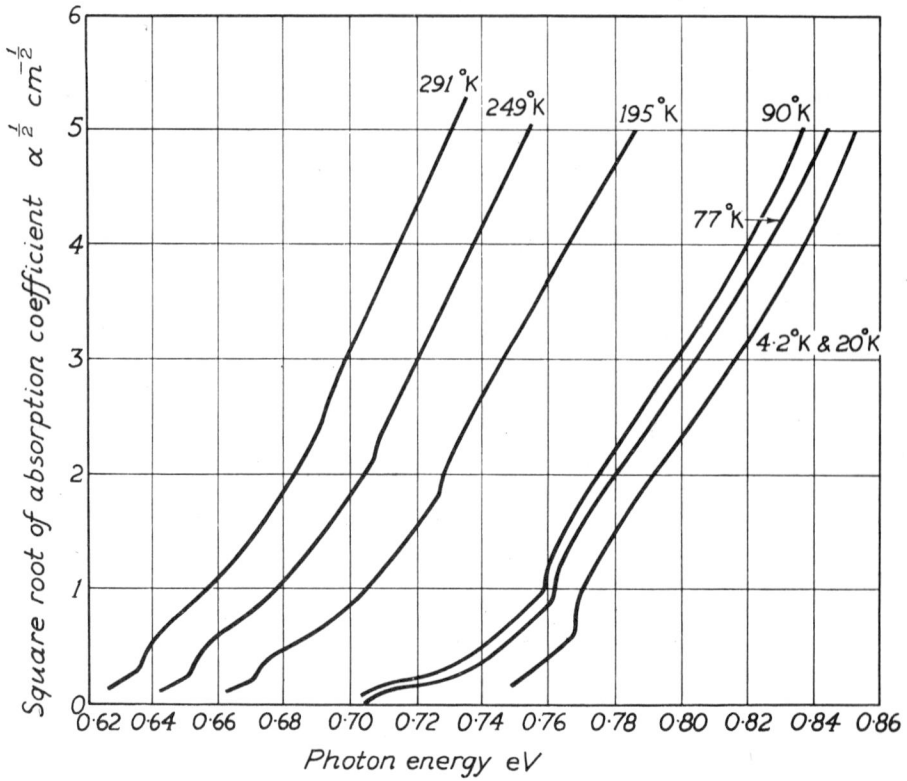

FIG. 7.15. Infra-red intrinsic absorption in high purity germanium showing the fine structure and the shift in the absorption edge with temperature due to the change in the energy gap (Macfarlane, McLean, Quarrington, and Roberts, 1957).

temperatures thermal expansion coefficients tend to zero and the lattice vibrational energy decreases as T^4 (1.11) one would expect that the gap energy should flatten off to a constant value as the temperature is reduced. As Fig. 7.15 shows, there is no difference between the absorption curves at 20 and 4·2° K for Ge. This constant value is the one which must be compared with that deduced from theoretical considerations. At high temperatures the energy gap $\mathscr{E}_g$, can be shown to be of the form (Fan, 1956)

$$\mathscr{E}_g(T) = \mathscr{E}_g(0+)\mathscr{E}'T, \tag{7.55}$$

where $\mathscr{E}_g(0)$ is the gap at $T = 0°$ K. Thus Hall and conductivity processes which are determined by a carrier concentration involving the exponential, $\exp\{-\mathscr{E}_g(T)/2kT\}$ (e.g. as in 7.22), will, because of (7.55), yield an apparent activation energy of $\mathscr{E}_g(0)$ rather than $\mathscr{E}_g(T)$. Quite good agreement is obtained between determinations of $\mathscr{E}_g(0)$ from Hall or conductivity measurements and those from infra-red absorption. Analysis of the Hall and conductivity measurements also enables $\mathscr{E}'$ to be calculated, again in quite good agreement with the optical measurements.

It will be noted from Fig. 7.15 that the absorption edge does not rise very sharply but that it has some fine structure. This is another consequence of the detailed band structure (section 7.14). For since in any interaction wave vector as well as energy must be conserved, we must have

$$\mathbf{k} = \mathbf{k}'+\mathbf{k}_p, \tag{7.56}$$

where $\mathbf{k}$ and $\mathbf{k}'$ refer to the wave vectors of the electron before and after excitation and $\mathbf{k}_p$ is the wave vector of the photon. However, because the photon wavelengths which we are considering are very long, $\mathbf{k}_p$ is small and can be neglected. Hence

$$\mathbf{k} \approx \mathbf{k}'. \tag{7.57}$$

We have, however, already seen that whilst the maximum energy in the valence band occurs at $\mathbf{k} = 0$, the minimum energy in the conduction band corresponds to larger values of $\mathbf{k}$. Thus if condition (7.57) was upheld, optical transitions could not occur across the minimum width of the energy gap; only 'direct' or vertical transitions could take place (Fig. 7.16). In practice it is found that the 'forbidden' transitions are observed because there is always sufficient interaction between the electrons and the lattice so that a phonon can be absorbed or emitted in order to maintain the wave vector balance; i.e.

$$\mathbf{k} = \mathbf{k}'\pm\mathbf{q}, \tag{7.58}$$

where $\mathbf{q}$ is the wave vector of this phonon. Such a process enables 'indirect' or diagonal transitions to take place from the maximum of the valence band to the minimum of the conduction band (Fig. 7.16). Some of the fine structure of the absorption curves shown in Fig. 7.15 can be ascribed to the change from indirect to direct transitions, but for a full interpretation the reader should consult the original paper.

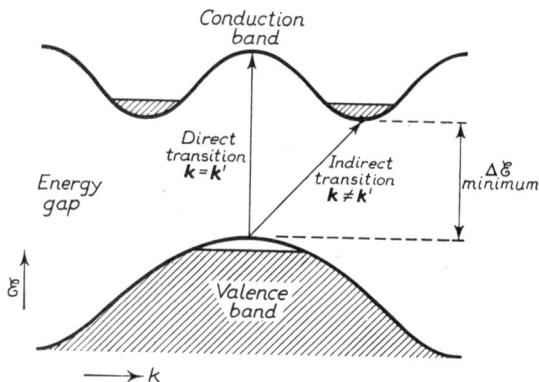

FIG. 7.16. Schematic diagram of the energy levels of a semiconductor illustrating direct transitions, i.e. in which $\mathbf{k} = \mathbf{k'}$, and indirect transitions in which the excitation energy is a minimum, but $\mathbf{k} \neq \mathbf{k'}$.

Very similar curves have been obtained by the same authors for the absorption edge of Si.

7.26. Carrier absorption

The wave vector rule, (7.57), should apply equally well to transitions of the carriers within a band as it does to interband transitions. Within a band, however, a given wave vector can only have a single energy associated with it and hence (7.57) implies that no transitions within the band, due to infra-red absorption, are possible. Once again, however, absorption *is* observed because electron-lattice interaction enables a phonon to be absorbed or emitted as in (7.58). At low temperatures, where electron-phonon interaction is small, the wave-vector balance can still be maintained because of the interaction between electrons and static defects such as impurity atoms. This type of interaction can be detected because it gives a different frequency dependence. (absorption proportional to $\nu^{-3.5}$) from that due to phonon interaction (absorption proportional to $\nu^{-1.5}$). Few investigations of specifically low temperature interest have, however, been made on this type of absorption.

7.27. *Absorption due to localized states*

The excitation of electrons to or from the shallow impurity levels by infra-red radiation can usually only be investigated at low temperatures. This is because at high temperatures these states are always completely excited by thermal activation and hence the infra-red will not be absorbed. Fig. 7.17 shows how the impurity absorption bands in boron-doped Si, which cannot be detected at 300° K, show up at 4·2° K. Since the energies involved are much smaller than that of the main

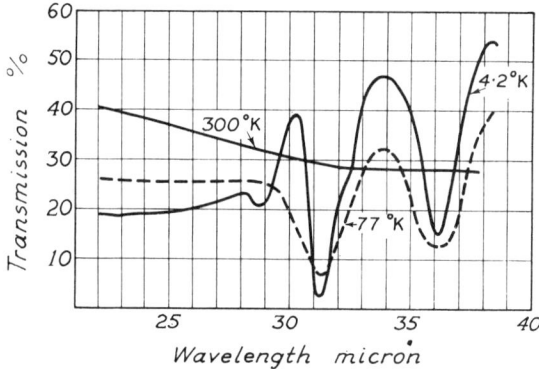

Fig. 7.17. Infra-red transmission spectra for boron-doped silicon (acceptor concentration 5×10^{15} cm^{-3}) showing how the absorption due to the localized states can only be detected at low temperatures (Burstein, Picus, Henvis, and Wallis, 1956).

energy gap, the absorptions always occur for wavelengths which are longer than those at which the intrinsic absorption edge (section 7.25) is observed. The shape of the absorption curve as a function of frequency can be calculated roughly by assuming as we did in section 7.2 that the impurity atom has energy levels similar to that of a hydrogen atom immersed in a matrix of high dielectric constant ϵ. Absorption of radiation will occur for frequencies which correspond to the energies of the various excited states of the atom—these should give rather sharp resonance peaks—and a continuous absorption band will begin at the ionization energy for which the electron is excited into the band. This absorption will tail off at higher frequencies. When the exciting frequency, ν, is very much greater than ν_i, the frequency corresponding to the ionization energy $\mathscr{E}_i$, then the absorption coefficient, α, is given by

$$\alpha = \frac{5 \cdot 26 \times 10^{-17} N}{r} \frac{m_e}{m_e^*} \frac{\mathscr{E}_H}{\mathscr{E}_i} \left(\frac{\nu_i}{\nu}\right)^{3 \cdot 5} \qquad (7.59)$$

(Fan, 1956) where r is the refractive index, N is the number of impurity atoms per unit volume, and $\mathscr{E}_H$ is the ionization energy of hydrogen. This formula is very approximate as it takes no account of the anisotropy of the effective mass nor of the other complications of the real band structure of semiconductors.

Fig. 7.18 shows the absorption of boron-doped Si at $4 \cdot 2° $ K plotted as a function of photon energy. The three peaks on the low-energy side of the main absorption band correspond to the excitation of the boron atom from the 1s ground state to the 2p, 3p, and 4p states respectively.

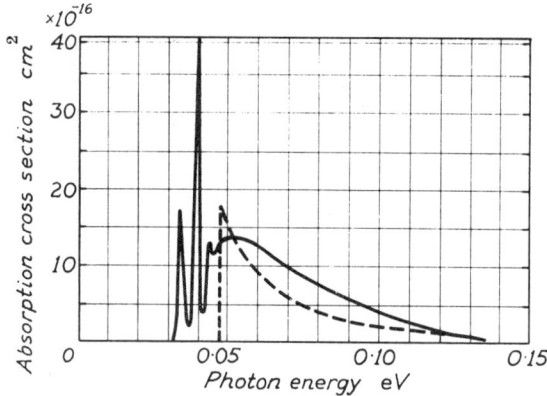

FIG. 7.18. Infra-red absorption in boron-doped silicon (as Fig. 7.17) showing the fine structure due to the excitation of the boron atom. The dotted curve is the theoretical expression for the absorption (7.59) (Burstein, Picus, Henvis, and Wallis, 1956).

The dotted curve indicates the absorption to be expected using the simple hydrogen model of (7.59). The overlapping valence bands (section 7.14) make it difficult to make a more accurate calculation of the activation and ionization energies of acceptor levels. For group V impurities in Si, where the excitation is to the neighbourhood of the conduction band, the energy levels from infra-red absorption determined by Picus, Birstein, and Henvis (1956) are in good agreement with theoretical values of Kohn and Luttinger (1955). The ionization energy determined from the infra-red absorption also correlates quite well with the thermal ionization energy deduced from conductivity and Hall effect measurements (see Table 7.4), although the optical values always tend to be higher than the thermal values.

In Ge the usual group III and V impurities have levels which are only about 10^{-2} eV from the band edge and thus investigations of the excited states and ionization energy must extend to wavelengths beyond 120 micron. Some experiments have now been made in this region at $4 \cdot 2° $ K

(Fan and Fisher, 1959), and Fig. 7.19 shows the excitation peaks and main ionization band for n-type Ge. The detailed structure of the peaks varies from one sample to another and with increasing impurity concentration some broadening and overlapping occurs. The results give

TABLE 7.4

Ionization energy of shallow impurity states

	Silicon (Picus, Burstein, and Henvis, 1956)		Germanium (Fan and Fisher, 1959)	
Impurity	*Optical ionization energy eV*	*Thermal ionization energy eV*	*Optical ionization energy eV*	*Thermal ionization energy eV*
As	0·0533	0·049	0·014	0·0127
Bi			0·0125	0·012
Sb	0·0426	0·039	0·0098	0·0096
P	0·0503	0·044	0·0128	0·0120

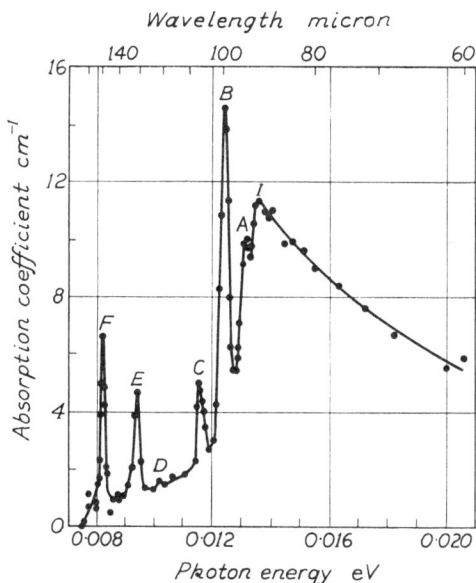

FIG. 7.19. Absorption spectra in the far infra-red, to show excitation and ionization of the shallow impurity states in n-type germanium (arsenic doped) (Fan and Fisher, 1959).

good agreement with the excitation levels calculated by Kohn (1957) and, as is shown in Table 7.4, the ionization energies for different impurities are similar to those calculated from Hall and conductivity measurements. As in Si the latter are always smaller.

Many other experiments have been made on Si and Ge doped with other impurities, in particular, Cu, Au, and Zn. These have multiple levels, some of which are a considerable distance from the band edges, and the absorption spectra therefore occur at shorter wavelengths where the experimental techniques are not so difficult. For details of these experiments and those in which the absorption is due to the defects produced by bombardment with high-energy particles, the reader is referred to the review article by Fan (1956).

7.28. *Lattice absorption*

In addition to the absorption mechanisms which we have described in the preceding sections, some other bands have been observed in the range 5 to 35 micron in Ge and Si which do not vary from one sample to another and are independent of the carrier concentration. It has been suggested that these bands are due to the interaction of the lattice vibrations with the photons. The amount of absorption is temperature dependent and appears to be proportional to the mean square displacement of the atoms. This would suggest that the most satisfactory investigations on the structure of these bands should be made at high temperatures and whilst it is true that they can readily be observed at room temperature their structure is very often masked by other absorption mechanisms, particularly that associated with the free carriers (section 7.26). To decrease the number of free carriers the specimens are cooled, although this will itself change the phonon spectrum.

A careful analysis of the spectra (Johnson, 1959) enables the energy associated with the various phonon modes of maximum (or minimum, in the case of optical modes) wave number to be determined.

7.29. *Photoconductivity*

The ionization of impurity atoms by infra-red absorption increases the number of free carriers in the semiconductor and these will enhance the electrical conductivity of the material. Thus the change in conductivity as a function of wavelength should be similar in form to the absorption curves themselves and this is borne out by experiment. An example is shown in Fig. 7.20, where the relative photoconductive response and the absorption coefficient is given for copper-doped Ge at 5° K. Since the electrical conductivity of semiconductors is usually quite small at low temperatures, it will be appreciated that with the addition of suitable impurities, they can be made into very sensitive detectors of infra-red radiation down to quite long wavelengths. In fact this sensitivity can be rather troublesome if the dark electrical conductivity is to be measured as it is difficult to stop radiation of 100 micron wavelength or more from entering the cryostat.

Photoconductivity can arise from radiation corresponding to (a) the intrinsic gap energy or to (b) the ionization energy of an impurity atom. In (a) both electrons and holes are produced in the conduction and valence bands respectively, whereas in (b) only one type of carrier will be produced. The measured value of the photocurrent will depend not only on the rate at which the carriers are excited but also on the time

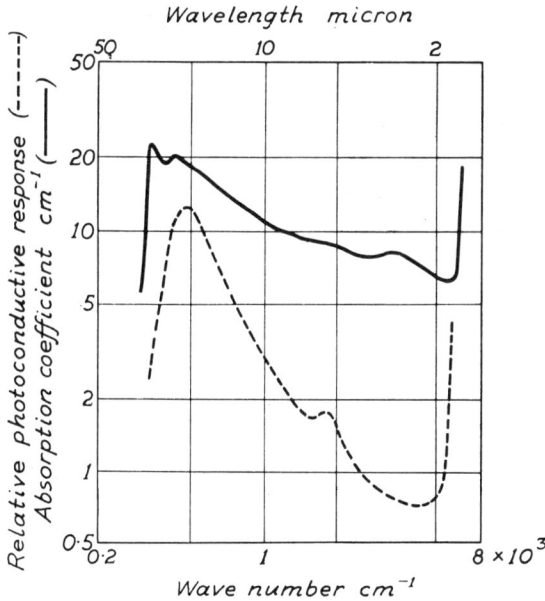

FIG. 7.20. The infra-red absorption coefficient (solid curve) and the relative photo-conductive response (broken curve) for a p-type copper-doped germanium crystal at 5° K (Kaiser and Fan, 1954).

for which they are free. A longer lifetime will give a larger photocurrent. Experiments show that for both types of photoconduction the trapping of the carriers always occurs at impurity atoms and other defects which have energy levels well within the energy gap. Electrons and holes never appear to recombine directly because this would lead to lifetimes (of the order of a second in Ge and Si) which are much longer than those which are observed (of the order of 10^{-3} to 10^{-6} second). The group III and V impurity levels are not very efficient as traps (except perhaps at helium temperatures) because they are so close to the band edges that the release of a trapped carrier can occur very easily.

Studies of photoconductivity can be made for three main purposes: (a) to investigate energy levels associated with various impurities, (b) to determine the lifetime of excited carriers, and (c) the development of sensitive detectors of radiation for various wavelength ranges. Low-temperature experiments are necessary, as they were with the infra-red absorption measurements, in order to avoid thermal activation and ionization of the levels which are being investigated. It has also been found that under some circumstances the free-carrier lifetime increases very much at low temperatures, thus enabling specimens of very high photosensitivity to be produced.

Apart from the fact that the onset of photoconductivity is a much more sensitive method of determining the ionization energy of various impurities than that which is available by the direct measurement of the infra-red absorption, the results which are obtained are similar to those described in sections 7.25 and 7.27. It should, however, be noted that the measurement of the energies of excited *bound* states cannot be made by photoconduction because free carriers are not produced.

7.30. *Carrier trapping and recombination*

Investigations on the trapping and recombination of carriers by various impurity atoms have been the basis of many experiments. Pulses of radiation a few microseconds long with a frequency corresponding to the intrinsic gap energy, are applied to the specimen and the build-up and decay curves of the photocurrent are observed on an oscilloscope. For Ge the impurity atoms which can act as traps can be, amongst others, Au, Cu, Zn, Cd, Mn, Ni, Co, or Fe. These atoms all have at least two energy levels within the intrinsic gap and all these levels (except the lowest one for Au) are acceptor states (Newman and Tyler, 1959; Johnson and Levinstein, 1960). It is believed that the impurity atom first traps an electron and then becomes negatively charged so as to attract a hole (or vice versa) and in this way electron–hole recombination occurs. The rate of trapping will be more efficient at lower temperatures when thermal activation of the trapping centres becomes less likely. Thus the decay times decrease exponentially. A full statistical theory has been given by Shockley and Read (1952).

At sufficiently low temperatures the decay times tend to become constant, being dependent on the number of trapping centres which are available. This behaviour is illustrated in Fig. 7.21 in which curves of the decay time against $1/T$ are plotted for some specimens of high-purity Ge to which no impurities were added intentionally.

Other experiments have been made in which trapping impurities have been added to Ge which has also been doped in the usual way with group III or V impurities to make it low-resistance p or n-type. Fig. 7.22 shows how the decay time varies with $1/T$ for Fe-doped Ge. For p-type material the decay time is approximately constant at low temperatures and is similar to that shown in Fig. 7.21. For n-type specimens it rises

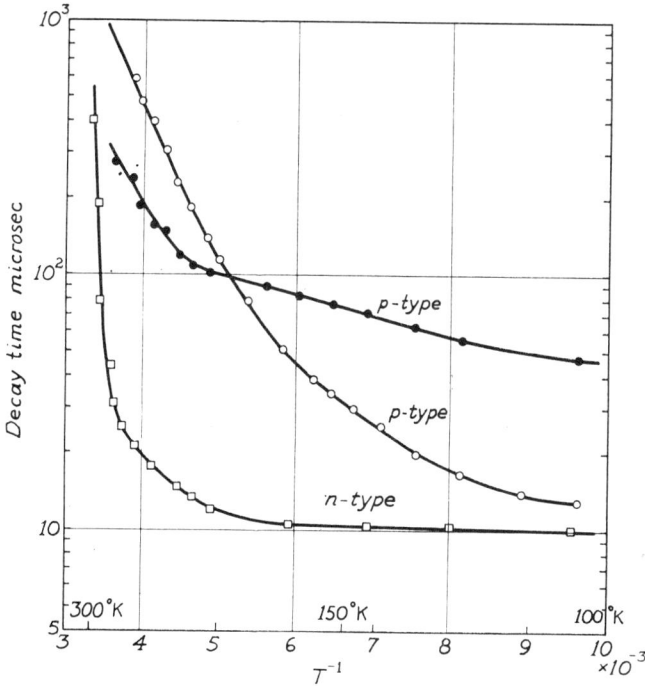

FIG. 7.21. Photoconductive decay time as a function of the inverse temperature for high-purity germanium samples not doped deliberately. All samples were intrinsic at 300° K with decay times of $\sim$ 1000 μsec (Newman and Tyler, 1959).

by several orders of magnitude. The interpretation of this effect is as follows. In p-type material the trapping atoms (which it will be recalled have acceptor states and which are situated towards the middle of the band gap) will be neutral and the situation at low temperatures is similar to that which we have already described in the preceding paragraph for high-purity material. In n-type material, however, the donor levels (if there are sufficient of them) will doubly ionize the trapping atoms. They will then be negatively charged and so they can attract a positive hole, but after this they will still have a single negative charge and hence they will repel a free electron. Thus recombination of the hole and the electron can only occur with the assistance of some thermal activation to enable the electron to overcome the potential barrier around the trap.

This mechanism will become much more difficult as the temperature is reduced and it leads to the exponential increase in lifetime which is shown in Fig. 7.22. Confirmation for a mechanism of this kind has been given in some experiments on Te-doped Ge. The Te atoms give traps

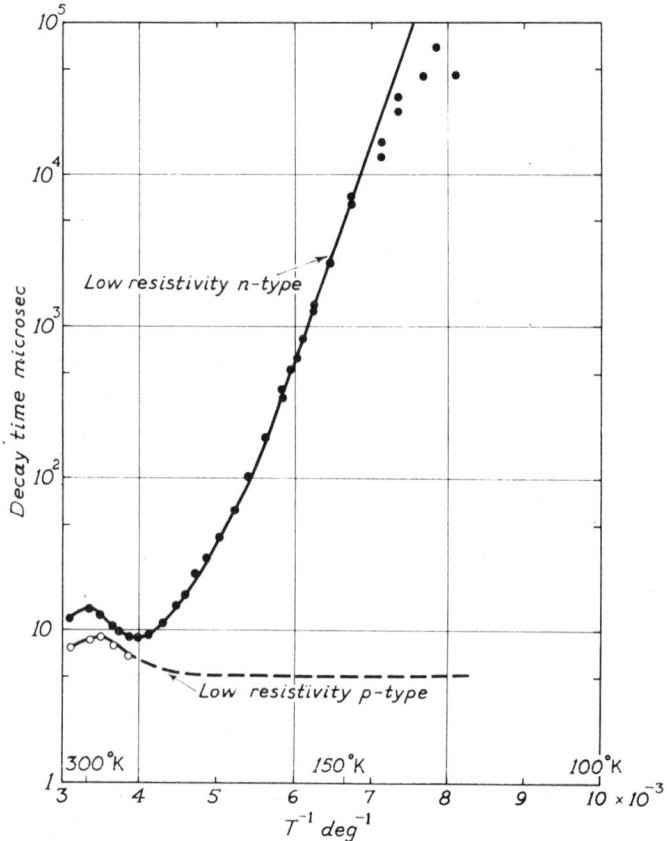

FIG. 7.22. Photoconductive decay times for Fe-doped germanium of low resistivity, showing the very marked difference between the behaviour of *n*- and *p*-type specimens (Newman and Tyler, 1959).

which are *donor* levels. It is then found that it is the decay time of the *p*-type material which increases rapidly at low temperatures, which is what one would expect if the model is correct.

A full account of trapping and recombination mechanisms would be out of place in this book. It is an exceedingly complicated subject because of the interplay of the various energy levels which can take part in the mechanism. At liquid-helium temperatures for example, due to the lack of thermal activation, the group III and V impurity levels themselves can also act as trapping centres. The reader is referred to the article by Rose (1957) for further information on the subject.

7.31. *Infra-red detectors*

We should make some remarks on the use of semiconductors as detectors of radiation. With the very large amount of information which is now available, it is possible to select a semiconductor with suitable doping, which will be photoconductive for a particular part of the spectrum. Either intrinsic or impurity activation can be used, depending on the wavelength to be investigated. A sensitive detector is one which has a large ratio of dark to illuminated current. It is clear that for this to occur one criterion† must be that the energy difference between the levels which are going to be excited by photons must be large compared with kT, otherwise the thermal activation will give a large 'dark' current. In practice, this means that detection of visible light and infra-red down to ~ 4 micron can be made at room temperature, whereas sensitive infra-red detectors for 4 to about 10 micron must be operated at liquid-nitrogen temperature or below. For the detection of very long wavelengths in which excitation of shallow impurity states is utilized it is necessary to go to helium temperatures in order to achieve adequate sensitivity.

Beyen, Bratt, Davis, Johnson, Levinstein, and MacRae (1959) describe two types of Au-doped germanium photo-cells. One was made using plain Au-doped Ge which is p-type. It operates by the excitation of electrons from the valence band to one of the Au levels with an energy corresponding to 9·5 micron. It was found that this detector was four times more sensitive at $60°$ K than it was at $80°$ K due to the reduction in the thermally activated dark current. The second cell was made with n-type (Sb-doped) germanium which also had some Au doping. The Sb donors ionize the Au levels and the electrons on these Au ions are then excited to the conduction band by the photons. The excitation energy for this process corresponds to that of 6 micron photons. In this case there was no change in sensitivity on cooling from 80 to $60°$ K, since even at $80°$ K the dark current was small. Considerations on the design of suitable photo-detectors are given by Wright (1958) and by Roberts and Wilson (1958).

7.32. 'Hot' carriers and impact ionization

In our discussion of mobility in section 7.8 ff. we have assumed implicitly that the carriers are in thermal equilibrium with the lattice.

---

† This is not the only criterion. Other factors which must be considered are mobility, recombination time, and the absorption coefficient.

When an electric field is applied and the carriers are accelerated, a drift velocity is superimposed on their random thermal motion, but owing to phonon collisions the mean energy remains constant. This will only be true if the interactions are effective enough to dissipate the energy which the carriers receive on being accelerated by the electric field. Owing to the small effective mass of the carriers their velocity is quite high and only a small amount of their excess energy can, in fact, be lost at each collision. The outcome of this is that the drift velocity of the carriers will no longer be proportional to the applied electric field if it is above a certain value, i.e. Ohm's law is no longer valid. A simplified analysis of the problem is given by Gunn (1957). If the field is sufficiently large, the energy of the carriers can become high enough for ionizing collisions to occur which result in the formation of a free electron and a hole. These new carriers are then themselves accelerated by the field, further ionization occurs, and hence an avalanche breakdown of the material can arise. At room temperature departures from Ohm's law and breakdown tend to occur in relatively high fields of a few hundreds or thousands of volt cm^{-1}, but at low temperatures phonon interaction is so small that fields of less than a volt cm^{-1} are sufficient to accelerate the carriers up to the point where they are not in thermal equilibrium with the lattice. These are called 'hot' carriers.

At low temperatures the 'hot' carriers can cause another type of avalanche breakdown by the ionization of neutral impurity atoms, particularly those of groups III and V, which, as we have already seen, have very small ionization energies. This breakdown, which is usually called impact ionization, results in the increase in current through the specimen by several orders of magnitude when the applied field reaches a certain critical value, $\mathscr{E}_c$. For the III–V levels $\mathscr{E}_c$ can lie between about 4 and 40 volt cm^{-1} depending on the purity of the sample. This type of breakdown does not occur at room temperature because the impurity atoms will have been already ionized by thermal activation. The phenomenon can be observed in Si as well as in Ge, but the breakdown fields are much higher, 170 volt cm^{-1} being a typical figure.

A general investigation of impact ionization in Ge has been given by Sclar and Burstein (1957) and by Koenig and Gunther-Mohr (1957). They have made conductivity and Hall measurements which show that at breakdown the Hall coefficient decreases very rapidly, i.e. the number of carriers increases, whilst the mobility actually decreases slightly. A typical series of results on n-type Ge is shown in Fig. 7.23. On the graph is also drawn the extrapolation of the current-field curve at low fields where Ohm's law is obeyed. It can be seen that deviations from Ohm's law occur at very low fields, of the order of 3 per cent of $\mathscr{E}_c$. The rise and fall in the mobility is explained as follows. In low fields the mobility is dominated by ionized impurity scattering (μ_i, section 7.10). As the field is increased the energy of the carriers increases and the effectiveness of this type of scattering *decreases* (7.37). Thus μ_i becomes

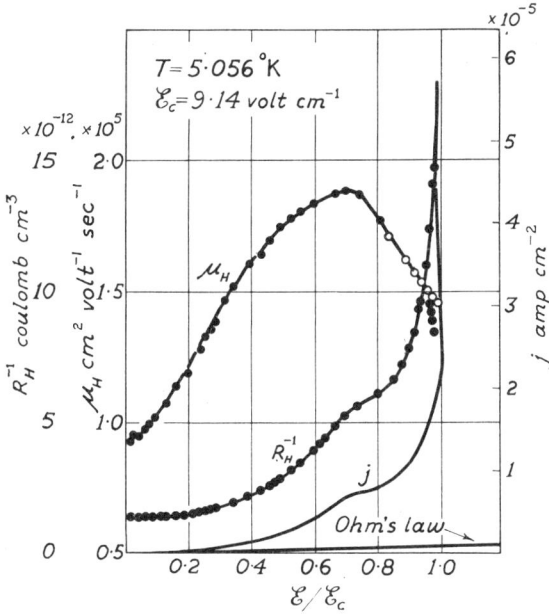

FIG. 7.23. Impact ionization. Current density, j, reciprocal Hall coefficient, R_H^{-1}, and Hall mobility, μ_H, plotted against the electric field for n-type germanium (Sb doped; $N_D - N_A - 2\cdot2 \times 10^{14}$ cm^{-3}). The extrapolated Ohm's law behaviour is also shown (Koenig and Gunther-Mohr, 1957).

higher and at a certain field it will become equal to the lattice-scattering mobility (μ_g, section 7.9). Now μ_g is *inversely* proportional to (energy)$^{\frac{1}{2}}$ (7.33) so that with a further increase in field this mobility decreases and becomes the dominant process.

7.33. *Factors affecting the breakdown*

The change in current at breakdown is found to decrease as the temperature is raised because at higher temperatures more impurity atoms are already thermally ionized and so there are fewer left for impact ionization. In Ge at about 30° K all the III–V impurity levels are thermally excited and no breakdown effect can be detected above this temperature. The breakdown field for a given specimen does not change very much with temperature, although it is dependent on the density of impurities. It has been found that the breakdown field can either decrease (Koenig, 1959) or increase (Sclar and Burstein, 1957) with less pure specimens. This seemingly contradictory behaviour is apparently dependent on the amount of compensation which is present in the specimen (Koenig, 1959).

Impact ionization does not occur in specimens where band overlap occurs (in Ge 10^{18} impurity atoms cm^{-3} or more) because then the impurities are already ionized.

At breakdown the time taken for current build up is very short. Sclar and Burstein (1957) suggest that the rise time is $< 10^{-8}$ sec and this behaviour has been utilized in the development of fast switching devices for computer circuitry. Such devices are sometimes called 'cryosars' This rise time can be decreased if the specimen is illuminated with room-temperature radiation. This itself will excite carriers from the impurity levels and will increase the number of free carriers when the impact ionization process begins, making the process more efficient and rapid.

For specimens which are at temperatures where impurity conduction (section 7.20) occurs, it is still possible to observe impact ionization, although there is a change in mobility behaviour. Details are given by Koenig and Gunther-Mohr (1957).

7.34. The 'hot' carrier system

So far we have discussed the action of 'hot' carriers in producing ionization. Many other investigations have been made on the 'hot'-carrier system itself and in particular with the mechanism which determines the mobility of the 'hot' system. It is not necessary to apply a field to the specimen in order to produce 'hot' carriers. Experiments at helium temperatures by Honig and Levitt (1960) on n-type Si have shown that room-temperature radiation can excite carriers which are 'hot'. This method of excitation has been used by Rollin and J. M. Rowell (1960) in determining the mobility behaviour of 'hot' carriers in p-type Ge.

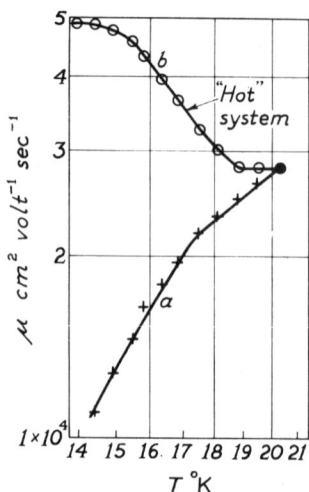

FIG. 7.24. 'Hot' carriers. The variation of mobility with temperature for p-type germanium (a) unilluminated specimen; (b) specimen illuminated with room-temperature radiation which produces a 'hot' carrier system (Rollin and J. M. Rowell, 1960).

Hall and conductivity measurements of specimens were made with and without illumination. The difference in the Hall mobilities is very striking (Fig. 7.24). From $21°$ K down to $17°$ K (below which impurity conduction (section 7.20) occurs and complicates the situation) μ is proportional to $T^{\frac{3}{2}}$ for the unilluminated specimen, and is presumably being limited by ionized impurity scattering (7.38). When the specimen is illuminated with room-temperature radiation, μ *increases* as the temperature is reduced. This is because the scattering by ionized impurities is much less for the more energetic 'hot' carriers (7.37) and it appears that phonon interaction becomes the dominant scattering mechanism (7.33).

7.35. Magnetic field effects

Apart from our description of cyclotron resonance (section 7.21) and the Hall effect (section 7.16) we do not intend to give a detailed description of the changes in the various properties which occur when measurements are made in a magnetic field H. Since the application of a field alters the electron trajectories, and also the energy levels, it follows that all the phenomena which we have discussed will to some extent be affected. In many cases the changes which are observed are quite large. Owing to the complicated energy surfaces at the band edges, the magnetic effects are very dependent on the direction of the field relative to the crystal axes and also relative to the specimen axis.

The phenomena which have been studied most intensively are as follows.

(*a*) The effect of a magnetic field on the electrical resistivity: the resistivity usually increases with H, although in the impurity conduction region a decrease can occur, e.g. in InSb (Fritzsche and Lark-Horovitz, 1955).

(*b*) The change in the thermopower (section 8.11) with H, and the observation of an extra thermal e.m.f., at right angles to both H and the temperature gradient (Nernst effect) (Herring, Geballe, and Kunzler, 1958, 1959, 1959*a*).

(*c*) The effect of H on the 'hot' carrier system (section 7.32). This results in increasing the breakdown field for impact ionization and also in a change in the behaviour in the non-ohmic region which occurs before breakdown (Sclar and Burstein, 1957).

It has proved more convenient in the development of this book to treat the following properties of semiconductors under the appended sections.

The specific heat—section 1.17.
The thermal conductivity—section 5.20.
The thermopower—sections 8.11 ff.
The de Haas–van Alphen effect—section 10.6.

8

THERMO-ELECTRICITY

8.1. Introduction

MANY measurements have been made of the e.m.f. which is developed within a closed circuit formed of two dissimilar metals when the junctions of the metals are kept at different temperatures. Whilst according to the third law of thermodynamics this e.m.f. should tend to zero as $0°$ K is approached, it is found that, even at liquid-helium temperatures, potentials of several micro-volts deg^{-1} can be obtained. Besides the utility of such an effect as a secondary thermometer, the phenomenon itself is one which must be explained in any satisfactory theory of conduction. Unfortunately, even more than in other electronic properties which we have discussed, the magnitude and sign of the thermo-electric effects are very sensitive to the deviations of the electron distribution from that which we assume in the free electron approximation.† This means that each material must be considered separately before even a qualitative description of its thermo-electric properties can be given.

Whilst the detection of the thermal e.m.f. in a system containing two dissimilar junctions is the most usual type of experiment there are, in fact, three types of thermo-electric behaviour (which are all interrelated) which can be investigated. These effects are usually described by the following coefficients.

(a) Seebeck coefficient. This is the e.m.f. F_s produced when two conductors A and B are joined together as in Fig. 8.1 with their two junctions at different temperatures, T_1 and T_2, under the condition that no current flows in the circuit. It is the thermal e.m.f. which is measured in an ordinary thermocouple arrangement. The thermo-electric power is the e.m.f. produced per unit temperature difference between the two junctions, i.e. dF_s/dT.

(b) The Peltier coefficient Π_{AB} is derived from the inverse of the Seebeck effect. It is the heat evolved per unit time at a junction between two conductors A and B when unit current passes across the junction.

(c) The Thomson coefficient σ_t relates to a single metal. It is the rate at which heat is evolved when unit current flows through a conductor when the ends of the conductor have a temperature difference between

---

† A very useful textbook by MacDonald (1962) has been published recently.

them of 1 degree, the direction of the conventional current being from the lower to the higher temperature.

These coefficients are not independent and by simple thermodynamical arguments (or more rigorously, and not really with very much greater difficulty, by the methods of the thermodynamics of irreversible pro-

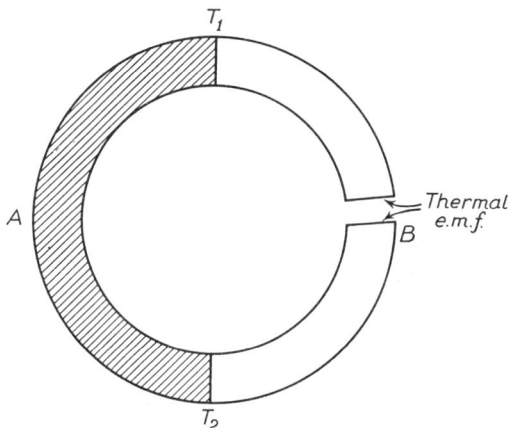

FIG. 8.1. A schematic diagram showing the arrangement of two metals A and B with their junctions at different temperatures, in order to produce a thermal e.m.f. between them.

cesses, see Ziman, 1960, p. 273) the following relationships may be derived:

Peltier coefficient,
$$\Pi_{AB} = T \int_0^T \frac{\sigma_{tB} - \sigma_{tA}}{T} \, dT. \tag{8.1}$$

Seebeck coefficient,
$$F_s = \int_{T_1}^{T_2} \frac{\Pi_{AB}}{T} \, dT. \tag{8.2}$$

Thermo-electric power,

$$dF_s/dT = \int_0^T \frac{\sigma_{tB} - \sigma_{tA}}{T} \, dT. \tag{8.3}$$

From these relations it can be seen that the fundamental quantity for each conductor which may be used to determine all the other coefficients is

$$\mathscr{S} = \int_0^T \frac{\sigma_t}{T} \, dT. \tag{8.4}$$

$\mathscr{S}$ is called the absolute thermo-electric power or the *thermopower* of a conductor. If $\mathscr{S}$ is known for any conductor A, then a determination of thermo-electric power (8.3) for a pair of junctions of that conductor with a second conductor, B, will enable one to calculate the thermo-power of B.

8.2. Low-temperature technique

The experimental determination of $\mathscr{S}$ at low temperatures has been made simpler by the fact that the thermopower of a superconductor is zero (section 8.10) and thus, as (8.3) shows, a measurement of the thermo-electric power of a metal against a superconductor gives a direct value for $\mathscr{S}$. The superconductor which has been used most commonly in such experiments has been lead, because this has a relatively high transition temperature ($T_c = 7\cdot22^\circ$ K) and hence direct measurements of the thermopower may be taken right up to T_c. For measurements above this temperature lead is still used as the reference junction and carefully determined values of the thermopower of lead are used. The most satisfactory values from the transition temperature up to 18° K, are those obtained by Christian, Jan, Pearson, and Templeton (1958). These are given in Table 8.1. They utilized a junction of lead and the superconducting alloy Nb$_3$Sn ($T_c \approx 17\cdot9^\circ$ K) in order to determine $\mathscr{S}$ directly for lead.

TABLE 8.1

Absolute thermo-electric power of lead

(from Christian, Jan, Pearson, and Templeton, 1958)

temp. (° K)	$\mathscr{S}$ (μV/deg)	temp. (° K)	$\mathscr{S}$ (μV/deg)	temp. (° K)	$\mathscr{S}$ (μV/deg)
7·25	$-0\cdot20_4$	11·0	$-0\cdot51_6$	15·0	$-0\cdot74_6$
7·5	$-0\cdot22_1$	11·5	$-0\cdot55_6$	15·5	$-0\cdot76_0$
8·0	$-0\cdot25_7$	12·0	$-0\cdot59_3$	16·0	$-0\cdot77_1$
8·5	$-0\cdot29_7$	12·5	$-0\cdot62_8$	16·5	$-0\cdot77_7$
9·0	$-0\cdot34_3$	13·0	$-0\cdot65_8$	17·0	$-0\cdot78_1$
9·5	$-0\cdot39_0$	13·5	$-0\cdot68_3$	17·5	$-0\cdot78_3$
10·0	$-0\cdot43_4$	14·0	$-0\cdot70_6$	18·0	$-0\cdot78_{45}$†
10·5	$-0\cdot47_5$	14·5	$-0\cdot72_8$	19·0	$-0\cdot78_{45}$†
				20	$-0\cdot78_4$†

† Calculated from interpolated Thomson heat measurements.

Above 18° K no superconductors are available and $\mathscr{S}$ must be found by direct measurements of the Thomson coefficient. These have been made for lead and tin by Borelius, Keesom, Johansson, and Linde (1930).

8.3. Elementary theory

It is quite easy to see why a thermopower exists. Let us consider a current flowing through a metal rod, the ends of which are maintained at temperatures T_1 and T_2 (Fig. 8.2), ($T_1 < T_2$). Then because of the temperature gradient, the Fermi distribution at one end of the specimen is different from that at the other, as is shown in the lower part of the figure. Thus an electron must be given extra energy as it is transported from T_1 to T_2 and this it will absorb from the phonons, giving rise to the Thomson effect. This extra energy may be calculated from the electronic specific heat per unit volume c_e. If there are n electrons per unit volume,

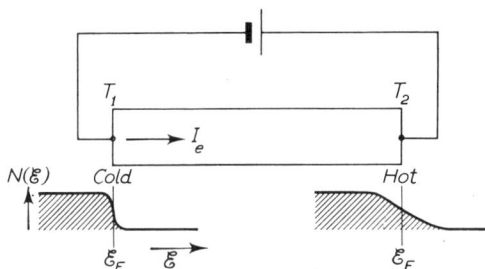

FIG. 8.2. The Thomson heat. Heat is absorbed as the electron current, I_e, flows from the cold to the hot end of the specimen. The electron energy distribution at either end of the specimen is shown in the lower part of the figure.

the mean specific heat of each is c_e/n and hence if T_2-T_1 is unity the heat which is absorbed per unit charge is $c_e/(ne)$,

i.e.
$$\sigma_t = c_e/(ne). \tag{8.5}$$

We have already seen (1.22) that for a free electron metal

$$c_e = \pi^2 k^2 n T/(2\mathscr{E}_0)$$

and so from (8.5) the thermopower is

$$\mathscr{S} = \int\limits_0^T \frac{c_e(T)}{neT}\, dT = \frac{\pi^2 k^2 T}{2e\mathscr{E}_0} = -3{\cdot}7\times 10^{-2}\frac{T}{\mathscr{E}_0\,(\mathrm{eV})}\mu\mathrm{V}\ \mathrm{deg}^{-1}. \tag{8.6}$$

From this expression we should expect that the thermopower should be (a) proportional to the absolute temperature, (b) dependent on the sign of the electronic charge, e, i.e. it should be negative, and (c) it should be rather small in magnitude; of the order of a few tenths of a $\mu\mathrm{V}\ \mathrm{deg}^{-1}$ in the liquid helium region. Whilst at temperatures down to, say, 100° K, the temperature dependence and magnitude of $\mathscr{S}$ are in fair agreement with the theory, nevertheless, even for the monovalent metals, not all these predictions are verified. The thermopower is sometimes positive, it is usually much larger than that given by (8.6) and it is not always linear with temperature.

8.4. *Modifications for a real metal*

The simple argument leading to (8.6) neglects the fact that when the electrons go from T_1 to T_2 they have to change their energies by various scattering processes (e.g. with impurities or with phonons, as has been discussed in section 4.5), and a more general expression† may be derived (see, for example, Mott and Jones, 1936, p. 310). This is

$$\mathscr{S} = \frac{\pi^2 k^2 T}{3e} \left. \frac{\partial(\log_e \sigma)}{\partial \mathscr{E}} \right|_{\mathscr{E}=\mathscr{E}_F} \qquad (8.7)$$

where the differential means the change in the electrical conductivity, σ, as the Fermi energy, $\mathscr{E}$, changes, evaluated at the actual Fermi energy, $\mathscr{E}_F$, of the metal. In order to appreciate the significance of (8.7) we should recall that the simple kinetic equation (4.3) shows that σ depends on $\mathscr{E}$ via the mean free path, l, and the quantity n/v_F. Thus the differential in (8.7) may be written as the sum of two terms

$$\frac{\partial(\log_e l)}{\partial \mathscr{E}} + \frac{\partial\{\log_e(n/v_F)\}}{\partial \mathscr{E}}. \qquad (8.8)$$

The first term in (8.8) seems reasonably certain to be positive since we should expect that a more energetic electron would not be scattered so easily as one with lower energy; hence l should increase with $\mathscr{E}$. The second term might be positive or negative. In a free electron metal where $\mathscr{E}$ is proportional to k^2, n is proportional to k^3 (1.18) and v_F is proportional to k (4.1). Thus n/v_F is proportional to k^2, i.e. to $\mathscr{E}$. Hence the second term of (8.8) will also be positive and will, in fact, be equal to $1/\mathscr{E}_F$. It should be noted that if we insert this term alone into equation (8.7) (i.e. if we neglect the effect of any scattering of the electrons) then we obtain the simple formula (8.6) except that the factor 2 in the denominator has been replaced by 3. If, however, the free electron approximation is not a valid one for the metal we are considering, it is possible that the term in n/v_F might be negative. This may come about as follows. We have just seen that in the free electron case n/v_F is proportional to k^2, i.e. to the area of the Fermi surface. In a more general treatment (see Ziman, 1960, p. 397) the second term of (8.8) can in fact be written as $\partial(\log_e \mathscr{A})/\partial \mathscr{E}$, where $\mathscr{A}$ is the *effective* area of the Fermi surface (i.e. the area which is *not* in contact with a zone boundary, see section 4.4). If, therefore, we have a metal in which the Fermi surface touches the zone boundary, $\mathscr{A}$ can *decrease* with $\mathscr{E}$ and hence the second term in (8.8) would be negative. This might overshadow the effect of the first term thereby producing a positive thermopower.

† This expression is not valid in the region where small-angle scattering by phonons is the dominant resistive mechanism (see section 5.5).

Experimental results

8.5. *The noble metals*

Above about 40° K a positive thermopower is indeed observed for copper (Fig. 8.3 (a)) where we have very good evidence (section 4.26) that

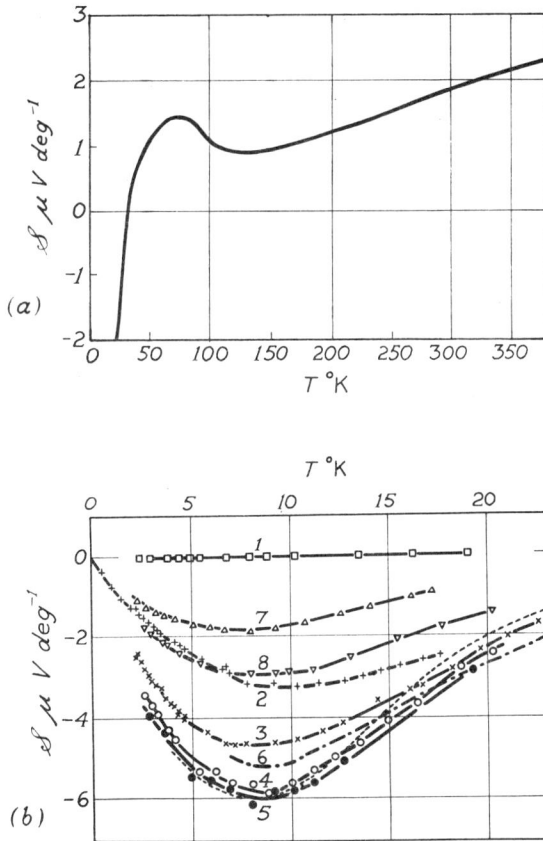

FIG. 8.3. The thermopower of high-purity copper (a) showing the positive values, linear in T at higher temperatures, which change to negative values below about 40° K; (b) low-temperature results for several specimens of high-purity copper showing the marked variations between samples from different sources (Gold, MacDonald, Pearson, and Templeton, 1960).

the Fermi surface does touch the zone boundary. It is also observed in silver and gold, where the evidence for a touch is also very strong. Unfortunately, at lower temperatures (e.g. Fig. 8.3 (b)) the thermopower of all three metals is found to be negative (MacDonald, Pearson, and Templeton, 1958; Gold, MacDonald, Pearson, and Templeton, 1960) and this would not be expected if the positive thermopower found at slightly higher temperatures was merely a Fermi surface effect. In addition to this the magnitude of the thermopower below 1° K can be quite high—for gold it is 4 μV deg^{-1} at 0·5° K. This is much more than one would expect from the simple theory which we have described in section 8.3.

8.6. *The alkali metals*

The thermopower of the alkali metals is also quite complicated. A long series of experiments has been made by MacDonald, Pearson, and

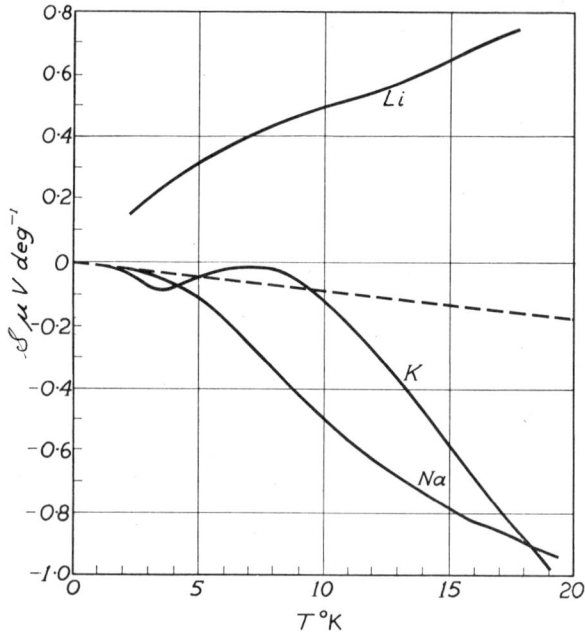

FIG. 8.4. The thermopower of lithium, sodium, and potassium. The dashed line is that of (8.7) for a free electron metal (Macdonald, Pearson, and Templeton, 1958a).

Templeton (1958a, 1960), in which measurements have been taken down to about 0·05° K. Some results above 2° K are shown in Fig. 8.4. Both sodium and potassium have negative thermopowers, but whilst that for sodium is approximately linear with T there are substantial differences between specimens having, nominally, the same purity. For potassium the thermopower passes through a local minimum and maximum. For lithium the thermopower is linear with T but is uncompromisingly positive, whereas rubidium and caesium whilst both positive (Rb does go negative at the lowest temperatures) both have very pronounced maxima in their curves.

There have not been any recent experiments on the thermopower of other pure metals. Borelius, Keesom, Johansson, and Linde measured the thermo-electric power of Fe and Pd to 16° K (1930) and Pb, Sn, and Pt down to liquid-helium temperature (1932)—all against a 'silver-normal' alloy (Ag+0·37 at per cent Au). All these metals exhibited maxima in their thermopower similar to that shown in Fig. 8.3 (a).

8.7. Phonon drag

This complicated behaviour, which seems to be observed in nearly all the metals which have been investigated, cannot be explained away merely in terms of the caprices of the hills and dales of the Fermi surface. These would not, by themselves, produce maxima and minima; nor do they explain why the magnitude of the thermopower is often so very much larger than the theory we have described would lead us to suppose. It has become customary to explain these large values of the thermo-power, particularly in the alkali metals, as an effect due to 'phonon drag'. The basis of this mechanism, which was first suggested by Gurevich (1945, 1946), can be understood when one considers a system where both an electrical potential gradient and a thermal gradient are present. We have always assumed implicitly that under such conditions the phonon distribution was in equilibrium when any interaction with the electron system was being considered. This will not be the case, however, if the relaxation time (or the mean free path) of the phonons, for interaction with other phonons, impurities, or boundaries is much longer than the relaxation time for phonon-electron interaction. For then, if we have an electric current flowing, momentum will be transferred from the electrons to certain phonons and this momentum will not be able to be dissipated into the whole phonon system before further electron inter-action occurs to increase the momentum still more. Hence these phonons will not be able to come into thermal equilibrium with the rest of the system. We have already noted in our discussion on phonon heat conduction how phonon-phonon and phonon-impurity interactions become very small as the temperature is reduced; this was shown by the very rapid rise in phonon conductivity at low temperatures (section 3.6). The phonons to which momentum has been transferred will there-fore be swept along by the electron flow and the temperature gradient will be changed. If, conversely, and as usually happens in most experi-mental determinations, the temperature gradient is kept fixed, some momentum will be transferred from the phonons to the electrons with a consequent change in the current flow. This phenomenon is usually

referred to as phonon drag. A proper analysis of the problem (e.g.
Ziman, 1960, p. 407) shows that whilst the electrical (but not the thermal)
conductivity can be slightly affected by these considerations, the
influence of the phonon drag adds an extra term to the thermopower
(in order to provide an e.m.f. to counteract the extra current flow)
which can be quite considerable. We give here a simple argument,
first presented by MacDonald (1954), which shows the order of magnitude of the phonon-drag effect.

Let us consider a metal of unit cross-sectional area along which there

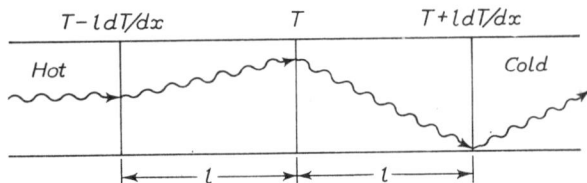

FIG. 8.5. Phonon drag. Phonons enter the section with an energy corresponding to
$T-l\,dT/dx$ and· leave it with an energy corresponding to $T+l\,dT/dx$. The energy
absorbed in the section is transferred to the electrons. This gives rise to the phonon
drag thermopower.

is a temperature gradient, dT/dx, and in which the mean free path of
the phonons is l. There will be a flow of energy down the temperature
gradient and we study a small section of the metal of thickness $2l$; this
can be considered to have a mean temperature, T. If the phonons make
a collision just before entering the section from the high-temperature
side they will have an energy density, U, appropriate to a temperature
of $T-l\,dT/dx$ (Fig. 8.5) (dT/dx is negative, so this temperature is higher
than T). The phonons which leave the section do so with an energy
corresponding to a temperature of $T+l\,dT/dx$. The energy absorbed
in the section will therefore be

$$2l\{U(T-l\,dT/dx)-U(T+l\,dT/dx)\}, \tag{8.9}$$

which to a first approximation is equal to $-(2l)^2\,(dT/dx)(dU/dT)$. But
dU/dT is, of course, the specific heat of the lattice per unit volume, c_v,
and so the energy absorbed per unit volume is

$$-2lc_v\,dT/dx. \tag{8.10}$$

This absorbed energy will give rise to a kind of radiation pressure or
momentum flow in the x direction which under normal conditions would
be transmitted to the other phonons. If, however, the phonon-electron
interactions are more important than the phonon-phonon interactions,

then a large proportion of this pressure will be transferred to the electrons in the form of extra momentum. Let us assume for simplicity that it is all transferred to the electrons. If no current is to flow, an opposing field E must be set up such that the force on the $2ln$ electrons is equal and opposite to that from the phonons, i.e.

$$2lEen = 2lc_v \, dT/dx. \tag{8.11}$$

Since the thermo-electric power is equal to $E \, dx/dT$, that due to the phonon drag will be

$$\mathscr{S}_p = \frac{c_v}{en}. \tag{8.12}$$

At high temperatures c_v is equal to $3nk$ (if we assume that there is 1 atom per electron) and so we might have a value for $\mathscr{S}_p$ which is of the order of

$$k/e = -86 \, \mu\mathrm{V} \, \mathrm{deg}^{-1}. \tag{8.13}$$

This is one or two orders of magnitude greater than that obtained from (8.6). In fact whereas the ordinary thermopower is of the order of c_e/ne, that due to the phonon drag is c_v/ne. Since at low temperatures the phonon-electron interaction tends to become more important than the phonon-phonon interaction we should still expect to observe a phonon-drag contribution, although since c_e is proportional to T (1.20) whereas c_v at low temperatures is proportional to T^3 (1.13), the phonon drag thermopower should become insignificant at very low temperatures. It will be noted that if n is small (as in semiconductors) then (8.12) can become very much larger than the value for metals. We shall discuss this in section 8.12.

It is obvious that a detailed calculation of the phonon drag involves a full consideration of all the phonon scattering mechanisms. In particular we need to know the electron-phonon interaction. This we have touched upon in section 4.20 where we noted that, in addition to the normal electron-phonon scattering, we could also have umklapp scattering. The probability of this second type of interaction was very closely bound up with the shape of the Fermi surface and, in particular, with the proximity of the Fermi surface to the zone boundary. These complications raise their ugly heads in the present problem. Ziman (1959) and Bailyn (1958) have shown that for umklapp scattering the phonon drag thermopower is positive, whereas for normal processes it is negative. The difference is essentially due to the fact that the phonon emitted after a u-process is in approximately the opposite direction to one emitted after an n-process. Since the u-processes become frozen out at very low temperatures, one can very well obtain the bumps in the curves which are observed in potassium, rubidium, and caesium. At temperatures higher than the maximum in the curves, umklapp scattering dominates the scene whereas at lower temperatures the phonon-drag effect is dictated by normal processes. Ziman (1959) has shown that this theory, combined with the assumption of various amounts of distortion of the Fermi surface, can yield behaviour which is in reasonable agreement with the experimental curves. It seems very likely that the linear positive thermopower of lithium is due to the fact that in this metal the Fermi surface does touch the zone boundary. The behaviour for sodium agrees very well with the phonon drag to be expected for a metal with a spherical Fermi surface. So far there has been no quantitative explanation of the behaviour of the noble metals using these ideas and it has been suggested (Gold, MacDonald, Pearson, and Templeton, 1960) that the change in sign at low temperatures and the large values of $\mathscr{S}$ which are observed might be due to small amounts of iron impurity rather than to the phonon drag mechanism.

8.8. Effect of impurities and defects on the thermopower

We have already seen in Chapter 4 that the addition of quite small amounts of impurity can cause profound changes in the electrical resistivity. Their effect on the thermopower is in general even more pronounced, since, as we have just described, $\mathscr{S}$ is exceedingly sensitive to distortions of the Fermi surface and these, of course, can be caused by impurities. None of the impurity effects can as yet be explained unambiguously. Alloys of lithium with up to 1 per cent magnesium show a marked reduction in the value of the thermopower (Fig. 8.6 (a)) and rubidium is also very sensitive to impurity content (MacDonald, Pearson, and Templeton, 1960). These last authors have shown, however, that in the case of sodium, the low-temperature (31° K) martensitic effect which causes a partial transformation from the body centred cubic to the hexagonal structure, does not appear to affect the thermopower, nor does there appear to be any size effect analogous to that observable in the case of the electrical resistivity (section 4.23).

A systematic study of the effect of impurities on the thermopower of potassium has been made by Guénault and MacDonald (1961)—an example from their work for specimens containing rubidium is shown in Fig. 8.6 (b). They have been able to account for the progressive change in the thermopower from the purer to the more impure specimens as being due to the quenching of the phonon drag effect by impurity scattering.

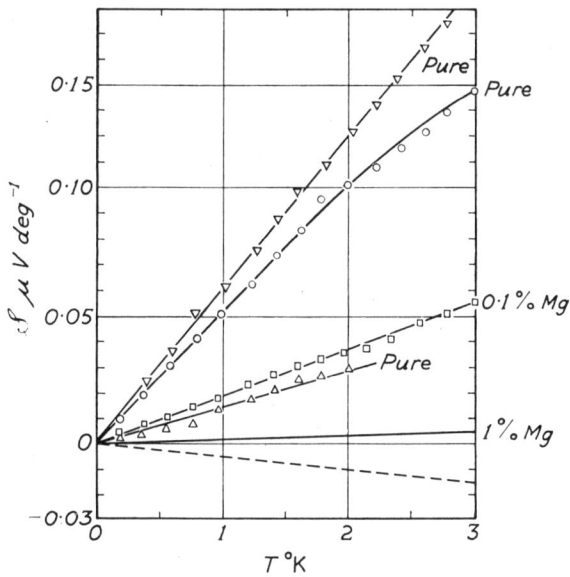

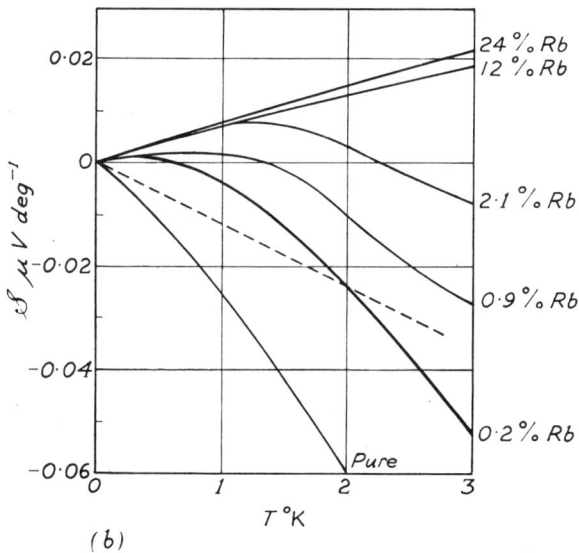

FIG. 8.6. The effect of impurity on the thermopower of (a) lithium (MacDonald, Pearson, and Templeton, 1960); (b) potassium (Guénault and MacDonald, 1961). The dashed line in each diagram is that of (8.7) for a free-electron metal.

For those alloys of copper in which there is a minimum in the electrical resistivity (section 4.15) very peculiar effects may be observed in the thermopower. As Fig. 8.7 shows, the addition of tin can make the normally positive thermopower of copper become negative and it can

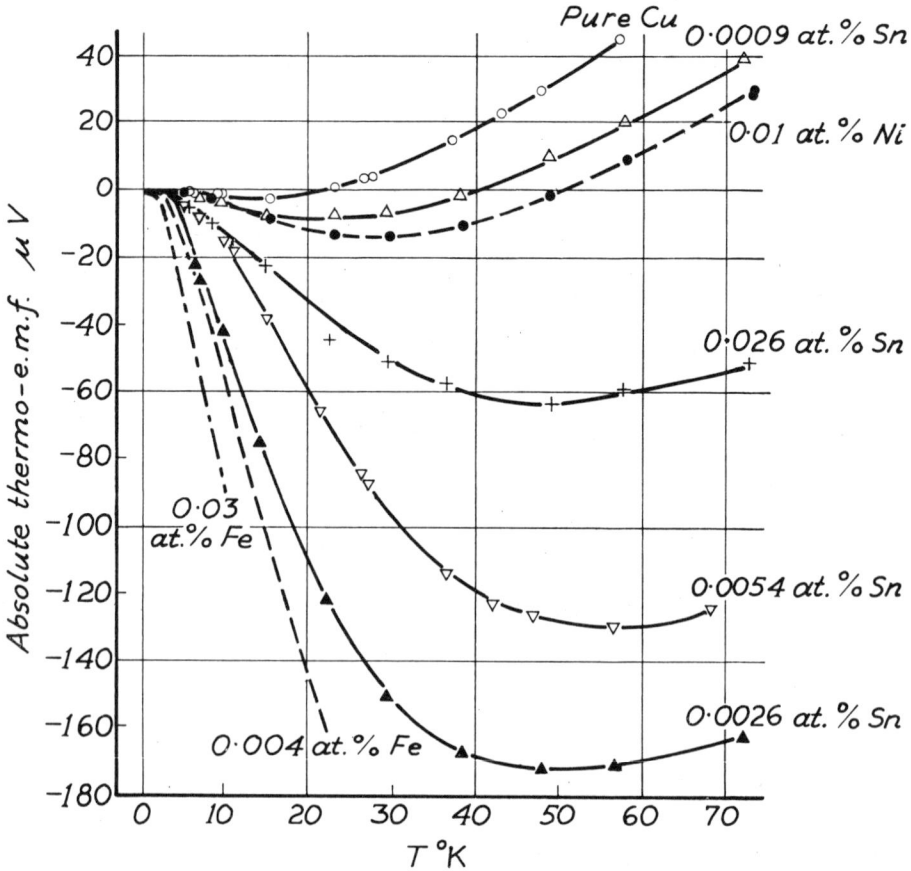

FIG. 8.7. The absolute thermoelectric force of copper and alloys showing the anomalies which occur with alloys which also exhibit an electrical resistance minimum. Note that this diagram does not plot $\mathscr{S}$, the thermopower; this is given by the slopes of the curves (MacDonald and Pearson, 1953).

also cause minima to appear in the curves. Note that this figure shows the absolute thermo-e.m.f. $\mathscr{S}$ is the slope of these curves. A large amount of equally complicated data on alloys has been obtained in the early work of Borelius, Keesom, Johansson, and Linde (1930a, 1932, 1932a).

The effect of cold work on the thermopower of pure silver has been determined by Pullan (1953). He found that whilst both the annealed and cold-worked samples had a positive thermopower, the temperature dependence of the annealed specimen was $T^{1\cdot79}$ whereas that for the cold-worked sample was directly proportional to T. The reproducibility of copper-constantan thermocouples can also be affected by cold working the constantan (MacDonald and Pearson, 1953).

We give no explanation of any of these effects. It is obvious that with a quantity such as the thermopower, which is so dependent on the shape and bounds of the Fermi surface, and the various scattering mechanisms of the electrons and the phonons both with one another and with other scattering centres in the crystal, that even though a qualitative description of the observed results might be possible, at the present time a quantitative theory would be very difficult and exceedingly speculative.

8.9. *Low-temperature thermocouples*

The thermocouple is a very convenient method of measuring both temperatures and temperature differences and it is therefore important that there should be some which can operate at low temperatures. Unfortunately, as we have seen in preceding sections, the thermopower tends to zero as the temperature is reduced and this makes it difficult to find a thermocouple which is very sensitive. The formula for the thermoelectric power (8.3) shows that this has its largest value when the thermopowers of the two metals comprising the junction are as different as possible from one another, particularly if they can have opposite signs. For this reason sensitive thermocouples are usually made of pairs of metals or alloys which are quite dissimilar. The usual combination is a monovalent with either a transition metal or an alloy containing a transition metal. A large number of experiments were made by Borelius, Keesom, Johansson, and Linde (1932) to investigate suitable junctions. At low temperatures, two kinds of thermocouples have been commonly used. Copper: constantan and gold+2·1 atomic per cent cobalt: silver+0·37 atomic per cent gold (Keesom and Matthijs, 1935). Typical curves of the thermoelectric force of these couples when one junction is at 4·2° K are shown in Fig. 8.8. Detailed tabulations are given by Powell, Bunch, and Corruccini (1961). Whilst the thermoelectric power of the copper-constantan couple is only about half that of the gold-cobalt: silver-gold couple at the lowest temperatures (about 3 μV per degree compared with about 5 μV per degree at 5° K) the ready availability of the copper-constantan makes it very convenient unless high sensitivity is required. The gold-cobalt:silver-gold couple does have the advantage that since both components of the junctions are alloys, their heat conductivity is low (section 5.11) and there is therefore very little heat flow down the connecting leads. This is one reason why the silver is alloyed with a little gold. Otherwise the silver-gold wire can be replaced by copper. It is also possible that some of the

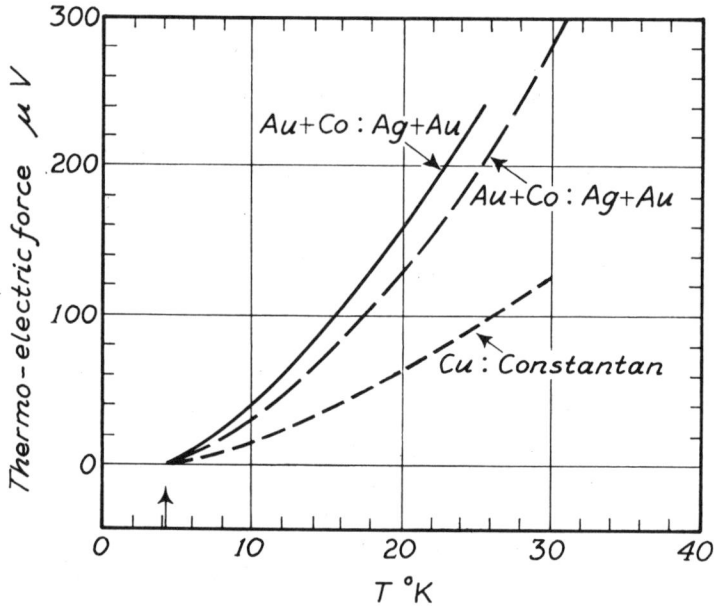

FIG. 8.8. The thermoelectric force for couples of Au+Co:Ag+Au and Cu:constantan. The reference junction is at 4·2° K (after White, 1959).

dilute alloys which have such anomalous thermopowers might be useful as thermocouples; e.g. Pearson and Templeton (1961) quote a value of 11 μV deg^{-1} for the thermopower of Au+0·02 per cent Fe in the liquid helium range. This would make it very useful as a thermometer. In order that any of these thermocouples should be reliable and reproducible, it is essential that the alloys should be homogeneous throughout their length and that there should be no mechanical strains in any of the leads.

8.10. Thermopower of superconductors

We have already mentioned (section 8.2) that the zero value of the thermopower of superconductors is utilized in making measurements of the thermopowers of other metals. The early experiments on superconductors were made by Meissner (1927), Borelius, Keesom, Johansson, and Linde (1931), and Daunt and Mendelssohn (1938, 1946). Fig. 8.9 shows the results of a particularly careful series of measurements (Pullan, 1953) on the thermopower of tin. It will be seen that at the super-conducting transition temperature there is a very rapid drop to zero.

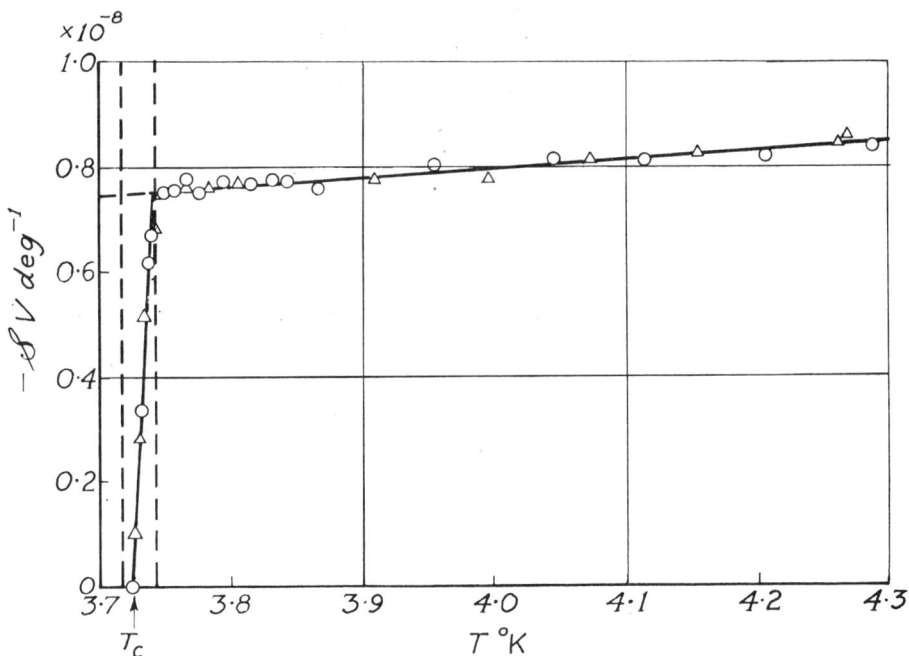

FIG. 8.9. The thermopower of tin, showing the rapid decrease to zero at the super-conducting transition temperature (Pullan, 1953).

If a magnetic field is applied which is strong enough to destroy the superconductivity, then the thermopower reverts to finite values and these fall on an extrapolation of the higher temperature thermopower curve.

This effect of zero thermopower is consistent with the concept of the zero entropy of the superconducting electrons which is assumed in the two-fluid model (section 6.7). In the terms of our simple explanation of the thermopower (section 8.3) no extra energy will need to be given to a superconducting electron when it goes from a part of the specimen at one temperature to a part at a higher temperature. Therefore we should expect the thermopower to be zero.

8.11. Thermo-electric effects in semiconductors

We have already described how the low-temperature thermopower of metals can only be accounted for if an additional term due to the 'phonon drag' effect is included (section 8.7). In semiconductors this effect is even more pronounced and it provides the major part of $\mathscr{S}$ at low temperatures.

We shall first consider, in simple terms, the modifications to our elementary derivation of $\mathscr{S}$ (section 8.3) which become necessary when we deal with semiconductors. The essential difference is that in a semiconductor the electrons are usually non-degenerate (section 7.17) and they have a Maxwellian energy distribution. The specific heat per electron is therefore $\frac{3}{2}k$. This is much greater than the specific heat of the electrons in a metal (1.22). The expression for $\mathscr{S}$ (8.6) will then read

$$\mathscr{S} = \int_0^T \frac{c_e}{neT}\,dT = \int_0^T \frac{3k}{2eT}\,dT. \tag{8.14}$$

Our simplicity now lets us down because on integrating we are left with an infinity at $T = 0°$ K, but of course this is avoided in reality because as $T \to 0$ the electron distribution becomes degenerate and so the specific heat $\to 0$ as in the case of electrons in metals. The main term for $\mathscr{S}$ will therefore be of the form

$$\mathscr{S} \approx \frac{3k}{2e} \log_e T. \tag{8.15}$$

This should be compared with the correct expression (see, for example, Wilson, 1953, p. 232)

$$\mathscr{S} = \frac{3k}{2e} \log_e T + \frac{k}{e} \big[2 - \log_e n - \log_e \{ \tfrac{1}{2} h^3 (2\pi m_e k)^{-\frac{3}{2}} \} \big]. \tag{8.16}$$

The two logarithmic terms on the extreme right-hand side cancel each other if n is about 6×10^{15} cm^{-3} and $m_e =$ the free electron mass. At room temperature $\mathscr{S}$ will then be of the order of k/e (86 μV deg^{-1}) whereas for a metal, since $kT/\mathscr{E}_0$ can be 10^{-2} at $T = 300°$ K, $\mathscr{S}$ from (8.6) will be of the order of $10^{-2}k/e$. Hence we see that for a sample with a carrier concentration which would be quite typical of that present in many semiconductors $\mathscr{S}$ will be very much greater for the semiconductor than for a metal. In practice $\mathscr{S}$ for germanium can approach 10^{-3} volt deg^{-1} at room temperature whereas for a metal it is only a few microvolts deg^{-1}.

Equation (8.16) predicts a slow decrease of $\mathscr{S}$ as the temperature is

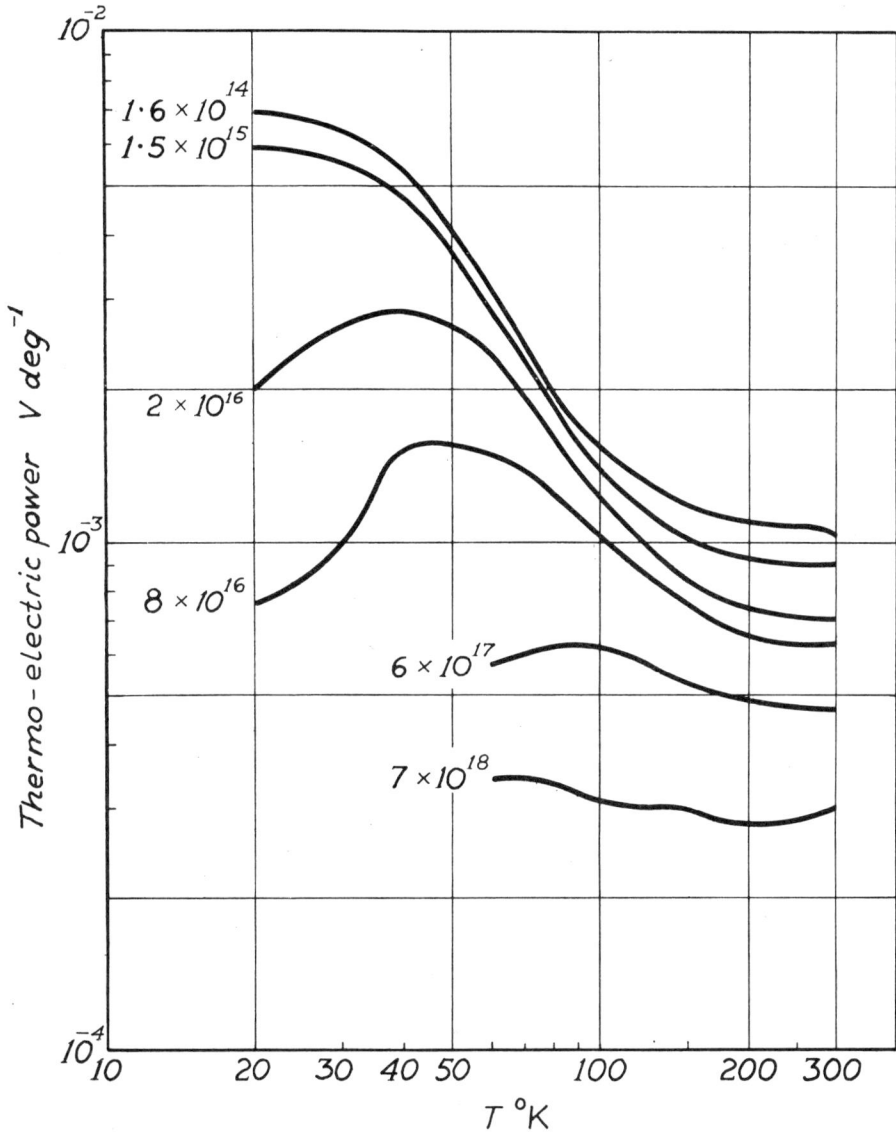

FIG. 8.10. The thermoelectric power of p-type (gallium doped) germanium, showing the anomalous rise at low temperatures due to the phonon-drag effect. The gallium concentration is marked against each curve (after Geballe and Hull, 1954).

reduced if n remains constant. For specimens with large n, $\mathscr{S}$ should be small. Experiments first made by Frederikse (1953) and by Geballe and G. W. Hull (1954) do not entirely bear out these predictions. Fig. 8.10 shows the thermoelectric power of some samples of p-type Ge.

Whilst the dependence on n is satisfactory, the magnitude and the temperature dependence are not in good agreement with the theory except around room temperature. At low temperatures, $\mathscr{S}$ rises rapidly. If the theoretical value (8.16) of $\mathscr{S}$ is subtracted from the experimental data there remains an anomalous thermopower which has a temperature

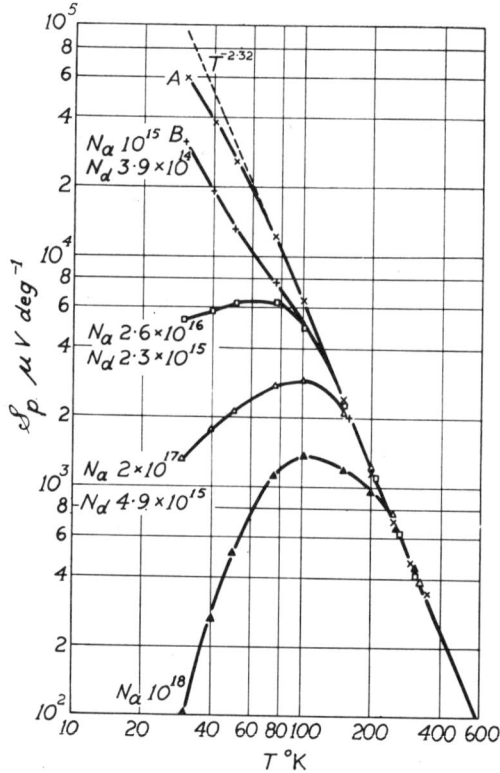

FIG. 8.11. The anomalous thermopower for various samples of p-type silicon (boron doped). N_a and N_d are the densities of acceptor and donor atoms respectively. The size effect is demonstrated by sample A which had the same impurity concentration as sample B, but whereas B had a cross-section of $2 \cdot 5$ mm $\times 1 \cdot 8$ mm, that of A was 5 mm $\times 5$ mm (Geballe and Hull, 1955).

dependence of about $T^{-3 \cdot 2}$ for the purest specimens. Similar behaviour has been observed in Si with a temperature dependence of $T^{-2 \cdot 3}$ (see Fig. 8.11) (Geballe and G. W. Hull, 1955) and in p-type InSb (Frederikse and Mielczarek, 1955).

8.12. *Phonon drag in semiconductors*

It is now generally agreed that the rapid rise in $\mathscr{S}$ to a value at the maximum which is about twenty times that which is predicted by

(8.16) is due to a phonon-drag effect. As in the case of the ordinary thermoelectric power, the phonon-drag contribution is much greater for semiconductors than it is for metals. The reason for this is fairly easy to understand. As we have already pointed out in section 8.7, phonon drag becomes important when the phonon relaxation time for phonon-electron interactions becomes much shorter than the relaxation time for all other phonon interactions. In semiconductors, there are few carriers and these will have low energies of the order of kT, hence their wave vectors $\mathbf{k}$ will also be small. When an electron interacts with a phonon it can only do so if both energy and wave vector are conserved, but because the energy of an electron with a given wave vector is much higher than that of a phonon with the same wave vector, the conservation relations (4.19) and (4.20) can only be satisfied if the electrons interact with phonons of very small wave vector, i.e. of long wavelengths. In order to dissipate the electron momentum these phonons must interact with other phonons. Once again, in order to obey the wave vector and energy conservation laws, these phonons can only interact with other phonons of small wave vector, q, and since the density of states is proportional to q^2 there will not be many phonons available for small q. Thus with long wavelength phonons, the phonon-phonon relaxation time is long and the conditions are favourable for a large phonon drag effect.

The expression (8.12) which we derived for the phonon drag thermopower, whilst it gives the order of magnitude to be expected is not as detailed as we might like because the phonon relaxation time τ_p for all other interactions except those with electrons, is assumed to be infinite. Since in actual fact we would expect that τ_p would be temperature dependent, it is useful to derive an expression for the phonon drag in which τ_p appears explicitly. This we do, following the arguments of Herring (1958). Let us assume that the phonons with which the electrons interact move through the crystal with a drift velocity v_d. If the energy density of these phonons is U', then the energy flux will be $v_d U'$. This flux may be equated to the heat flow through the material by using the kinetic theory expression for the thermal conductivity κ' (3.4). We then obtain

$$v_d U' = \kappa' \, dT/dx \qquad (8.17)$$

$$= \tfrac{1}{3} c' l v \, dT/dx, \qquad (8.18)$$

where c' is the specific heat per unit volume of the phonons we are considering. Since these are of long wavelength, we can assume that they behave classically and hence

$$U' = c'T. \tag{8.19}$$

Thus
$$v_d = \frac{lv}{3T}\frac{dT}{dx}. \tag{8.20}$$

If the electrons interact only with these phonons, then, in the absence of an electric field, they will eventually acquire the phonon drift velocity, v_d. In order to stop this current flow a field E must be applied so as to produce a velocity equal but opposite to v_d. From the usual definition of the mobility, μ, this field will be $E = \pm v_d/\mu$ ($+$ for electrons; $-$ for holes). The phonon-drag thermopower, $\mathscr{S}_p$, is then equal to $E/(dT/dx)$ and from (8.20) will be

$$\mathscr{S}_p = \frac{lv}{3\mu T}, \tag{8.21}$$

or, since $l = \tau_p v$,
$$\mathscr{S}_p = \frac{\tau_p}{3\mu}\frac{v^2}{T}. \tag{8.22}$$

We should note that $\mu = e\tau_e/m_e^*$ (7.30) and hence $\mathscr{S}_p$ is a function of τ_p/τ_e. This emphasizes the fact that it is only when τ_p is large compared with τ_e, that $\mathscr{S}_p$ becomes important.

We are now in a position to determine the temperature dependence of $\mathscr{S}_p$. In an anisotropic crystal, such as Ge, for longitudinal phonons of small wave vector, q (these are the phonons which have the longest τ_p and hence will be the most effective for producing drag), Herring (1958) shows that τ_p is proportional to $1/(q^2T^3)$.

We then assume that the phonons which can interact with the electrons will have values of q which vary with T in a manner similar to that in which the wave vector $\mathbf{k}$ of the electrons depends on T. Since $\mathbf{k}$ is proportional to the square root of the electron energy (i.e. $(kT)^{\frac{1}{2}}$) then q will also have this dependence. Thus q^2 will be proportional to T and hence τ_p is proportional to T^{-4}. If μ is determined by phonon scattering, then it is proportional to $T^{-\frac{3}{2}}$ (7.35) and thus, with the T^{-4} dependence of τ_p, we find using (8.22) that $\mathscr{S}_p$ is proportional to $T^{-3\cdot5}$. It will be recalled that the experiments of Geballe and G. W. Hull (1955) gave a dependence of $T^{-3\cdot2}$.

This very simple treatment neglects many factors which must be considered in a proper explanation of the experiments. We should, of course, consider the effect of all phonon modes in the determination of τ_p, the other electron-scattering processes which we dealt with in the sections on mobility (section 7.8 ff.) and we should also take account of the complicated band structure of semiconductors and of the effect of varying the density of carriers. A very extensive treatment has been given by Herring (1954) although his 1958 paper is recommended as an introduction to the subject.

8.13. *Boundary scattering*

One particular aspect, however, should still be discussed. As we have noted, the important phonons for producing drag are those of long wavelength, and in the previous paragraph we showed that τ_p for these phonons had a very rapid temperature dependence, being proportional to T^{-4}. This τ_p is the relaxation time for all interaction processes except those with electrons. As the temperature is reduced τ_p will eventually become constant because it will be determined by the scattering of the phonons by the specimen boundaries (section 3.7). This will become the most effective mechanism for dissipating momentum and the relaxation time will be proportional to the specimen size. Thus a larger value of $\mathscr{S}_p$ should be observed for large specimens than for small ones, and this has been verified experimentally by Geballe and G. W. Hull (1954) (see also curves A and B of Fig. 8.11). It should be noted that the size effect for phonon drag becomes evident at a much higher temperature than it does for thermal conductivity. This is because for phonon drag the most important phonons are those of long wavelength. These will be scattered by the specimen boundaries at higher temperatures than are the phonons of shorter wavelength, which since they carry more energy, are the ones which are most important for heat transport. The shorter wavelength phonons are still being scattered by u-processes or impurities at these temperatures.

A constant value of τ_p in (8.22) would lead to the prediction that $\mathscr{S}_p$ should be proportional to $T^{\frac{1}{2}}$ at very low temperatures. This has not yet been shown with any great certainty in experiments. To some extent this is because when one enters the impurity conduction range (section 7.20) the thermopower mechanism changes and $\mathscr{S}$ decreases very considerably, thereby masking the phonon drag effect. In addition to this, the considerations of the next section very often come into play.

8.14. *Effect of the carrier concentration*

The theory which we have outlined so far is only applicable to specimens for which the carrier concentration, n, is so small that it has no effect on the mean free path, l, of the phonons in (8.21). If, however, n is so large that phonon-electron scattering materially decreases the thermal conduction of the low-frequency phonon modes which are responsible for producing the drag, then l in (8.20) will be reduced and hence, for a given temperature gradient, v_d in the same equation will also be less. An inspection of (8.20) and (8.21) shows that if $\mathscr{S}_{p0}$ is the phonon drag thermopower for a very low carrier concentration, then $\mathscr{S}_p/\mathscr{S}_{p0}$ will vary in the same way as does the ratio of the drift velocities v_d in the two cases. A more complete treatment by Herring (1954, 1958), which, however, does not include the effects of degeneracy or impurity conduction, results in the expression

$$\frac{\mathscr{S}_p}{\mathscr{S}_{p0}} \approx \left\{ 1 + \frac{3\mathscr{S}_{p0}\,ne}{k}\left(\frac{h^2}{6\pi m_e^*\,kT}\right)^{\frac{3}{2}} \right\}^{-1}. \tag{8.23}$$

Thus $\mathscr{S}_p$ will be reduced for high values of n and the effect will become more pronounced as the temperature is reduced. The initial rapid increase in $\mathscr{S}_p$ as the temperature is lowered, is halted and the thermopower passes through a maximum and then decreases (Fig. 8.11). In non-degenerate specimens, however, complications can arise because as the temperature is reduced, the number of carriers which are thermally excited becomes less, and so n in (8.23) is no longer a constant. This decrease in the carrier concentration will tend to increase $\mathscr{S}_p$ and might overcome the effect of the decrease in T. This type of behaviour is shown in the two upper curves of Fig. 8.11. The very rapid decrease in $\mathscr{S}_p$ for the lowest curve may be due in some part to impurity band conduction (section 7.20).

9

MAGNETIC PROPERTIES

9.1. Introduction

ALL substances have an effective magnetic dipole moment m induced in them when they are placed in a magnetic field H. In general there are three main types of behaviour which are observed.

(a) Diamagnetism, when m is in a direction opposite to H and is proportional to the magnitude of H.

(b) Paramagnetism, when m is in the same direction as H and is proportional to it.

(c) Ferromagnetism, when m is parallel but not proportional to H; m increases very rapidly as H is increased and it then saturates at a value which is many orders of magnitude higher than any which is found in a paramagnetic substance. When the field is reduced to zero some of the magnetic moment remains, and if this is large enough, the material is said to be a permanent magnet.

Of these three effects, paramagnetism is the one which has the largest low-temperature interest because, as we shall see, in a paramagnetic, m increases very much as the temperature is reduced. The magnetic behaviour of diamagnetic and ferromagnetic materials does not change at low temperatures.

It would be out of place in this book to give a full explanation of the various types of magnetic behaviour, particularly for dia- and ferromagnetism, and we shall content ourselves here with giving only an outline of their theoretical explanation. A good elementary introduction to magnetic properties is given by Bleaney and Bleaney (1957), chapters 8, 20, and 21.

All the magnetic effects we shall consider are due to the interaction of a magnetic field with the electrons of the atoms of the substance which is being investigated. This interaction occurs because the electron, which is both spinning on its own axis and also circulating around the nucleus, can be considered as a circulating current. Since such a current always has a magnetic moment associated with it, it follows that the electron also possesses a magnetic moment which will tend to orient itself in a magnetic field.

The total magnetic moment of an electron can initially be discussed in terms of two components, m_l, due to its orbital motion and m_s which is associated with its spin. When a magnetic field is applied to an atom, then for each electron m_l and m_s will tend to align themselves with respect to H along one of the special directions which are permitted by quantum mechanics.† The spins are only allowed to have two such directions and these are such that the projections of m_s along H are equal but opposite. It is clear that if we just consider the spin moment then an atom with an odd number of electrons will always have a non-zero value of m_s when a field is applied. If the atom has an even number of electrons, however, there might be equal numbers of them with their m_s in opposite directions so that the resultant spin moment is zero.

9.2. *Diamagnetism*

Diamagnetism occurs when, as described in the previous paragraph, an atom or ion has no resultant spin dipole moment. The phenomenon is produced by the effect of an applied magnetic field on the *orbital* motion of the electrons. Using classical theory and recalling that the magnetic moment = current × area enclosed by circuit, the orbital magnetic moment is given by

$$m_l = -eG_l/(2m_e), \qquad (9.1)$$

where G_l is the orbital angular momentum, $-e$ is the charge, and m_e is the mass of the electron. At first sight one might think that the application of a magnetic field would tend to align all the orbits, which would yield a paramagnetic effect. This does not occur, however, because the interaction between the orbits within an atom is so very much stronger than the influence of any available magnetic field (10^4 to 10^5 oersted) that, to a first approximation, the general orientation remains unchanged.

The diamagnetic effect can be easily deduced from Lenz's law of electromagnetic induction. The change of flux through an orbit will induce an e.m.f. which will produce a current (i.e. it will modify the motion of the electrons) in such a direction that the field associated with it opposes the applied field. Since an electron in orbital motion about the nucleus does not experience any electrical resistance, the motion induced by the application of the field will continue until the field is

† In this chapter we discuss the magnetic properties in terms of the *alignment* of the orbit or the spin with respect to the magnetic field H. This gives an oversimplified picture of the process. In actual fact when H is applied, the orbital and spin magnetic moments *precess* with H as the axis of precession. Thus the components of the orbital and spin moments in the direction of H remain constant during the precession and it is these which we call m_l and m_s.

removed. Thus the effect of applying H is to induce a dipole moment in a direction opposite to that of the field. The extra angular momentum ΔG_l given to an electron when the flux due to a field H threads the orbit, whose projection perpendicular to H has a radius r, is

$$\Delta G_l = Her^2/2. \tag{9.2}$$

This can be shown very simply either from the Larmor precession theorem (Bleaney and Bleaney, 1957, p. 184) or directly from a calculation of the force on the electron due to a change of flux through the orbit. The contributions due to ΔG_l from each electron can be summed because they are all in the same direction. Thus from (9.1) and (9.2), the magnetic susceptibility per unit volume, χ_v, defined by

$$\chi_v = ZNm/H, \tag{9.3}$$

where Z is the number of electrons per atom and N is the number of atoms per unit volume, is given by

$$\chi_v = -NZe^2\langle r\rangle^2/(4m_e), \tag{9.4}$$

where $\langle r\rangle^2$ now relates to an average of all the electron orbits. Assuming spherical symmetry, $\langle r\rangle^2 = \frac{2}{3}\langle R\rangle^2$ where $\langle R\rangle^2$ is the mean square distance of the electrons from the nucleus. Thus

$$\chi_v = -NZe^2\langle R\rangle^2/(6m_e). \tag{9.5}$$

It will be noted that χ_v is essentially negative and is dependent on the number of electrons in the atom. Since $\langle R\rangle^2$ should alter only very slightly with temperature, χ_v will be practically independent of temperature. For solids and liquids χ_v is of the order of 10^{-6} e.m.u. cm^{-3}. The diamagnetism of the conduction electrons in a metal is dealt with in Chapter 10.

9.3. Paramagnetism

If each atom in an assembly does have a permanent spin dipole moment, then in a field H these dipoles will tend to align themselves parallel to H and hence paramagnetism can occur. It is not possible, however, to describe the phenomenon purely in terms of the spin moment, because the separation of the total moment into m_l and m_s is really an oversimplification. To discuss paramagnetic behaviour even qualitatively requires some familiarity with the manner in which the orbital and spin moments of the individual electrons in an atom interact and combine. We shall summarize the points which are important for our purpose with no attempt at justification (a full description is given by Herzberg, 1944). The motion of the electrons about a nucleus can be characterized by four quantum numbers. The energy is determined largely by the main quantum number, n; this number is integral and $n = 1$ relates to the smallest value of the energy. The orbital angular momentum of an electron is characterized by l and this can have integral values from 0 to $(n-1)$. Associated with the angular momentum is an orbital magnetic dipole moment, as in (9.1), and in a magnetic field H the dipole can orient itself in a limited number of positions relative to H. The energy of the dipole will depend on its orientation and it is determined by the values of a magnetic quantum number, m; this is also integral and can have values $-l,...,0,...,+l$. In addition to n, l, and m, the electron has a spin quantum number, s, for which there are only two possible values, $\pm\frac{1}{2}$. The electrons will tend to exist in states of the lowest energy compatible with the Pauli exclusion principle, i.e. that only one electron can be present in a state characterized by a given n, l, m, and s. Since s can only have the two values $\pm\frac{1}{2}$, we can also say that not more than two electrons can have the same values of n, l, and m.

If all the states for a given n, l are occupied, their resultant spin dipole moment will be zero because there will be as many electrons with $s = +\frac{1}{2}$ as with $s = -\frac{1}{2}$. They will also have zero orbital dipole moment in the presence of a field because for an electron in a state $+m$ there will be another in a state $-m$. Such an assembly of electrons form what is called a closed shell.

The only electrons which will contribute to the dipole moment will therefore be those which are in incomplete shells. In general these will be the so-called valence electrons. Now the alignment of electron spins within an atom and between neighbouring atoms is governed by a special type of interaction, known as the exchange interaction, whose nature is briefly discussed in section 9.16. The exchange interaction between the electrons in an incomplete shell is such that the energy is a minimum when the electrons line up with their spins parallel (provided that the Pauli principle is not violated) and thus most atoms will have a dipole moment and will be paramagnetic. It should, however, be emphasized that this is only true for *free* atoms and not for molecules. The reason for this is that the exchange energy of electrons of *neighbouring* atoms is usually a minimum when their spins are anti-parallel and hence the total dipole moment of the molecule will be zero. In ionic crystals the outer electrons of one atom are transferred to complete the shell of its neighbour so that both ions have closed shells of electrons around them and these also have no dipole moment. Thus, in general, only diamagnetism is observed.

Hence paramagnetism will usually only occur when the atom has an incomplete shell in addition to that occupied by the valence electrons. An inspection of a table of the electronic configuration of the elements will show that there are five groups in which this occurs. These are the iron group, with an incomplete $3d$ shell, the palladium group (incomplete $4d$), the lanthanides or rare earths (incomplete $4f$), the platinum group (incomplete $5d$), and the actinides (incomplete $5f$). Almost all the experimental work has been concerned with salts of the iron or of the rare earth groups and it is on these that we shall concentrate.

9.4. *Russell–Saunders coupling; Hund's rules*

We must now describe how the angular momenta and the magnetic moments of the individual electrons in an incomplete shell are combined. In general the orbital and the spin moments cannot be treated separately because the field due to one will interact with that due to the other. The ground state of the atom or ion is determined by combining the quantum numbers of the individual electrons in accordance with a set of instructions known as Hund's rules. These yield a composite quantum number J. The rules are as follows.

(*a*) The spins of the electrons are arranged so that as many of them as possible are parallel to one another without violating the Pauli principle (i.e. only two electrons to each value of m). Taking $s = +\frac{1}{2}$ or $-\frac{1}{2}$ for each electron, depending on the direction of spin, $\sum s$ is calculated. This sum is designated by S, the combined spin momentum.

(*b*) The electrons with the spins assigned as in (*a*) are divided amongst the possible values of m so that $\sum m$ is a maximum. This sum is denoted by L, the combined orbital momentum.

(*c*) The states of the atom or ion are characterized by a quantum number J which runs in integral steps from $L-S$ to $L+S$ (Fig. 9.1 (*a*)). The ground state is given by $J = L-S$ for a shell which is less than half full and by $J = L+S$ for a shell which is more than half full. This method of combination is called Russell-Saunders coupling. The state is often designated by using spectroscopic notation in which a capital letter signifies the value of L ($L = 0, 1, 2, 3, 4, 5,...$ being represented by $S, P, D, F, G, H,...$ respectively). A superscript preceding the letter gives the value of $(2S+1)$ and a subscript following the letter gives the value of J. Thus Sm^{3+} which has 5 electrons in the $4f$ shell ($l = 3$) will have them arranged with all spins parallel with the following m values:

m	3	2	1	0	-1	-2	-3
s	↑	↑	↑	↑	↑		

S will be $\sum s = \frac{5}{2}$ and L will be $\sum m = 5$. $J = L - S = \frac{5}{2}$. The ground state will therefore be designated by $^6H_{5/2}$.

The dipole moment m_J associated with J is given by

$$m_J = -g\beta\{J(J+1)\}^{\frac{1}{2}},\tag{9.6}$$

where β is the Bohr magneton ($\beta = 9\cdot27 \times 10^{-21}$ erg gauss^{-1}) and g is the Landé splitting factor, given by

$$g = \frac{3J(J+1)+S(S+1)-L(L+1)}{2J(J+1)}.\tag{9.7}$$

For a system which only has spin (i.e. $L = 0$), $g = 2$. When a magnetic field is applied there will be a tendency for the composite dipole (9.6) to rotate until it is parallel to the field because by so doing its energy will be reduced. However, in a similar manner to that by which a single electron can only orient itself at certain positions relative to H, depending on the value of m, so the whole electron assembly can only have special orientations. These are characterized by a magnetic quantum number M that runs from $+J$, $(J-1)$,... to $-J$ in integral steps. Thus there are $2J+1$ possible orientations. For a certain orientation the energy of the dipole in the field is given by an expression whose terms are, to second order in H,

$$U_M = Mg\beta H - \tfrac{1}{2}\alpha H^2...,\tag{9.8}$$

and since the projection of the dipole moment in the direction of H is given by $(-\partial U_M/\partial H)$ it will have the values

$$m_H = -Mg\beta + \alpha H... .\tag{9.9}$$

The coefficient α in the correction term depends on the spacing of the state we are considering from neighbouring states. The closer the states are together, the larger is α. Thus if we apply a magnetic field to a system, the atoms will orient themselves in the various positions relative to H which are determined by all the possible values of M. It will only be at higher temperatures, however, that there will be many atoms in an orientation associated with the larger values of M, because from (9.8) these will have a higher energy. Thus at high temperatures the dipole moments of the atoms will be spread over all possible orientations and so the total effective moment will be small. At low temperatures, all the atoms will tend to be in the lowest energy state with the same value of M, their moments will be parallel, and so the total moment will be much larger. A more detailed theory of this process is given in the next section.

9.5. *Paramagnetic susceptibility; Curie's law*

If the magnetic system is in thermal equilibrium with the crystal lattice, the distribution of the atoms amongst the various values of M will be governed by the Maxwell–Boltzmann statistics. The probability that an atom or ion will be in a state M will be proportional to

$$\exp(-U_M/kT)$$

and hence the dipole moment per unit volume will be

$$m_v = \frac{N \sum\limits_M m_H \exp(-U_M/kT)}{\sum\limits_M \exp(-U_M/kT)}. \tag{9.10}$$

Substituting for m_H, U_M from (9.8) and (9.9) and neglecting the terms in α,

$$m_v = \frac{N \sum\limits_M -Mg\beta \exp(-Mg\beta H/kT)}{\sum\limits_M \exp(-Mg\beta H/kT)}. \tag{9.11}$$

Putting $x = g\beta H/kT$ and expanding the exponential we obtain for $x \ll 1$ (i.e. except for high fields at low temperatures)

$$m_v = \frac{Ng\beta \sum\limits_M (-M+M^2x)}{\sum\limits_M (1-Mx)}. \tag{9.12}$$

The summation over M is zero because its values run from $-J$ to $+J$, and since the sum of the squares of the first M natural numbers is $M(M+1)(2M+1)/6$, we obtain

$$m_v = \frac{Ng\beta x J(J+1)(2J+1)}{3(2J+1)} \tag{9.13}$$

and so

$$m_v = \frac{Ng^2\beta^2 HJ(J+1)}{3kT}. \tag{9.14}$$

The susceptibility per unit volume is given by m_v/H, i.e.

$$\chi_v = \frac{Ng^2\beta^2 J(J+1)}{3kT}. \tag{9.15}$$

The restriction that $x \ll 1$ is usually not a very serious one as can be judged from the fact that for $g = 2$ and $T = 1°$ K, $x = 1$ when H is 7,400 oersted. Thus (9.12) to (9.15) will be invalid at very high fields or very low temperatures. Equation (9.11) must then be solved more accurately and it can be shown (Bates, 1951, p. 43) that

$$m_v = Ng\beta J\left[\frac{2J+1}{2J}\coth\{(2J+1)x/2\} - \frac{1}{2J}\coth(x/2)\right]. \tag{9.16}$$

The expression in square brackets is called the Brillouin function and is tabulated for $J = \frac{1}{2}, \frac{3}{2}, \frac{5}{2}$ by J. R. Hull and R. A. Hull (1941) and for

$J = \frac{7}{2}$ by Giauque, Stout, Egan, and Clark (1941). For large values of the argument coth tends to unity, and so for high fields at low temperatures m_v saturates to a value

$$m_v = Ng\beta J. \tag{9.17}$$

Equation (9.15) shows that under normal circumstances the susceptibility should be inversely proportional to the temperature. This relationship is known as Curie's law and is often quoted in the form

$$\chi_v = C/T. \tag{9.18}$$

This is the key susceptibility equation and experimental data are always compared with a relation of this type, where C is called the Curie constant. Experiments show that whilst the general form of the susceptibility is that of (9.18), a strict proportionality to $1/T$ is not usually found. The reasons for this we shall discuss in the following paragraphs.

9.6. *The internal electric field*

In a solid the paramagnetic ions lie on special sites in the crystal lattice. Experiments are usually made on crystals which are magnetically dilute, i.e. the paramagnetic ions are sufficiently far apart from one another for the magnetic interaction between them to be so small that the orientation of one does not influence that of its neighbour, except at very low temperatures. This dilution is often achieved by using double salts such as the alums, and the double nitrates. Many of these salts have a large number of molecules of water of crystallization which increases the dilution. In fact the immediate neighbours of the positive paramagnetic ions are usually the negative oxygen ions of the water molecules. Whilst, therefore, there is very little magnetic interaction between the paramagnetic ion and its surroundings, it is still under the influence of a strong electric field due to these oxygen ions. This field will affect the charge distribution of the electrons in the incomplete shell of the paramagnetic ion. There will be a tendency for the charge cloud to avoid regions which are near to the oxygen ions, i.e. certain charge configurations, or states, will have energies which differ from those of the free ion. The energy levels are said to be split by the field (Fig. 9.1 (b)). The effect of the internal electric field (it is sometimes called the crystalline field) is different in the iron and in the rare-earth groups of elements and whilst it introduces some complexity in the results, it is the study of the effect of these energy splittings, both by the internal field and by other interactions which has led to the intense interest and unending variety of paramagnetic investigations.

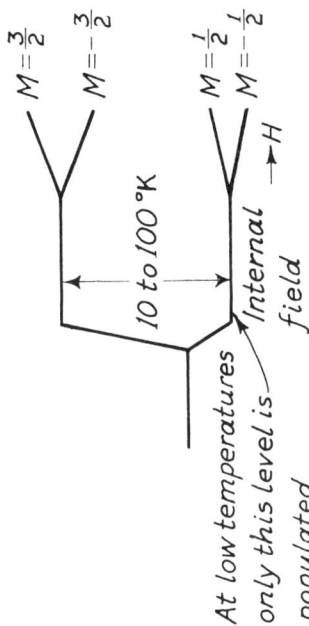

$J = \frac{9}{2}$ ——————— *10-fold degen.*

$J = \frac{7}{2}$ ——————— *8-fold degen.*

$J = \frac{5}{2}$ ——————— *6-fold degen.*

1000's °K

$J = \frac{3}{2}$

$M = \frac{3}{2}$
$M = \frac{1}{2}$
$M = -\frac{1}{2}$
$M = -\frac{3}{2}$

→ *H*

(*a*)

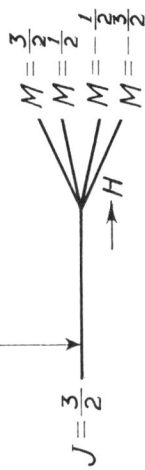

$M = \frac{3}{2}$

$M = -\frac{3}{2}$

10 to 100 °K

At low temperatures only this level is populated

Internal field

$M = \frac{1}{2}$
$M = -\frac{1}{2}$

→ *H*

(*b*)

Fig. 9.1. The energy levels of an ion with $L = 3$ and $S = \frac{3}{2}$. The different possible values of J running from $L - S$ to $L + S$ are separated by energies corresponding to several thousand degrees so that only the lowest level, $J = \frac{3}{2}$, is populated. (*a*) In the free ion the fourfold degeneracy of this level is removed by a magnetic field H. (*b*) In a salt the internal electric field will split this level into two doublets and at low temperatures only the lowest doublet will be populated. The application of a magnetic field splits the doublets as shown.

We shall delay further discussion on the nature of the energy splittings to a later section except to point out one important consequence. If the energy levels are split, then their population will change with temperature, even for $H = 0$. If this is taken into account in (9.10) then Curie's law is not obtained, except at temperatures which are sufficiently high compared with the splitting energy that equal occupation of the levels does still occur.

9.7. *Susceptibility of the iron group*

In these ions the partly filled $3d$ shell is responsible for the permanent dipole moment. This is the outermost shell of the ion and it is therefore very strongly influenced by the crystal field. The symmetry of this field tends to be imposed on the charge distribution. Now it is a general rule that the lower the symmetry of a field the more will be the tendency for states which had the same energy in the free ion, to split up and have different energies. This is readily seen, for example, if one considers the case of three similar mutually perpendicular orbits (*a*) in a cubic field, when they will all remain equivalent, and (*b*) in an axial field when at least one of them will now have a different energy from the other two.

Even in a cubic crystal there is, in fact, usually sufficient distortion from true cubic symmetry that there will be one orbital state which will have a lower energy than the others. There is a general theorem which states that if an atom or ion exists in a non-degenerate orbital state in the absence of a magnetic field, then its orbital angular momentum is zero. (A proof is given by Van Vleck, 1932, p. 273.) Thus the effect of the internal field is to make the orbital angular momentum zero and hence there is no dipole moment associated with it. The orbital moment is said to be 'quenched'. Essentially one can picture this as being due to the fact that there is an equal probability of an electron going round the atom in one direction as in the other.

There is, however, very little direct interaction between the electron *spins* and the crystal field and so when a magnetic field is applied, the spins will orient themselves in their permitted directions relative to H. Thus the spin-orbit coupling scheme, described in section 9.4, breaks down. L and S are not combined to give a J value. Instead the $2L+1$ orbital states are split by many thousands of degrees and each of them is $2S+1$ degenerate. Only the lowest orbital state (which will have zero orbital moment) will be occupied. This means that the dipole moment and the magnetic energy of the ion when in a salt are determined by the value of S and not by J. The permitted values of M in (9.11) will run therefore from $+S$, $(S-1)$, $(S-2)$,..., $-S$, rather than from $+J$ to $-J$. The expression for the susceptibility will be the same as (9.15) except that J must be replaced by S, and g will have the value 2 corresponding to a spin-only system. Table 9.1 gives a list of the elements of the iron group, and their spectroscopic ground states, together with experimental values of the Curie constant measured at room temperatures on double sulphates. These values should be compared with those calculated using (9.15) in which $g = 2$ and S is used instead of J. These are shown in the next column. In the last column values of J have been used and this

TABLE 9.1

The paramagnetic ions of the 3d group

(after Bleaney and Bleaney, 1957, p. 563)

No. of electrons in 3d shell	Ion	Ground state	S	L	J	C in units of $N\beta^2/(3k)$ (experimental)	$4S(S+1)$	$g^2J(J+1)$
1	Ti^{3+}, V^{4+}	$^2D_{3/2}$	1/2	2	3/2	2·9	3	2·4
2	V^{3+}	$^3F_2$	1	3	2	6·8	8	2·67
3	V^{2+}, Cr^{3+}	$^4F_{3/2}$	3/2	3	3/2	14·8	15	0·6
4	Cr^{2+}, Mn^{3+}	$^5D_0$	2	2	0	(23·3)	24	0
5	Mn^{2+}, Fe^{3+}	$^6S_{5/2}$	5/2	0	5/2	34·0	35	35
6	Fe^{2+}	$^5D_4$	2	2	4	28·7	24	45
7	Co^{2+}	$^4F_{9/2}$	3/2	3	9/2	24·0	15	44
8	Ni^{2+}	$^3F_4$	1	3	4	9·7	8	31·3
9	Cu^{2+}	$^2D_{5/2}$	1/2	2	5/2	3·35	3	12·6

All values of C are for double sulphates except that for Cr^{2+} which is for $CrSO_4 . 6H_2O$.

shows poor agreement with the experimental results. For most of the iron group, Curie's law is obeyed very well until temperatures and fields are reached at which (9.16) has to be used. This can be seen in Fig. 9.2 which shows a plot of $1/\chi$ against T for chromium potassium alum.

Agreement between theory and experiment, however, even using S instead of J, is not always entirely satisfactory. In some cases the spin-orbit coupling is very strong so that the quenching of the orbital moment does have an effect on the spin system. This will modify the effective value of S—reducing it in shells less than half full and increasing it in those more than half full. This effect is noticeable in Ni^{2+} and Cu^{2+}. The proportionality of χ_v to $1/T$ (Curie's law) is unaffected. In some cases (e.g. Co^{2+}) the quenching of the orbital moment is not complete and a complicated set of energy levels which are split by a few hundred degrees K are produced. The occupation of these levels changes with temperature and hence Curie's law is not obeyed. This type of behaviour also occurs in the rare earths with which we shall deal in the next section.

9.8. *Susceptibility of the rare-earth group*

In these elements the incomplete $4f$ shell is surrounded by the charge cloud of the complete $5s$ and $5p$ subshells. The $4f$ electrons are therefore partly screened from the effect of the internal field. In addition, due to the larger atomic number of the atoms, the radius of the $4f$ shell is

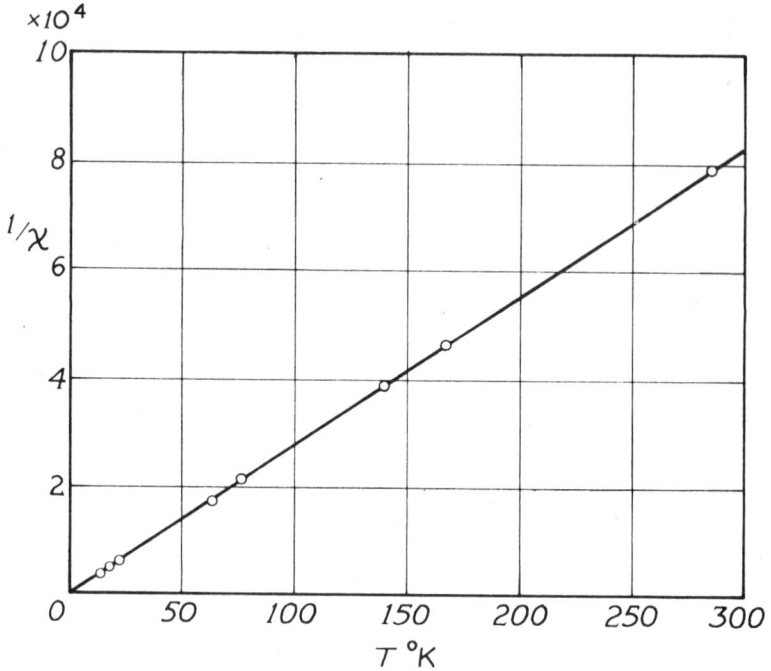

Fig. 9.2. The inverse susceptibility per gm of chromium potassium alum as a function of temperature, showing the very good agreement with Curie's law (after de Haas and Gorter, 1930).

actually smaller than that of the iron group $3d$ shell and hence the $4f$ electrons are farther away from the surrounding diamagnetic ions. For these reasons the orbital moment is not quenched and Russell–Saunders coupling (section 9.4) may be used. As Table 9.2 shows, except for Sm and Eu, the room-temperature experimental values of the Curie constant agree quite well with those calculated using (9.15). The reason for the lack of agreement in the case of Sm and Eu is that we have only been considering the dipole moment associated with the ground state J of the ion, where $J = L \pm S$ according to (c) of Hund's rules (section 9.4). In fact, as we have mentioned, J can have the values $(L-S)$, $(L-S+1)$, ..., $(L+S)$ which correspond to excited states, but these levels are usually separated by energies corresponding to thousands of degrees K, and so only the lowest level need be considered. For Sm and Eu, however, the excited states of J are not at such high energies (see Fig. 9.3 (b)) and their occupation must be taken into account. When this is done the figures in parentheses which are given in Table 9.2 are obtained and, as will be seen, good agreement with experiment is then achieved.

In contrast to the iron group the rare earths do not obey Curie's law down to very low temperatures. This is due to the splitting of the ground

TABLE 9.2

The paramagnetic ions of the rare earths $(4f)$ *group*

(after Bleaney and Bleaney, 1957, p. 568)

No. of electrons in $4f$ shell	Ion	Ground state	S	L	J	g	Average value of C in units of $N\beta^2/(3k)$ experimental	$g^2 J(J+1)$
1	Ce^{3+}	$^2F_{5/2}$	1/2	3	5/2	6/7	6	6·43
2	Pr^{3+}	$^3H_4$	1	5	4	4/5	12	12·8
3	Nd^{3+}	$^4I_{9/2}$	3/2	6	9/2	8/11	12	13·1
4	Pm^{3+}	$^5I_4$	2	6	4	3/5	..	7·2
5	Sm^{3+}	$^6H_{5/2}$	5/2	5	5/2	2/7	2·4	0·71 (2·5)
6	Eu^{3+}	$^7F_0$	3	3	0	..	12·6	0 (12)
7	Gd^{3+}	$^8S_{7/2}$	7/2	0	7/2	2	63	63
8	Tb^{3+}	$^7F_6$	3	3	6	3/2	92	94·5
9	Dy^{3+}	$^6H_{15/2}$	5/2	5	15/2	4/3	110	113
10	Ho^{3+}	$^5I_8$	2	6	8	5/4	110	112
11	Er^{3+}	$^4I_{15/2}$	3/2	6	15/2	6/5	90	92
12	Tm^{3+}	$^3H_6$	1	5	6	7/6	52	57
13	Yb^{3+}	$^2F_{7/2}$	1/2	3	7/2	8/7	19	20·6

The values in parenthesis are those calculated by Van Vleck allowing for the population of excited states at $T = 293°$ K.

state levels in zero field which we have already discussed. It has been found, however, that the effect of the changing population of the higher levels can be taken into account to some extent by means of a correction term in $1/T^2$. The Curie formula then becomes

$$\chi_v = \frac{C}{T}(1 - \Delta/T) \tag{9.19}$$

$$\approx \frac{C}{T(1 + \Delta/T)}, \tag{9.20}$$

or

$$\chi_v = \frac{C}{T + \Delta}. \tag{9.21}$$

The experimental results are represented quite well by (9.21) at temperatures which are not too low. A plot of $1/\chi_v$ against T will be linear with an intercept on the $1/\chi_v$ axis.

9.9. *Level splittings; Kramers theorem*

It is not possible to give any guidance about the value of Δ because the crystal field can split the $2J+1$ levels of the ground state in many different ways. There is one rule, however, which is quite general. If J is half integer (which will occur if there are an odd number of electrons in the shell) then the electric field can, at most, only split the $2J+1$ states into half that number of doublets (Fig. 9.1 (b)). This is the result of Kramers (1930) theorem. These doublets, which are often called Kramers doublets, are usually spaced unevenly and when a magnetic field is applied, each is split into the two levels $+M$ and $-M$ with dipole moments and relative energies given by (9.9) and (9.8).

If J is an integer (which can only occur if there are an even number of electrons in the shell) then, except in a field of high symmetry, the $2J+1$ states are usually split into singlet levels of different energy and each is characterized by one of the possible values of M. Even if some of the levels remain degenerate, the lowest level will be a singlet (the Jahn–Teller, 1937, effect). The levels are not usually spaced evenly and there is no rule as to which level has a particular value of M. It can even vary in different salts of the same ion.

In the rare earths the interaction of the internal field is such that there is an overall energy spread of between 10 and 100° K across the split levels (Figs. 9.1 (b) and 9.3 (c)). Since at room temperature this splitting is much less than kT, all the states will still be equally occupied at $H = 0$ and so Curie's law holds. The law breaks down at lower temperatures because the zero field population of the various levels will change and hence m_v will no longer be given by (9.11).

9.10. *The crystal symmetry*

The manner in which the levels are split by the internal field depends on the symmetry of the field, i.e. on the distribution of the oxygen ions around the paramagnetic ion. If the crystal has a high degree of

---

FIG. 9.3. Energy level diagrams. (a) The iron group showing how the $2L+1$ levels are split, first by a cubic field, and then by a rhombic field. The manner in which the $(2S+1)$-fold degeneracy of these states is removed by spin-orbit coupling and by an external field is shown for the lowest level. These schemes are not to scale (Gorter, 1947). (b) The energy levels of four rare-earth ions. These are to scale and the energy corresponding to kT at 293° K is also shown. Each of these levels is $(2J+1)$-fold degenerate (Van Vleck, 1932). (c) The manner in which the lowest level of two of the ions in (b) is split by internal fields and by an external magnetic field, H. The amount of splitting and the sequence of levels, even in a cubic field, can change in different materials. For $J = 4$ the effect of H is only indicated for the two extreme levels. The overall splitting might be of the order of 100° K. Not to scale.

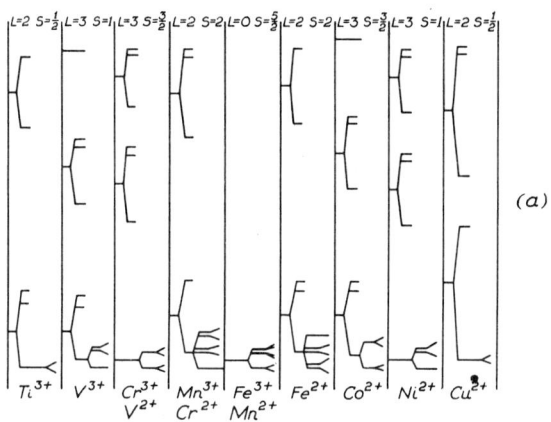

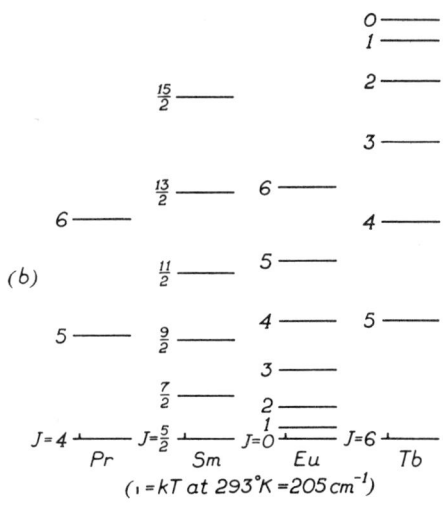

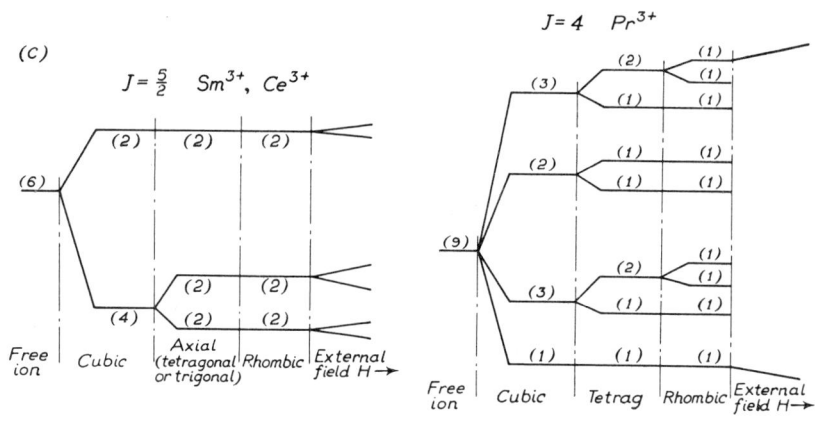

$L=2$ $S=\frac{1}{2}$ | $L=3$ $S=1$ | $L=3$ $S=\frac{3}{2}$ | $L=2$ $S=2$ | $L=0$ $S=\frac{5}{2}$ | $L=2$ $S=2$ | $L=3$ $S=\frac{3}{2}$ | $L=3$ $S=1$ | $L=2$ $S=\frac{1}{2}$

(a)

Ti^{3+} | V^{3+} | Cr^{3+} V^{2+} | Mn^{3+} Cr^{2+} | Fe^{3+} Mn^{2+} | Fe^{2+} | Co^{2+} | Ni^{2+} | Cu^{2+}

(b)

0 ——
1 ——
2 ——
$\frac{15}{2}$ ——
3 ——
$\frac{13}{2}$ —— 6 ——
6 —— 4 ——
$\frac{11}{2}$ —— 5 ——
5 —— $\frac{9}{2}$ —— 4 —— 5 ——
$\frac{7}{2}$ —— 3 ——
2 ——
1 ——
$J=4$ —— $J=\frac{5}{2}$ —— $J=0$ —— $J=6$ ——
Pr   Sm   Eu   Tb

$(\text{\i}=kT \text{ at } 293°K = 205 \text{ cm}^{-1})$

(c)

$J=\frac{5}{2}$ Sm^{3+}, Ce^{3+}

(6) | (2) | (2) | (2)
(4) | (2) | (2)
   | (2) | (2)

Free ion | Cubic | Axial (tetragonal or trigonal) | Rhombic | External field H →

$J=4$ Pr^{3+}

(9) | (3) | (2) | (1)
 | | (1)
 | (3) | (1) | (1)
 | (2) | (1) | (1)
 | | (1) | (1)
 | (3) | (2) | (1)
 | | (1)
 | (1) | (1) | (1)

Free ion | Cubic | Tetrag | Rhombic | External field H →

FIG. 9.3

symmetry, such as in a cubic crystal, then the levels need not be split to the full extent permitted by Kramers' theorem. In general, however, there are always departures from such high symmetry and these are nearly always sufficient to remove as much of the degeneracy as is permitted by the theorem. The distortions which are necessary to remove the degeneracy need only be very small and they cannot always be detected by X-ray diffraction measurements. Hence it is not possible, *a priori*, to predict what the detailed splittings will be. An introduction to crystal field splittings is given by van den Handel (1956).

However, in order to give a general idea of the type of level schemes which are encountered, some diagrams are given in Fig. 9.3. In Fig. 9.3 (a) are the levels for the ions of the iron group. It will be recalled that the orbital motion of these ions is quenched by the internal field, so that to a first approximation the $2L+1$ orbital states may be considered separately from the $2S+1$ spin states. The left-hand side of each diagram shows the manner in which the $2L+1$ orbital states are split, by the internal field, firstly by one with cubic symmetry and then by one with rhombic symmetry. Each of these levels will have a degeneracy of $2S+1$ due to the combined spin momentum S and the manner in which these levels are split by the spin-orbit interaction is indicated for the lowest of the L states. Finally the diverging lines show how these levels are affected by an external magnetic field. These diagrams are not to scale, but in general the splitting between the various L-states is of the order of thousands of degrees K and usually only the lowest doublet (or singlet, as the case may be) of the whole scheme is occupied.

Figs. 9.3 (b) and (c) show schemes for some rare earth ions. In Fig. 9.3 (b) are drawn the level diagrams for the $2S+1$ states of each ion (running between $J = L+S$ and $J = L-S$). These are to scale and the energy corresponding to kT at 293° K is also shown on the diagram. It will be noted that this is very small compared with the level spacings, except for Sm and Eu and indeed apart from these ions, only the lowest level need be considered in all the other rare earths. Each of the levels in Fig. 9.3 (b) is $(2J+1)$-fold degenerate and the manner in which the lowest level is split by the internal field and by an external magnetic field is shown in Fig. 9.3 (c) for two of the ions. The overall splitting of these levels is of the order of a hundred degrees K. It will be noted in both Figs. 9.3 (a) and (c) that those ions with an even number of electrons (S or J integral) have a singlet ground state, whereas those with an odd number of electrons (S or J half-integral) have a ground-state doublet whose degeneracy may only be removed by an external magnetic field. This is in accord with the theorems of Kramers and Jahn and Teller which we have mentioned in section 9.9.

9.11. *Susceptibility at very low temperatures; Van Vleck paramagnetism*

At very low temperatures we must expect a breakaway from (9.21) for a salt with Kramers doublets because the situation will eventually be reached in which all the ions are in the lowest energy state when the magnetic field is applied. Their moments will then be completely aligned and no further increase will be possible. When this behaviour is being approached the full expression (9.16) for the magnetic moment must be used in order to calculate χ_v and the magnetic moment will tend towards the saturation value given by (9.17).

For ions with an even number of electrons, in which the levels can split into singlets even in the absence of a magnetic field, we must also expect unusual behaviour. At high temperatures the upper levels are still sufficiently close together so that the extra splitting due to an applied magnetic field will change their populations. This gives rise to the temperature-dependent paramagnetism which we have discussed. In general, however, the lowest level is a singlet state separated from the rest (Jahn–Teller effect) and at low temperatures it will be the only one to be occupied. Since this state is a singlet the application of a magnetic field should not be able to provide any axis of preferred orientation where the energy is lower and hence one might expect the salt to be diamagnetic at low temperatures. From (9.9), however, we see that even for a state with $M = 0$ there will still be magnetic energy due to the correction term in αH^2. This will give rise to a dipole moment $N\alpha H$ and a temperature *independent* susceptibility $N\alpha$. This is called Van Vleck paramagnetism.† At higher temperatures there will be fewer atoms in the ground state and so this extra susceptibility decreases as the temperature is increased. It can be shown, for temperatures which are much greater than the splitting between the ground and the excited states, that the contribution to the susceptibility due to the αH term is proportional to $1/T$, i.e. it slightly modifies the value of the Curie constant.

Thus we would expect that the susceptibility of a substance having a singlet ground state would follow an equation of the form (9.21) at higher temperatures but would flatten off to a value of $N\alpha$ at low temperatures, whereas for a salt with a degenerate ground state, temperature-

† We should note that a ground-state singlet does not necessarily have to have $M = 0$ for Van Vleck paramagnetism to arise. It may be formed of admixtures of other states whose resultant is non-magnetic.

dependent paramagnetism should continue until $g\beta H \gg kT$. These two
types of contrasting behaviour can be observed. Fig. 9.4 shows a plot
of $1/\chi$ against T for ytterbium sulphate, $Yb_2(SO)_3 . 8H_2O$. Yb has an

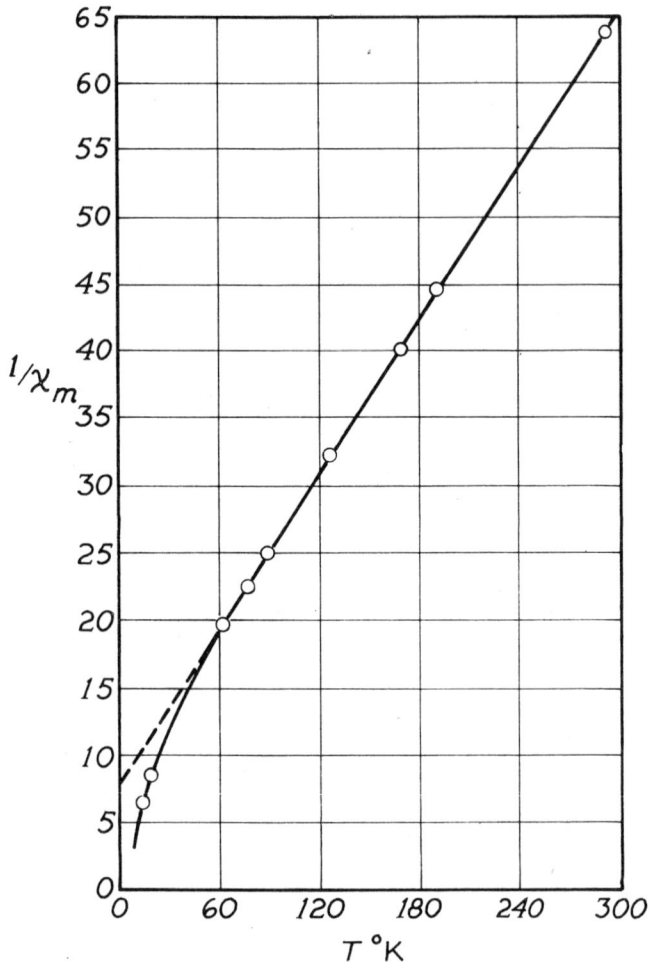

FIG. 9.4. The inverse molar susceptibility of ytterbium sulphate as a function of temperature showing the continuous increase in χ at low temperatures, characteristic of an ion
with Kramers degeneracy (Jackson, 1936).

odd number of electrons and hence according to Kramers' rule it should
be left with degenerate doublet states in zero field. It can be seen that
down to about 60° K a good straight line is obtained showing agreement
with (9.21) with $\Delta = 42°$ K. At lower temperatures, however, χ increases more rapidly than the simple theory predicts. Fig. 9.5 shows

a similar type of plot for praseodymium sulphate, $Pr_2(SO_4)_3.8H_2O$, Pr has an even number of electrons and the levels can be split into singlets. The flattening off of $1/\chi$ at low temperatures due to the Van

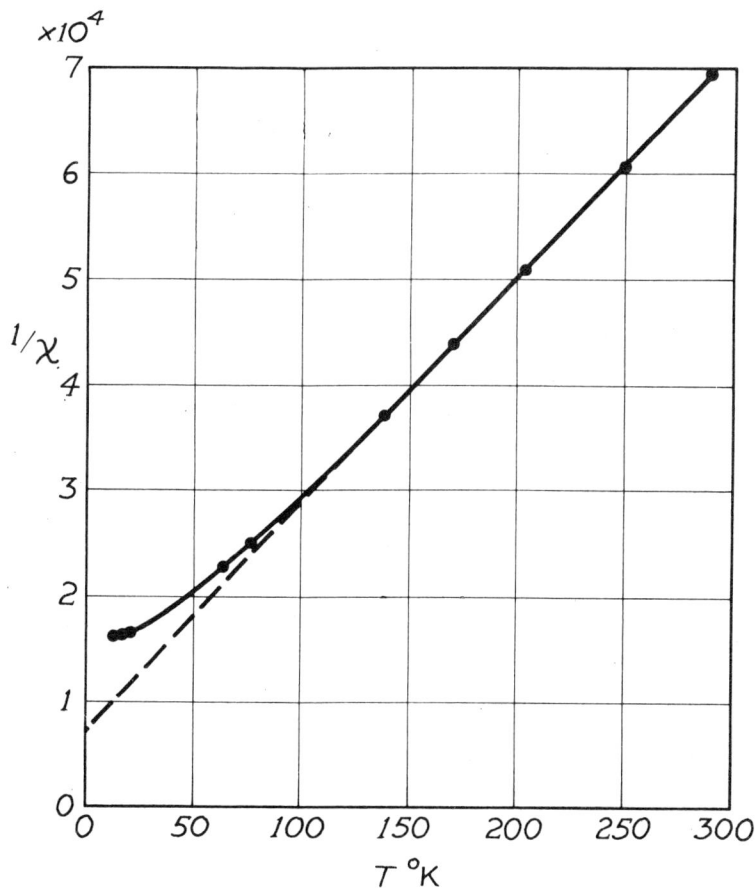

FIG. 9.5. The inverse susceptibility per gm of praseodymium sulphate as a function of temperature which shows a flattening off at low temperatures—Van Vleck paramagnetism—characteristic of an ion with a singlet ground state (Gorter and de Haas, 1931).

Vleck paramagnetism is quite evident, although at high temperatures (9.21) fits the results well.

It is impossible, without presenting data on individual substances, to give any more than a general picture of susceptibility behaviour. This is because the energy splittings which control the deviations from Curie's law are so very sensitive to the internal field that there are wide variations both in type and in magnitude. Hence every substance is a special case. Thus quite different behaviour can be obtained for different salts containing the same paramagnetic ion. An outstanding case is the change in behaviour between an ordinary salt and one in which the ion is in a covalent complex such as the complex cyanides. For example, the susceptibility of most of the salts of Fe^{3+} is isotropic and has a value corresponding to $S = \frac{5}{2}$ (see Table 9.1) whereas potassium ferricyanide has an anisotropic susceptibility which departs markedly from Curie's law and its magnitude is nearer to that expected for $S = \frac{1}{2}$ (e.g. see Baker, Bleaney, and Bowers, 1956). The cobalticyanides present an even more extreme example because they are diamagnetic.

In some salts the level scheme can be completely altered when the paramagnetic ions are diluted with diamagnetic ions of the same valency. For example, cerium ethylsulphate diluted 1:200 with lanthanum ethyl-sulphate has the $M = \pm\frac{1}{2}$ doublet below the $M = \pm\frac{5}{2}$ doublet. In the concentrated cerium salt the $M = +\frac{5}{2}$ doublet lies lowest (Bogle, Cooke, and Whitley, 1951).

9.12. *Salts commonly used in experiments*

Owing to the fact that slight departures from high symmetry can have quite a large effect on the energy levels, certain classes of salts have tended to be used in paramagnetic experiments because their crystal symmetry is well known. The most satisfactory crystals are those which have accurate cubic symmetry or those which have one main symmetry axis such as is present in a hexagonal structure. The field splittings can then be calculated accurately and this enables any other interactions which may be present (e.g. due to nuclear spin, exchange interaction, or interaction with neighbouring dipoles) to be estimated with a greater degree of assurance. It is also an advantage in the interpretation of the results to have only one paramagnetic ion per unit cell, or if there is more than one, that the orientation of the crystal field around each ion should bear a simple relation to the orientations of the others in the cell.

In order to try and fulfil some of these conditions, experiments have been made on a few special types of salt. In this section we list the ones which have been most commonly used, together with a few remarks on their characteristics. For further details the reader is referred to Bowers and Owen (1955). In the chemical formulae, X refers to a diamagnetic ion and M to a paramagnetic ion. It does not follow that salts of the types listed can be prepared for all the ions of a particular group (except the rare earth ethylsulphates).

Iron group

(a) Alums. $X^+M^{3+}(SO_4)_2.12H_2O$. These are used for trivalent ions of the group. The crystal field is approximately cubic although there is some trigonal distortion. There are four ions per unit cell. These are not magnetically equivalent.

(b) Tutton salts. $X_2^+M^{2+}(SO_4)_2.6H_2O$. These are used for divalent ions of the group. The crystal has low symmetry. There are two ions per unit cell. These are not magnetically equivalent (see section 9.32).

(c) Fluosilicates. $M^{2+}SiF_6.6H_2O$. These have a hexagonal structure and show axial (trigonal) symmetry. There are six magnetically equivalent ions per unit cell.

(d) Cyanide complexes. For trivalent ions, $X_3^+M^{3+}(CN)_6$; for divalent ions, $X_4^+M^{2+}(CN)_6.3H_2O$. Whilst these salts tend to have low symmetry properties they are of interest because they show the effect of strong covalent bonding compared with the ionic bonding of the other salts.

(e) Double nitrates. See section (c) in rare-earth group.

Rare-earth group

(a) Sulphates $M_2^{3+}(SO_4)_3.8H_2O$. These were the first salts of this group to be investigated, but the structure is not accurately known. Apart from the gadolinium salt they are not now used.

(b) Ethylsulphates $M^{3+}(C_2H_5SO_4)_3.9H_2O$. These are hexagonal crystals and have axial (trigonal) symmetry. There are two magnetically equivalent ions per unit cell. Isomorphous salts exist for all the rare earths. This series of salts is the one which has been investigated most extensively.

(c) Double nitrates. $X_3^{2+}M_2^{3+}(NO_3)_{12}.24H_2O$. These have a complicated structure with axial symmetry and one ion per unit cell. Some double nitrates of the divalent ions of the iron group have also been investigated, the X^{2+} in the formula now being the paramagnetic ion.

(d) Trichlorides $M^{3+}Cl_3$. These, like the ethylsulphates, are hexagonal but they are sometimes to be preferred because the complex around the paramagnetic ion is so much simpler.

Platinum group

These have been investigated using complex halides of the form $X_2^+M^{4+}Y_6$, where Y is a halogen. They show strong covalent bonding.

The most important salts are the chloro-iridates which appear to have exact cubic symmetry. There are four equivalent ions per unit cell.

The actinides

These have been investigated with the actinide in the uranyl complex $M^{4+}O_2$, usually in a double nitrate with rubidium, $M^{4+}O_2Rb(NO_3)_3$. They are hexagonal crystals.

9.13. *Anisotropy*

The foregoing list of salts shows that several of those which are investigated have a crystal field which has axial symmetry rather than a cubic (isotropic) field. With such crystals the magnetic properties are also anisotropic. This can be understood quite easily in a general way. An axial crystal field will tend to distort the electron cloud around the paramagnetic ion so that it, too, has a similar axial symmetry. Due to the spin-orbit coupling, the spins will also be influenced by this orbital motion. If an external magnetic field is applied, the spins will tend to orient themselves along a resultant of the two fields of force which they experience. The susceptibility will therefore be anisotropic. There will usually be two main axes for the susceptibility—one parallel to the crystal field axis and the other in any direction which lies in a plane at right angles to that axis. These two susceptibilities are denoted by $\chi_\parallel$ and $\chi_\perp$. The anisotropy can be very large. Fig. 9.6 shows $1/\chi_\parallel$ and $1/\chi_\perp$ for hexagonal cerium ethyl sulphate plotted against the temperature. It should be noted that the individual susceptibilities obey a modified Curie law (9.21). In some salts at low temperatures the anisotropy can be reversed and larger than is shown in Fig. 9.6; e.g. in cerium magnesium nitrate $\chi_\perp/\chi_\parallel$ can be as high as 100 at 1° K. Such crystals can be very useful as the cooling agents in experiments utilizing adiabatic demagnetization (section 9.38) if after demagnetization it is necessary to apply a magnetic field to the specimen which has been cooled by the paramagnetic salt (e.g. in the determination of the critical field of a superconductor). Normally the application of the field would raise the temperature of the

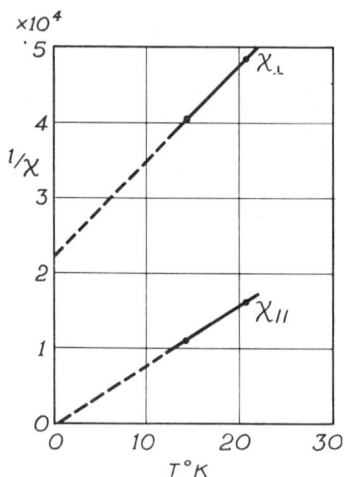

FIG. 9.6. The anisotropy of the susceptibility, per gm, of cerium ethylsulphate (Fereday and Wiersma, 1935).

salt and specimen. If, however, the salt used for cooling has a very anisotropic susceptibility then very little heating will be caused if the magnetic field applied to the specimen is in the direction of minimum susceptibility (Cooke, Duffus, and Wolf, 1953).

9.14. Interactions at very low temperatures

Besides the effect of the crystal field there are several other interaction effects which can further alter the spacing of the energy levels or can remove any degeneracies which may remain after the crystal field splitting. These interactions are usually very small, being in the energy range of about 10^{-2} to a few degrees, but at low temperatures they can become very important. Susceptibility measurements are not always the best method of investigating these effects—the methods of paramagnetic resonance are very often much more powerful and accurate (section 9.29 ff.) but for completeness we describe these other perturbations on the energy spectrum in this section.

9.15. *Dipole-dipole interaction*

We have so far assumed that the paramagnetic ions are sufficiently diluted that the field due to one of them has no effect on its nearest neighbour. However dilute the crystal might be, this can never be quite true. There will always be a small magnetic field present which could split a degenerate Kramers doublet. In the salts which we have discussed, when they are undiluted by diamagnetic ions, the paramagnetic ions are 6 or 7 Å apart. The field at a distance d from a dipole of strength m is of the order of m/d^3, hence since m is of the order of a Bohr magneton ($9 \cdot 27 \times 10^{-21}$ erg gauss^{-1}), the field at a neighbouring ion will be about 100 gauss. This will give splittings of about 10^{-2} degrees.

9.16. *Exchange interaction*

These interactions are a consequence of the quantum nature of the system and they have no classical counterpart. If the wave functions of the electrons of two atoms overlap, then the electrons of atom 1 are to some extent also associated with atom 2 and vice versa. Thus there must be some interaction between the two groups of electrons because they can exchange their roles in this manner. In particular there is a correlation between the dipole moments of the electrons. In general, the system has a lower energy if the magnetic moments of neighbouring atoms are anti-parallel. It is only in ferromagnetic materials that the parallel orientation has a lower energy. The magnitude of the exchange interaction will depend on the amount of overlap of the electron wave functions and hence on the paramagnetic dilution. High dilution, however, is not sufficient to avoid exchange effects because an indirect or 'super-exchange' mechanism can also take place utilizing exchange interaction with a neighbouring diamagnetic ion as an intermediary between the paramagnetic ions. A simple example of this effect is given by Dekker (1957), p. 490. In the rare earths, where the incomplete shell is well shielded by the outer electrons, exchange effects can be neglected. For the iron group they give interactions of the order of 10^{-2} degree but in covalent compounds, such as the $3d$ cyanides, the indirect exchange mechanism can give interactions of the order of 10^{-1} degree. The $4d$ and $5d$ covalent salts may have exchange splittings of several degrees.

9.17. *Nuclear interactions; magnetic hyperfine structure*

The nucleus of a paramagnetic ion can possess a spin characterized by a quantum number I and associated with this there will be a nuclear magnetic moment m_N. The nucleus can orient itself in $2I+1$ different positions relative to any magnetic field which it experiences. The greater part of this field will be that arising from the electron dipole. As a consequence there will be an interaction between the nuclear and the electron dipoles of the order of $m\,m_N/r^3$, where r is the radius of the electron orbit. In an external magnetic field the electron levels will each be split into $2I+1$ components with an energy spread of between $\sim 10^{-3}$ and $\sim 10^{-1}$ degree. This extra splitting is called hyperfine structure (HFS).

9.18. *Nuclear quadrupole interaction*

This is a type of nuclear interaction which is caused by the fact that the positive charge distribution on the nucleus is not spherically symmetrical but has an ellipsoidal shape. Such a distribution can be considered as an electric quadrupole and it will interact with the *gradient* of an electric field. This will tend to give the nucleus a preferred orientation. The main electric-field gradient which it experiences is that due to the incomplete electron shell around the nucleus. Thus the quadrupole interaction is another method by which a coupling between the nuclear moment and the electron moment may be achieved. The splitting due to this interaction is less than that caused by the HFS coupling, but it is often sufficient to modify the splitting produced by the HFS.

The effect of all these interactions is that at sufficiently low temperatures there will always be a region where the degeneracy of the lowest energy levels is removed, although if the interaction was very small an exceedingly low temperature might be necessary. The concept of the absolute zero as being a state of complete order would also require that all degeneracies be removed as $0°$ K is approached. The outcome of these effects, when we consider the magnetic susceptibility, is that however well Curie's law (9.18) or its modification (9.21) might hold at higher temperatures, a region will always be reached, as the temperature is reduced, where such relationships are no longer valid. It is clear from the number of different kinds of interaction which are possible, that there will be considerable differences in behaviour between the various types of crystals which have been investigated, and therefore once again, as with all paramagnetic effects, no very general rules can be given. From the order of magnitude of the splittings which have been given in the preceding sections, we might expect to observe anomalies in the magnetic behaviour in the range between 1 and 10^{-2} ° K; e.g. a flattening off in the value of the susceptibility and a trend towards the type of temperature-independent paramagnetism which is shown in Fig. 9.5. Usually, however, the anomalies are more marked and some reasons for this are given in the next section.

9.19. Ferromagnetism

In section 9.16 we introduced the concept of exchange forces. It is now generally agreed that these forces, although they cannot be calculated in any detail, are responsible for the phenomenon of spontaneous magnetization in ferromagnetic materials (e.g. see Bleaney and Bleaney 1957, chapter 21). In these substances it is thought that the exchange energy is a minimum when the spin systems of neighbouring atoms are parallel to one another. The exchange interaction on a particular ion can be considered as being produced by an internal magnetic field H_i at the ion which is proportional to the magnetic moment of its neighbours.† Thus H_i can be written

$$H_i = \lambda m_v, \tag{9.22}$$

where λ is a constant and m_v is the magnetic moment per unit volume of the material. It should be emphasized that H_i is a field which is additional to that produced by dipole-dipole interaction (section 9.15) and it is quite often very much greater. The effect of H_i on the

† Whilst it is customary to discuss the effect of the exchange forces in terms of an internal *magnetic* field it should be noted that the interaction is in reality due to the coulomb (i.e. *electrostatic*) potential between the electrons.

susceptibility can be found by substituting $H+H_i$ for H in the Curie law expression (9.18),

i.e.
$$m_v = C(H+\lambda m_v)/T, \tag{9.23}$$

and hence the susceptibility defined by m_v/H is

$$\chi_v = C(H+\lambda m_v)/(HT) \approx C/(T-\lambda C). \tag{9.24}$$

Hence
$$\chi_v = C/(T-\theta_w) \tag{9.25}$$

where
$$\theta_w = \lambda C. \tag{9.26}$$

This is the Curie–Weiss law and θ_w is the Weiss constant. At temperatures greater than θ_w the substance has a temperature dependent paramagnetic susceptibility as shown by (9.25). Below θ_w the exchange interaction is able to overcome the disorder due to the thermal energy of the ions and the neighbouring spins will tend to line up parallel with one another even in the absence of an external magnetic field. This will increase the value of m_v thereby making H_i greater (9.22) and this will make the exchange interaction even stronger, thereby producing a more efficient alignment of spins. Thus a co-operative reaction is set up (section 1.23) resulting in a complete alignment of spins at temperatures not very much less than θ_w. The temperature of the transition, as distinct from θ_w, is called the Curie temperature.

9.20. *Anti-ferromagnetism*

The exchange energy is very sensitive to the spacing of the paramagnetic ions and it is only for a very small range of separations that the energy is a minimum when the neighbouring spins are parallel. In most cases the exchange energy is smaller when neighbouring spins are anti-parallel. The alignment of the spins in an anti-parallel array is also a co-operative effect which occurs at a temperature known as the Néel temperature (T_N). This phenomenon is known as anti-ferromagnetism. An anti-ferromagnetic material can be considered to be arranged on two interpenetrating lattices, each lattice having its ions with their spins parallel, but with the spins on one lattice being anti-parallel to those on the other. A diagram of the arrangement in a simple cubic lattice is shown in Fig. 9.7. An internal field treatment, similar to that which we have described for ferromagnetism, may be used, in which it is assumed that an ion on lattice A is acted on by its neighbours (which will be on lattice B) by means of an internal field H_{iB}. If m_B is the volume magnetization of lattice B we can then write for the total field H_A on A

$$H_A = H+H_{iB} = H-\lambda m_B \tag{9.27}$$

and similarly
$$H_B = H+H_{iA} = H-\lambda m_A. \tag{9.28}$$

The negative signs are used because the exchange interaction will tend to destroy the alignment parallel to H. Above T_N we can calculate the susceptibility for each lattice separately, using H_A and H_B instead of H in (9.23).

Thus
$$m_A = \tfrac{1}{2}C(H-\lambda m_B)/T \quad \text{and} \quad m_B = \tfrac{1}{2}C(H-\lambda m_A)/T. \tag{9.29}$$

Then
$$m_v = m_A+m_B = \tfrac{1}{2}C\{2H-\lambda(m_A+m_B)\}/T. \tag{9.30}$$

Hence
$$\chi_v = m_v/H = C\{1-\lambda(m_A+m_B)/2H\}/T \tag{9.31}$$
$$\approx C/(T+\theta') \tag{9.32}$$

where
$$\theta' = \lambda C/2. \tag{9.33}$$

Thus above the Neel temperature an anti-ferromagnetic substance obeys a Curie–Weiss law but with θ' of opposite sign to θ_w.

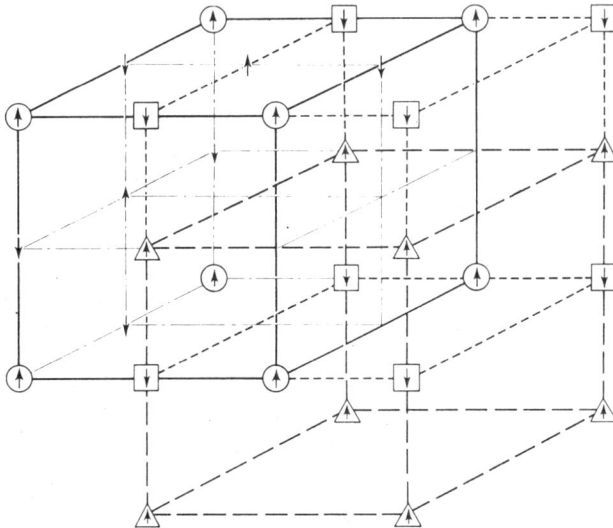

FIG. 9.7. Anti-ferromagnetism in a simple cubic lattice. The spins of the ions at the corners of the small cubes are arranged so that they form a series of interpenetrating cubic lattices with double the cell size. Three of these large cubes are shown, in heavy outline (ions denoted by circles), dashed line (ions denoted by triangles), dotted line (ions denoted by squares). The ions at the corners of any one large cube have all their spins parallel.

The Néel temperature may be deduced from (9.27) and (9.28) by setting the external field H equal to zero. When anti-ferromagnetic alignment occurs, a non-trivial solution of the equations must be obtained. On substituting for λ from (9.33)

$$\left.\begin{array}{l} H_A = -2\theta'm_B/C \\ H_B = -2\theta'm_A/C. \end{array}\right\} \tag{9.34}$$

If we assume that each sublattice obeys Curie's law, i.e.

$$m_A = \tfrac{1}{2}H_A C/T \qquad (9.35)$$

(the factor $\tfrac{1}{2}$ arises because one half of the atoms are on each sublattice)
then
$$\left.\begin{array}{l} H_A = 2Tm_A/C \\ H_B = 2Tm_B/C \end{array}\right\}. \qquad (9.36)$$

Thus (9.36) becomes

$$\left.\begin{array}{l} (T/C)m_A = -(\theta'/C)m_B \\ (T/C)m_B = -(\theta'/C)m_A \end{array}\right\}. \qquad (9.37)$$

If we make it a condition that $m_A, m_B \neq 0$, then the solution of (9.37) is $T = \theta'$. Thus the Néel temperature, T_N, should be θ'.

TABLE 9.3

Anti-ferromagnetic substances

Most of these data have been taken from the compilation by Nagamiya, Yosida, and Kubo (1955).

Substance	T_N °K	θ' °K	Substance	T_N °K	θ' °K
CoO	293	280	FeF$_3$	394	
CuO	173		MoF$_3$	185	
FeO	198	570	KCoF$_3$	114	
MnO	122	610	KFeF$_3$	115	
NiO	492–647		KMnF$_3$	88	
Cr$_2$O$_3$	311	1070	KNiF$_3$	275	
Fe$_2$O$_3$	∼950	2000	CoCl$_2$	25	−38·1
MnO$_2$	84		CrCl$_2$	40	149
V$_2$O$_4$	343	720	CuCl$_2$	70	109
CrS	165	528	FeCl$_2$	24	−48
FeS	613	857	NiCl$_2$	50	−68·2
MnS	154		TiCl$_3$	∼100	
MnTe	323	690	VCl$_3$	30	301
CoF$_2$	37·7	52·7	CuBr$_2$	193	
CrF$_2$	53		MnCl$_2$.4H$_2$O	∼1·7	
FeF$_2$	79	117	MnBr$_2$.4H$_2$O	∼2·2	
MnF$_2$	68	113·2	CuCl$_2$.2H$_2$O	4·31	
NiF$_2$	73·2	115·6	Cr	475	
CoF$_3$	460		α-Mn	100	
CrF$_3$	80				

However, whilst we have identified θ' and θ_w in (9.32) and (9.25) with the temperatures at which materials become anti-ferromagnetic and ferromagnetic respectively, this is not borne out exactly by experiment. The values of θ' and θ_w obtained from (9.32) and (9.25) using high-temperature susceptibility measurements are not exactly the same as the transition temperatures at which the co-operative effects are observed to begin. For anti-ferromagnetics, in particular, the actual value of T_N is much lower (by a factor of perhaps 2 or 3) than θ'. Some data are given in Table 9.3. This discrepancy can be explained if the interactions of the ions with their next-nearest neighbours are also taken into account. These, being on the same lattice as the ion we are considering, will tend to produce an interaction which will upset the anti-parallel arrangement and hence the Néel temperature will be reduced.

Whilst many substances become ferro- or anti-ferromagnetic at room temperature or above (chromium metal, for example, has a Néel temperature of about 200° C) there is considerable low-temperature interest in these transitions. The reason for this is that if we have a system of magnetic ions, then, however small the exchange interaction may be, there will be some temperature which is low enough for a co-operative alignment to be able to take place; i.e. at a sufficiently low temperature all paramagnetic crystals should become either ferro- or anti-ferromagnetic (or ferrimagnetic, see section 9.22). The general trend is towards anti-ferromagnetic behaviour.

9.21. *Susceptibility of an anti-ferromagnetic substance below the Néel temperature*

In the previous section we have shown that the susceptibility of an anti-ferromagnetic above T_N will follow a type of Curie–Weiss law

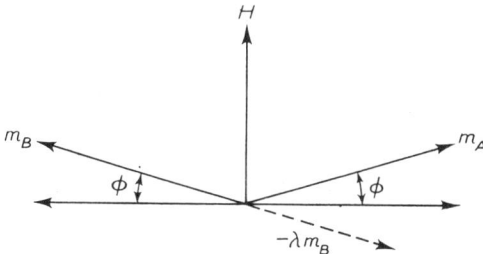

Fig. 9.8. The magnetization of an anti-ferromagnetic when the field is applied at right angles to the spin orientation. Each sublattice rotates through a small angle ϕ, yielding a net magnetic moment.

(9.32). In a cubic crystal it should be isotropic. This is not the case below T_N. We shall first consider a simple treatment of the susceptibility when the applied field is small and is perpendicular to the spin orientation in zero field. At absolute zero we shall assume that the sub-lattices are aligned accurately anti-parallel but that in a small field H their spins rotate by an angle ϕ with respect to their original directions (Fig. 9.8). The value of ϕ will be determined by the fact that the couple on each dipole should be zero. For this to occur there must be no resultant field perpendicular to a dipole. The interaction of lattice B on lattice A may be considered as being due to a field $-\lambda m_B$. This will act in a direction opposite to m_B as is shown. The component of this field perpendicular to m_A will therefore be $-\lambda m_B 2\phi$ and for equilibrium this must be equal and opposite to the component due to the external field H. Since ϕ is small, this component will be approximately equal to H. Hence

$$H = \lambda m_B 2\phi. \qquad (9.38)$$

The change in the magnetic moment of the two lattices due to the orientation produced by H will each be $m_A\phi$ and $m_B\phi$ and since these are equal to one another the total magnetic moment is

$$m_v = 2m_B \phi \qquad (9.39)$$

and the susceptibility, using (9.38), will be

$$\chi_\perp = m_v/H = 1/\lambda. \qquad (9.40)$$

From (9.32) and (9.33) we can see that at $T = \theta'$, χ_v is also equal to $1/\lambda$, and hence from these simple arguments we would expect that below the transition temperature $\chi_\perp$ should be approximately constant with a value of about $1/\lambda$.

If $\dot{H}$ is applied in a direction parallel to the spins a very different type of situation develops. At $T = 0°$ K the spins in the two sublattices are in perfect alignment and hence the application of a parallel field will produce no resultant moment on them. Thus $\chi_\parallel$ will be zero. As the temperature is raised the alignment is upset slightly and so H will be able to produce a small rotation of the spins and hence a small susceptibility. As T_N is approached, $\chi_\parallel$ will become larger tending towards $1/\lambda$.

Thus when a substance becomes anti-ferromagnetic at low temperatures the susceptibility of a single crystal will be strongly anisotropic (Fig. 9.9). This anisotropy can be distinguished from that which results from the axial symmetry of the crystalline field (section 9.13), because the latter will not show a discontinuous transition to anisotropy such as that which occurs at the Néel temperature of an anti-ferromagnetic.

If a powdered specimen of an anti-ferromagnetic is measured, the susceptibility below T_N will be an average of $\chi_\parallel$ and $\chi_\perp$ and is given by

$$\chi_{\text{powder}} = (\chi_\parallel + 2\chi_\perp)/3. \qquad (9.41)$$

Thus at $T = 0°$ K χ_{powder} will be equal to two-thirds of its value at T_N. The specific heat of anti-ferromagnetics is discussed in section 1.23.

9.22. *Ferrimagnetism*

In some crystals of a more complicated structure, the magnitude of the magnetic moments associated with each of the two sublattices (which we have suggested are responsible for anti-ferromagnetism) are not exactly the same. Thus when spontaneous anti-parallel alignment occurs, the material, instead of having zero magnetic moment, has a permanent magnetic moment. This phenomenon is known as ferri-magnetism. The most common material which shows this behaviour is magnetite, Fe_3O_4, which has a spinel structure. This has two sub-lattices, one containing positions for Fe ions surrounded by four oxygen

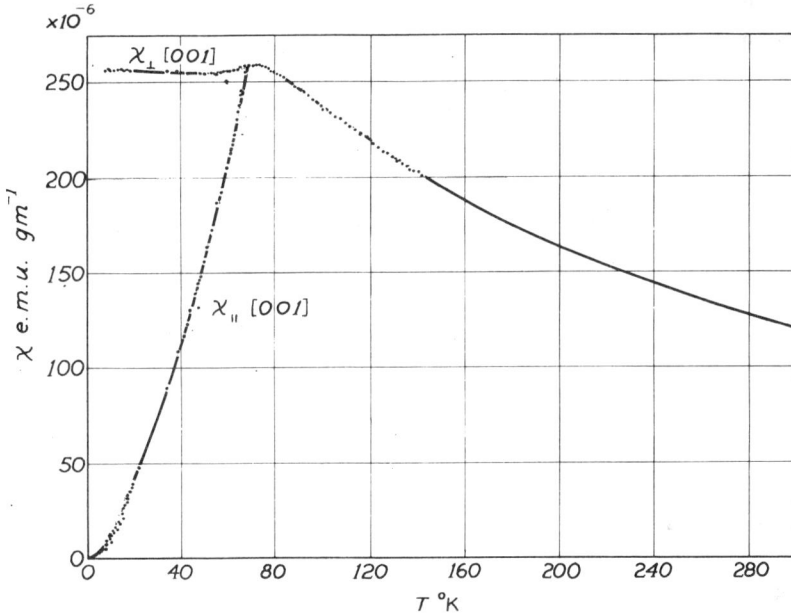

FIG. 9.9. The susceptibility of MnF_2 parallel and perpendicular to the [001] axis of the crystal. This shows the very large anisotropy of the susceptibility of an anti-ferro-magnetic below the Néel temperature (data by courtesy of Dr. S. Foner).

ions (tetrahedral sites) and the other containing twice the number of positions in which they are surrounded by six oxygen ions (octahedral sites). Trivalent Fe ions occupy all the tetrahedral sites and equal numbers of di- and trivalent Fe occupy the octahedral sites. Thus the sublattices corresponding to the octahedral and the tetrahedral sites are magnetically inequivalent and after anti-parallel alignment there is a net magnetic moment. The divalent Fe can be replaced by many other divalent metal ions to give one of the many types of mixed ferrites which are now available. The magnetization depends very much on the nature of the divalent ions which are added.

Ferrites are of great technical importance, because they have a high electrical resistivity. They can therefore be used as cores in inductive elements at frequencies which are very much higher than those for which iron cores are employed and they are also used as directional elements in micro-wave equipment. Nevertheless, from the fundamental point of view their magnetic properties are difficult to interpret in any detail because the crystal structure is so complicated that it is not possible to calculate the energy splittings with any accuracy.

For this reason considerable interest is now centred on another type of crystal which exhibits ferrimagnetism. These are crystals with the structure of the garnets. They have the formula $P_3^{2+}Q_2^{3+}R_3^{4+}O_{12}$, when they occur naturally, R being silicon. Whilst they have an apparently complicated crystal structure with a large unit cell containing 160 atoms, there is in fact a high degree of symmetry and the positions of corresponding ions are related to one another by simple reflection and rotation. Within the sub-units of the unit cell, the electric field has cubic symmetry and hence, compared with the spinels, their properties are much simpler to understand. In the crystals which are synthesized for research the di- and quadrivalent ions P and R are usually replaced by two trivalent ions. The substance of this type which has been most extensively investigated is yttrium iron garnet (YIG). It has the formula $Y_3^{3+}Fe_2^{3+}Fe_3^{3+}O_{12}^{2+}$, i.e. $Y_3Fe_5O_{12}$. Ferrimagnetics in which the diamagnetic yttrium is replaced by a rare-earth ion have also been investigated.

Since these crystals appear to be very suitable ones in which to investigate exchange interactions many experiments have also been made on garnets which are paramagnetic. These are usually made with Ga_5^{3+} or Al_5^{3+} replacing the Fe_5^{3+} ions and with some (to get a high degree of dilution) or all of the Y_3^{3+} being replaced by a rare-earth ion. Lu is also used as a diluent instead of Y.

The susceptibility of a ferrimagnetic material above the Néel temperature may be calculated by using the concept of internal magnetic fields in a manner similar to that which we adopted for anti-ferromagnetics. The difference between the two treatments lies in the fact that the magnetic moment of each sublattice is different. It can be shown (Dekker, 1957, p. 493) that under these circumstances the susceptibility does not follow a Curie–Weiss law, but rather that a graph of $1/\chi$ against T is concave towards the T axis.

Below the Néel temperature the situation is even more complicated. As in a ferromagnet, a domain structure exists and when an applied magnetic field is increased from zero, these domains will be rotated and so the magnetic moment of the specimen will increase until with sufficient field the domains will line up and the moment will saturate. Thus there will be a field-dependent susceptibility. This behaviour will, of course, depend very much on the specimen. The quantity which is physically more significant is the saturation moment itself. In a ferromagnet this has a fairly straightforward behaviour. As the temperature is reduced below the Curie point, the alignment of the spins improves and the magnetic moment increases and eventually flattens off at a value which

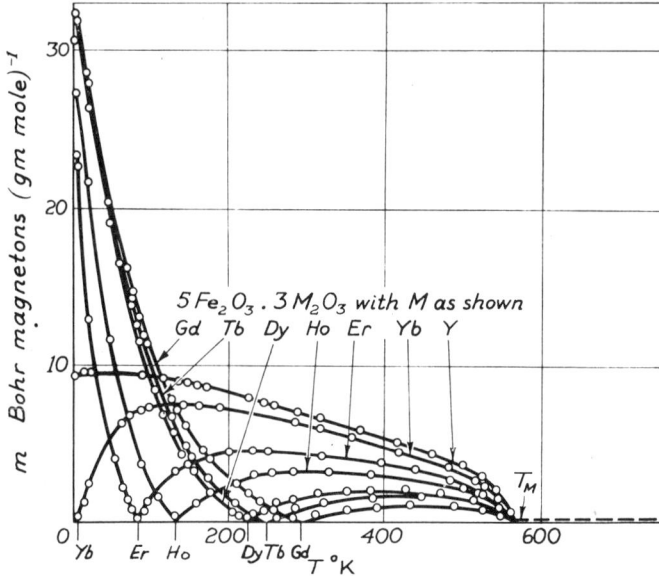

Fig. 9.10. The spontaneous magnetization curves for several rare-earth iron garnets ($5Fe_2O_3.3M_2O_3$). The position of the compensation point for each crystal is marked on the temperature axis by the symbol for its rare-earth ion. Whilst the scale of this diagram does not permit it to be shown, it should be noted that as $T \rightarrow 0°$ K, m flattens off to a constant value (after Pauthenet, 1958).

corresponds to the greatest permissible spin alignment. In a ferrimagnet similar behaviour can occur, but since the magnetic moment is the resultant moment of two anti-parallel and inequivalent sublattices more complicated behaviour may be observed. In particular, the magnetic moments of the sublattices will not each have the same temperature dependence and it is possible for them to have equal and opposite magnetic moments at one temperature so that the total magnetic moment passes through a zero point corresponding to a true anti-ferromagnetic transition. Such behaviour is shown in many specimens (Fig. 9.10) and the temperature of zero magnetic moment is called the compensation point. Below this point the magnetic moment again rises and it generally flattens off to a constant value as T tends to $0°$ K

For further information and full references on ferrimagnetism the reader is referred to the review article by Wolf (1961).

9.23. Susceptibility of gases

Whilst the most important paramagnetic materials are those of the types which we have discussed in the preceding sections of this chapter, we should point out that two common gases are also paramagnetic—oxygen and nitric oxide, NO. The latter has an odd number of electrons and hence it must have a resultant spin, and in the oxygen molecule there are two parallel spins which give an effective $S = 1$ (a simple explanation for this is given by Heitler, 1945, p. 122). It has been possible to make susceptibility measurements on both of these molecules down to low temperatures while they are still in the 'gaseous' state. This has been achieved by incorporating the molecules in a clathrate compound. The clathrates have a structure consisting of a series of cages formed by the molecular arrangement of the β-quinol lattice. An O_2 or an NO molecule can be put into each cage where it apparently behaves very much as if it were free. Fig. 9.11 shows the susceptibility of the oxygen clathrate, plotted against $1/T$ for the range 1 to 20° K. Curie's law is obeyed very accurately from 20 to 10° K but at very low temperatures the susceptibility flattens off and becomes constant (Van Vleck paramagnetism, section 9.11) indicating that the three molecular levels are split so that the ground state is a singlet. These experiments give very strong confirmation to the belief that in these compounds the molecules of O_2 behave like a gas down to very low temperatures. Results of a similar nature have also been obtained for the NO clathrate (Cooke and Duffus, 1954).

9.24. A summary of susceptibility behaviour

The preceding sections show that the dependence of the susceptibility on temperature falls into two ranges—above and below the temperature of magnetic ordering. At higher temperatures where the material is paramagnetic, a plot of $1/\chi_v$ against T is linear, except for ferrimagnetics (section 9.22) and so follows the Curie–Weiss law, $\chi_v = C/(T+\theta)$. It is important to note that this might be a 'genuine' Curie–Weiss law, in which the sign and magnitude of θ are related to the nature of the co-operative transition, or it might be spurious, as, for example, in some of the rare-earth salts. In these the ground state is not always well separated from the higher states and then the changing occupation numbers of these states as the temperature is altered gives rise to a non-zero value for θ. This θ is not connected with ferro- or anti-ferromagnetism as in (9.32), but it is actually the Δ of (9.21). Thus care must be taken before predicting any co-operative transition from high-temperature behaviour. For the iron group salts and others which have the ground state well separated from the excited states, the Curie–Weiss law usually holds very well and θ tends to be small.

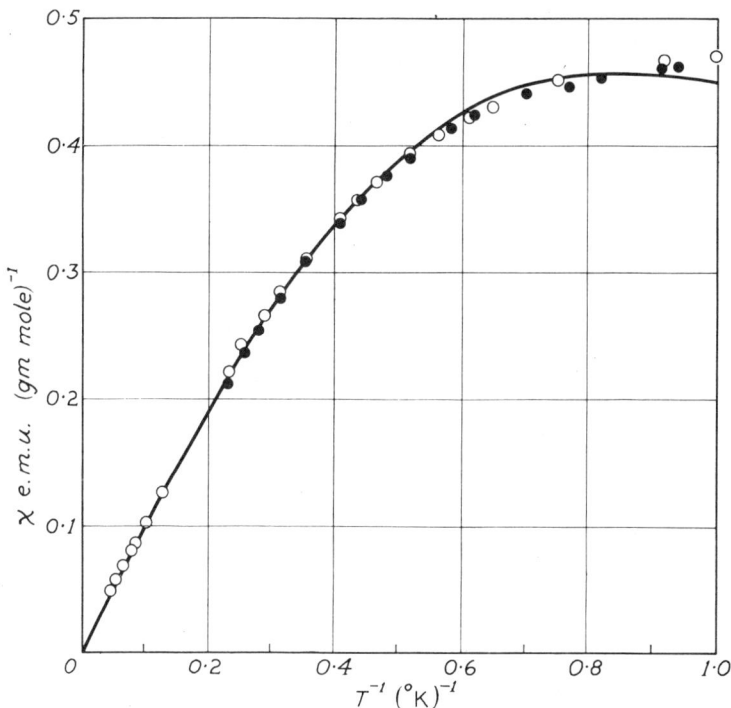

FIG. 9.11. The susceptibility of oxygen in β-quinol clathrate as a function of $1/T$. Above $10°$ K Curie's law is accurately obeyed, but at low temperatures χ tends to a constant value (Van Vleck paramagnetism). The full line is the calculated susceptibility of an ideal gas of normal oxygen (Cooke, Meyer Wolf, Evans, and Richards, 1954). Measurements have since been taken down to $0.25°$ K (Meyer, O'Brien, and Van Vleck, 1958).

Salts which have a singlet ground state will show temperature independent paramagnetism when the temperature is sufficiently low for this state to be the only one which is occupied (section 9.11).

At very low temperatures anomalies sometimes occur in the susceptibility due to the small energy splittings produced by (a) nuclear interactions (HFS and quadrupole, sections 9.17 and 9.18) and (b) interactions with surrounding paramagnetic ions (exchange, section 9.16, and dipole-dipole, section 9.15).

At a sufficiently low temperature magnetic ordering will occur. This is usually anti-ferromagnetic in nature. If it is not masked by the anomalies referred to in the preceding paragraphs the nature of the transition can be determined from the sign of θ; a positive value indicates anti-ferromagnetism whilst a negative value gives evidence for ferro-magnetism. The actual transition to the anti-ferromagnetic state will be shown by a maximum in the susceptibility at the Néel temperature and a strong anisotropy below it (section 9.21). Below the transition temperature ferro- and ferrimagnetic materials will give complicated, field-dependent, values of the susceptibility. At low temperatures in very high fields paramagnetics will also have a field-dependent suscepti-bility and for large H/T the magnetic moment will saturate, correspond-ing to a complete alignment of the spins as in (9.17).

9.25. Paramagnetic relaxation

In the preceding sections we have tacitly assumed that the suscepti-bility was measured in a static magnetic field. If an alternating field is used some rather interesting frequency-dependent effects are found. Details of the experiments and the techniques are given by Gorter (1947) and Cooke (1950). When a field is applied to a paramagnetic crystal, the dipoles can lower their energy by reorienting themselves but they will take a finite time, τ, to get rid of their excess energy and come to equilibrium. If the period of the applied field is much greater than τ, the dipoles can follow the variations of the field quite easily and the susceptibility under these conditions will be the same as that measured in a static experiment. If, however, the frequency is increased, the orientation of the ions will tend to get out of phase with that of the field. The susceptibility can then be treated as a complex quantity of the form

$$\chi_v = \chi' - i\chi'' \tag{9.42}$$

where χ'' is the out of phase component. Both χ' and χ'' will be frequency dependent. Following the relaxation theory of Debye (1929) χ' and χ'' are given by

$$\chi' = \frac{\chi_T}{(1+\omega^2\tau^2)}, \qquad \chi'' = \frac{\omega\tau\chi_T}{(1+\omega^2\tau^2)}, \tag{9.43}$$

where χ_T is the susceptibility in a static field, ω is the angular frequency of the applied field, and τ is the time characteristic of the relaxation process. Since in one cycle the absorption of energy from the field is given by $\oint \mathbf{H}\,d\mathbf{m}$ the presence of χ'' will give rise to a net absorption of energy, and the specimen will therefore heat up. In fact the first investigations on paramagnetic relaxation were made by Gorter (1936) who actually incorporated his salt inside a gas thermometer and in this way measured the heating effect directly (at a frequency of the order of 10 Mc/s). If H_0 is the amplitude of the r.f. magnetic field, then the rate of absorption of energy is equal to

$$\tfrac{1}{2}\omega\chi''H_0^2. \tag{9.44}$$

More interest in this work was stimulated when it was found that the effect was much more pronounced and occurred at lower frequencies if

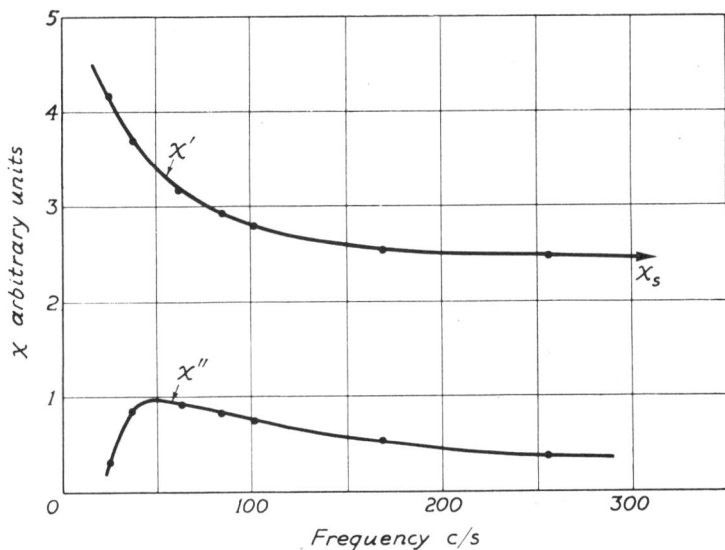

FIG. 9.12. The values of χ' and χ'' for chromium potassium alum in a field of 775 oersted at 2·04° K. The adiabatic susceptibility χ_s is given by the value of χ' at high frequencies (after Casimir, Bijl, and du Pré, 1941).

a strong constant magnetic field was applied which was parallel to the alternating field. In the helium region χ'' was greatest at audio frequencies. A typical series of results at 2·04° K is shown in Fig. 9.12 in which both χ' and χ'' are plotted as a function of the frequency. It will be noted that χ' decreases with frequency and flattens off to a constant value at high frequencies whereas χ'' has a marked maximum. In the next section we shall describe why different effects are observed with and without a magnetic field and we shall also give a qualitative explanation for the behaviour of χ' and χ''.

9.26. Relaxation processes

Let us consider a system of spins initially in zero external field. Owing to the effects of neighbouring spins there will always be some internal field H_i present (of the order of a few hundred oersted in a concentrated paramagnetic salt) and so a particular spin will always have some defined orientation axis. If a *small* external field H_e is now applied ($H_e \ll H_i$) it will cause the effective field on the spins to be altered slightly in direction, but not in magnitude. They will therefore reorient themselves accordingly, but there will be little change in energy. The initial reorientation of the spins will, however, change H_i and so the orientations will have to be readjusted a little more. This will continue until eventually a new equilibrium configuration will be established throughout the whole spin system. All this will take a finite time to occur. This is called the spin-spin relaxation time. It is usually very short—of the order of 10^{-10} sec—and it is independent of the temperature. Thus except at very high frequencies χ'' will be very small and there will be little energy absorption.

A different situation arises when $H_e \gg H_i$. H_i can be neglected and H_e defines the orientation axis. If H_e is now suddenly increased by a small amount, the magnitude of the effective field on the dipoles is altered by, say, dH, but its direction is unchanged. The energy of the atomic states giving rise to the paramagnetism will therefore increase by $\sim g\beta \, dH$ and so as we have seen in section 9.5, the populations of the states will change until their occupation is again in accord with the Maxwell–Boltzmann statistics for a temperature T. In order for this to occur, some of the atoms whose states have been raised in energy by the application of dH, will have to reorient themselves and go into a lower state. To do this they emit a phonon whose energy is equal to the difference in energies of the two states. Whilst it is possible for phonons to be emitted spontaneously, the probability for this to occur is very small; in general there must be some coupling between the spins and the phonon system. This is achieved because the lattice vibrations give rise to a perturbation of the internal field which will influence the spins and vice versa. The larger the perturbation (e.g. at high temperatures), the stronger will be the coupling and the probability of a spin transition is then increased. Thus a spin-lattice relaxation time, τ_{sl}, can be defined which is a measure of the period required for the spins to give up their energy and come to thermal equilibrium with the crystal lattice, after H_e has been changed. This time is a rapidly varying function of the temperature. At helium temperature it can range from 10^{-2} sec to over a minute, whilst at room temperature it might be 10^{-6} sec.

If instead of changing H_e suddenly, we superimpose on·to H_e a small alternating field, then χ'' will be controlled by τ_{sl} and the energy absorption will be much greater than was the case in our first example in which the main relaxation process was through spin-spin interactions. As is apparent from (9.43) χ'' is a maximum when $\omega\tau_{sl} = 1$ and this condition is easily attainable. There will then be sufficient time during each cycle for the excess energy to be given to the crystal lattice and for thermal equilibrium to be maintained. At higher frequencies, however, there is less time available for the energy transfer and the spin system will not be in thermal equilibrium with the lattice. The spins will become an isolated system in which the 'temperature', as defined by the actual population of the energy levels, is different from that of the lattice. It should, however, be noted that whilst the value of χ'' decreases at high frequencies, the absorption of energy remains constant. From (9.43) we see that for $\omega\tau \gg 1$, χ'' is proportional to ω^{-1} and this just cancels out the frequency dependence in (9.44), thereby yielding a constant energy absorption. At low frequencies, such that $\omega\tau_{sl} \ll 1$, the spin system will have time to come to equilibrium and all the necessary energy will be able to be transferred to the lattice, but since ω is small, the actual rate of energy dissipation will be reduced as is shown by (9.44).

The behaviour of the in-phase component of the susceptibility χ' is easily understood. At low frequencies there is sufficient time for energy transfer to the lattice so that $\mathbf{m}$ can keep in phase with $\mathbf{H}$. χ' will have the same value as that determined in a static field. At high frequencies (taking the simplest type of ion with J or $S = \frac{1}{2}$ as an example) the populations of the parallel and anti-parallel states will not change very much with the alternations of the field, since in order, say, to flip from the anti-parallel to the parallel orientation as the field increases, energy must be transferred to the lattice. There is only a very short time in which this can be done and hence the populations of the two states change very little. The susceptibility therefore decreases to a value considerably below its static value. Since the spins can be considered to be an isolated system, this high-frequency susceptibility is very often referred to as the adiabatic susceptibility, χ_s. The importance of this will be indicated in section 9.28.

9.27. *The spin-lattice relaxation time*

Measurements of τ_{sl} as a function of temperature and field have been very difficult to interpret. Qualitatively one can appreciate that as the temperature is raised there will be a greater perturbation due both to the increase in numbers and to the higher energy of the phonons. Since, however, these only produce changes in the internal electric field they do not act directly on the spin system but only with the orbital motion of the electrons. Thus the phonons are only able to influence the spins via the spin-orbit coupling. Hence there are two links in the rather mysterious chain reaching from spins to phonons. One from the spins to the orbital motion and another from the orbital motion to the phonons. So far, however, it has not been possible to take account of the spin-orbit interaction quantitatively and hence, whilst recognizing its existence, we do not explicitly consider it in the following paragraphs.

There are two distinct mechanisms whereby energy can be transferred

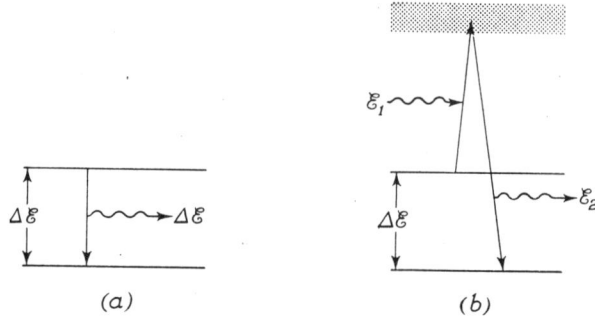

(a)   (b)

FIG. 9.13. (a) A direct relaxation process. The ion goes to the ground state by emitting a phonon with energy $\Delta\mathscr{E}$. (b) A Raman process. The ion relaxes by first absorbing a virtual phonon of energy $\mathscr{E}_1$ and then emitting another of energy $\mathscr{E}_2$ such that

$$\mathscr{E}_2 - \mathscr{E}_1 = \Delta\mathscr{E}.$$

The intermediate excited state is not well defined because the process occurs so rapidly.

between the spin and the phonon systems. The first is known as a direct process. When this occurs the spin system reduces its energy by an amount $\Delta\mathscr{E}$ in the most obvious manner; it emits a phonon with energy $\Delta\mathscr{E}$ (Fig. 9.13 (a)). There is, however, another way in which this transition can be achieved. The spin system could first *absorb* a phonon of energy $\mathscr{E}_1$ so that it was excited to a still higher energy state and then emit another phonon with energy $\mathscr{E}_2$, such that $\mathscr{E}_2 - \mathscr{E}_1 = \Delta\mathscr{E}$, to bring it down to the required final state (Fig. 9.13 (b)). If this absorption-emission process is very rapid so that the ion is only in the excited state for a very short time then by the uncertainty principle its energy is ill defined (it is called a virtual state), and does not need to correspond to a real excited state of the ion. (We met a similar type of virtual interaction in the theory of superconductivity, section 6.29.) Thus there is no limitation on the energy of the phonon which is absorbed and hence the whole spectrum can take part. This mechanism is called a Raman process.

Preliminary calculations for these two processes were first made by Kronig (1939), and Van Vleck (1940) made full computations for τ_{sl} for some iron group salts. At relatively high temperatures (e.g. $90°$ K) high-energy phonons are available and Raman processes are dominant. According to Van Vleck† these should give a relation of the form

$$\tau_{sl} = \text{constant} \times (\Delta\mathscr{E}_2)^6/T^9, \tag{9.45}$$

where $\Delta\mathscr{E}_2$ is the energy difference between the original state and the first *real* excited state. Results at $90°$ K on chromium potassium alum show that in order to get agreement with the theory $\Delta\mathscr{E}_2$ should only be about $100°$ K, instead of being the several thousands of degrees which we expect in the iron group (section 9.10). Unfortunately experiments to determine the actual temperature dependence in this region have not been carried out.

For direct processes Van Vleck† shows that

$$\tau_{sl} = \text{constant} \times (\Delta\mathscr{E})^4/TH^4. \tag{9.46}$$

Experiments at liquid-helium temperature indicate that τ_{sl} varies as $T^{-\frac{3}{2}}$ or T^{-2}, which is vaguely correct, but it does not exhibit an H^{-4} dependence. Thus the theory is not at all satisfactory.

There is one recent series of experiments in which a much better agreement with theory has been achieved (Finn, Orbach, and Wolf, 1961). These have been made on a rare-earth salt, cerium magnesium nitrate, whose first excited state is much closer to the ground doublet than is found in the iron-group salts; i.e. $\Delta\mathscr{E}_2$ is not too much greater than kT. Under these conditions the authors show that a type of Raman process can occur in which a *real* excitation to the excited state occurs, i.e. the phonon which is absorbed *does* have the energy $\Delta\mathscr{E}_2$. τ_{sl} is then of the form

$$\tau_{sl} = \text{constant} \times \exp(\Delta\mathscr{E}_2/kT). \tag{9.47}$$

Experiments between 3 and $1\cdot9°$ K do show that τ_{sl} has this exponential dependence down to about $2°$ K (Fig. 9.14) and that $\Delta\mathscr{E}_2$ has the value $34k$. This agrees very well with the value of the splitting found by specific heat measurements of the Schottky anomaly (section 1.19) which give $\Delta\mathscr{E}_2 = 34\pm1k$ and also from susceptibility measurements which give $\Delta\mathscr{E}_2 = 37\pm3k$.

Whilst this mechanism has so far only been demonstrated experimentally for the Ce ion, it should be applicable to many rare-earth ions since several of them have fairly low-lying states close to the ground state which could act as a 'bridge' state in the relaxation process.

† (9.45) and (9.46) are for Kramers salts only. For non-Kramers salts τ_{sl} is proportional to T^{-7} and $H^{-2}T^{-1}$ respectively.

9.28. *Thermodynamics of paramagnetic substances*

In section 1.19 we described how a specific-heat anomaly of the Schottky type arose when a system has low-lying excited states. It was caused by the changing populations of those states at temperatures where kT was of the order of the energy separation. As we have seen in the earlier sections of this chapter, this is just the kind of situation which arises in paramagnetics. We now wish to show that it is possible

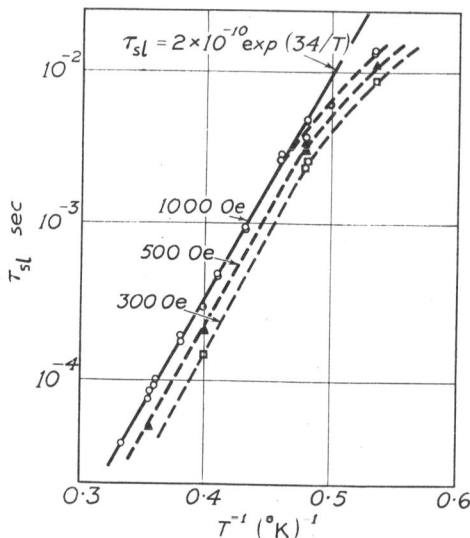

FIG. 9.14. A plot of the measured relaxation time for Ce^{3+} in cerium magnesium nitrate against $1/T$ for three field strengths, 300, 500, and 1,000 oersted. The solid line is a plot of $\tau_{sl} = 2 \times 10^{-10} \exp(34/T)$ against $1/T$ (Finn, Orbach, and Wolf, 1961).

by purely magnetic measurements to determine the anomalous specific heat.

We first note the fact which is shown in most texts on thermodynamics (e.g. Pippard, 1957 a, chapter 6) that for a magnetic substance one may write (ignoring any change in volume)

$$dU = H\,dm + T\,dS, \tag{9.48}$$

where S is the entropy and U is the internal energy plus the energy of the magnetic field. From this relationship one may derive a series of expressions analogous to Maxwell's thermodynamic equations in which the pressure and volume in Maxwell's equations are replaced by H and $-m$ respectively. We now define an isothermal susceptibility

$$\chi_T = \partial m / \partial H |_T$$

and an adiabatic susceptibility $\chi_S = \partial m/\partial H |_S$. These differentials may be written as

$$\chi_T = -\frac{\partial m}{\partial T}\bigg|_H \frac{\partial T}{\partial H}\bigg|_m \quad \text{and} \quad \chi_S = -\frac{\partial m}{\partial S}\bigg|_H \frac{\partial S}{\partial H}\bigg|_m , \text{ respectively,}$$

(9.49)

and hence we obtain

$$\frac{\chi_T}{\chi_S} = \frac{\partial m}{\partial T}\bigg|_H \frac{\partial S}{\partial m}\bigg|_H \bigg/ \frac{\partial S}{\partial H}\bigg|_m \frac{\partial H}{\partial T}\bigg|_m .$$

(9.50)

The denominator of this expression is $\partial S/\partial T |_m$ and this is equal to c_m/T where c_m is the specific heat at constant magnetic moment.

Thus

$$\frac{\chi_T}{\chi_S} = \frac{T}{c_m} \frac{\partial m}{\partial T}\bigg|_H \frac{\partial S}{\partial m}\bigg|_H .$$

(9.51)

We must now find an expression for the right-hand differential, since this cannot be manipulated by Maxwell's equations. We write

$$dS = \frac{\partial S}{\partial m}\bigg|_T dm + \frac{\partial S}{\partial T}\bigg|_m dT,$$

(9.52)

and hence

$$\frac{\partial S}{\partial m}\bigg|_H = \frac{\partial S}{\partial m}\bigg|_T + \frac{\partial S}{\partial T}\bigg|_m \frac{\partial T}{\partial m}\bigg|_H .$$

(9.53)

The first differential on the right-hand side may then be written using Maxwell's relations as $-\partial H/\partial T |_m$ and this is equal to $\partial H/\partial m|_T \, \partial m/\partial T|_H$. Thus (9.53) becomes

$$\frac{\partial S}{\partial m}\bigg|_H = \frac{\partial H}{\partial m}\bigg|_T \frac{\partial m}{\partial T}\bigg|_H + \frac{c_m}{T} \frac{\partial T}{\partial m}\bigg|_H .$$

(9.54)

Putting this expression into (9.51) we obtain

$$\frac{\chi_T}{\chi_S} = \frac{T}{\chi_T c_m} \frac{\partial m}{\partial T}\bigg|_H^2 + 1,$$

(9.55)

and hence

$$c_m = T \frac{\partial m}{\partial T}\bigg|_H^2 \frac{\chi_S}{\chi_T(\chi_T - \chi_S)} .$$

(9.56)

The reader will doubtless be able to find other derivations which are either more or less tortuous than the one given. $\partial m/\partial T |_H$ is taken at that value of H which will produce the constant moment m at a temperature T. Equation (9.56) shows that we can calculate the magnetic contribution to the specific heat from purely magnetic measurements. χ_T and χ_S may both be determined from susceptibility experiments at low and high frequencies respectively, as has been described in section 9.26, and the values of $\partial m/\partial T |_H$ can be found from static or low-frequency magnetic moment or susceptibility measurements. It can easily be shown for an 'ideal' paramagnetic, i.e. one for which m is of the form $m = f(H/T)$, that c_m is independent of the field and so it may therefore be taken as the specific heat in zero field. If Curie's law is obeyed, $m = CH/T$ and hence

$$\partial m/\partial T|_H = -CH/T^2 = -\chi_T H/T. \qquad (9.57)$$

Thus
$$c_m = CH^2\chi_S/\{T^2(\chi_T - \chi_S)\}. \qquad (9.58)$$

It might be thought that this expression contradicts the statement in the previous paragraph that c_m should be independent of field. However, in section 9.26 we have already described how the energy change on slightly increasing the field depends on the value of the original field. From this it follows that, even in an ideal paramagnetic, χ_S will decrease as H increases. This does, in fact, make c_m independent of H.

It should be emphasized that this method of determining the specific heat only yields the magnetic contribution. Since this is usually the quantity of interest, the necessity of making allowance for the lattice specific heat, c_g, which is necessary in the calorimetric method, is avoided.

The reader may well ask, since specific heats can be measured in this rather neat and sophisticated manner, whether straightforward calorimetric determinations are no longer necessary. Unfortunately, they are. At very low temperatures, say, below $1°$ K, the heating effects produced during a high-frequency magnetic measurement are so great that the method has to be abandoned and resort must again be made to calorimetric methods. This, however, is not so disappointing as it may sound, because the advantage of the magnetic method of measurement lies in the higher temperature ranges where a large correction for c_g would normally be necessary. At very low temperatures c_g is so small that almost the entire specific heat which is measured calorimetrically will be that due to the magnetic anomaly and very little correction is needed.

9.29. Paramagnetic resonance

Of all the experiments which have been developed for the investigation of low-lying energy states of paramagnetic ions, the resonance technique (Zavoisky, 1945) is the most powerful. There are very many research groups active in the field and a considerable literature has grown up. An introductory account of the experimental techniques and of the general experimental work is given in the book by Ingram (1955) whilst a more fundamental account of the experiments and of their interpretation is given by Bleaney and Stevens (1953). More details of experimental results have been given in the series of review articles by Bowers and Owen (1955), Bagguley and Owen (1957), and Orton (1959). A comprehensive account of the subject is also given by Low (1960). Much of the work, however, has little specific low-temperature interest and we shall limit ourselves in this section to describing the basic principles of the experiments and the main fields of investigation.

Paramagnetic resonance is generally observed in substances which have a degenerate ground state which can be split into two separate energy levels by an external magnetic field. For simplicity let us consider a degenerate state with zero orbital angular momentum ($L = 0$)

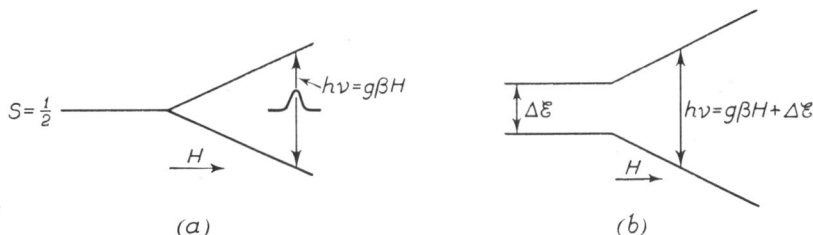

FIG. 9.15. The splitting of a doublet by a magnetic field (a) with no zero field splitting, (b) when the levels are already split by some other interaction. Paramagnetic resonance may be detected when $h\nu = g\beta H$. In (b) it is clear that $h\nu$ must be greater than $\Delta\mathscr{E}$ before resonance will be achieved.

and with $S = \frac{1}{2}$. The application of a magnetic field H will split these levels so that they have an energy separation of $g\beta H$, (9.8). This is shown in Fig. 9.15 (a). These states will be populated in accordance with the Maxwell–Boltzmann distribution function appropriate to the temperature of the specimen, and at equilibrium we may say that, due to interactions with phonons, the number of transitions from the lower to the higher energy state with the *absorption* of a phonon is the same as the number of transitions from the higher to the lower state with the *emission* of a phonon. If, now the substance is placed in a radio-frequency field which has a frequency ν such that

$$h\nu = g\beta H, \tag{9.59}$$

then transitions will be induced from the upper to the lower state and vice versa, with the emission or absorption of energy $h\nu$ respectively. These transitions have equal probabilities. Since, in general, the lower state will initially have the higher population, there will be more transitions to the upper state than to the lower, and so the populations will tend to equalize. This does not usually happen (except in strong r.f fields) because after excitation to the higher state the ions tend to relax to the lower state by *phonon* emission in order to return to the Maxwell–

Boltzmann distribution (see Andrew (1955) p. 13 for further discussion). Thus to a first approximation we can say that the ions are excited by interaction with the r.f. field but they relax by interaction with the phonon system. There will thus be a net absorption of r.f. energy and this can be detected as a drop in the r.f. power which is transmitted through the system.

In practice frequencies of the order of 10^4 Mc/s or higher are used,

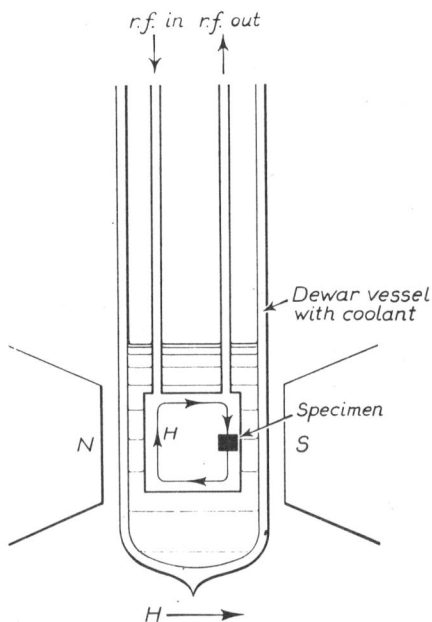

FIG. 9.16. A schematic representation of a paramagnetic resonance experiment. The specimen is placed in a microwave resonant cavity in a position where the r.f. magnetic field is at right angles to the static magnetic field. The actual position of the specimen depends on the cavity geometry and need not be as shown here.

corresponding to wavelengths in the cm or mm range. Most experiments are in fact made using 3-cm equipment ($\sim 9.5 \times 10^3$ Mc/s). This is a standard radar frequency, the equipment is readily available, and the geometry and size of the apparatus is convenient. The specimen is placed in a microwave resonant cavity which is designed so that the radio-frequency magnetic field is at right angles to the steady field H (Fig. 9.16). The frequency is always kept constant and H is varied until an absorption is detected in the resonant circuit. If $g = 2$, then for 3-cm waves a resonance will be observed in a field of about 3,400 oersted. Whilst smaller fields could be used if the experiments were made at lower frequencies, the population of the levels would be more nearly equal and the net absorption would be small. Due to the low frequency the rate of excitation would also be very much less and so the absorption would be even more difficult to observe. In addition to this, if there is any splitting of the levels in zero field, then clearly $h\nu$ must be at least as great as this splitting for resonance to occur (Fig. 9.15 (b)). Inhomogeneities in the internal field (section 9.6) can broaden the levels by a few hundred oersted and hence greater accuracy can be achieved in high H. It is sometimes for these reasons that experiments may be made at frequencies of 36×10^3 Mc/s (8-mm wavelength) or even higher.

9.30. *Low-temperature experiments*

There are two main reasons why low temperatures are very often essential before resonance can be observed. Firstly, there is the effect of the spin-lattice relaxation time, τ_{sl}. We have already seen in section 9.27 that τ_{sl} is very short at higher temperatures. Because of this strong interaction with the lattice vibrations the energy levels become broadened to such an extent that instead of a resonance peak being detected sharply over a small range of H, it is smeared out over a wide region of H (of the order of a few thousand oersted) so as to be almost or entirely undetectable. Very often sharp resonances may be obtained by liquid-hydrogen temperature (20·4° K) but sometimes experiments must be made at liquid-helium temperatures for good resolution. It may also be necessary to use low temperatures for another reason: viz. if there are other energy states which lie fairly close to the one which we wish to investigate, then at higher temperatures these states would also be occupied and might give resonance peaks which would overshadow those due to the lower state. Indeed it is sometimes possible, from the change in the height of the peaks as a function of temperature, to estimate the separation of the sets of states. In some cases resonances may be obtained which are only a few oersted wide, but this requires carefully prepared specimens and of course a very uniform magnetic field must be used in order to attain such high resolution.

9.31. *Methods for determining energy splittings and for identifying the levels*

We now consider the type of information which can be obtained from the experiments. Once the field which produces resonance in a simple doublet state (Fig. 9.15 (a)) has been determined, then g, the splitting factor, may be calculated from (9.59). This, by itself, is important but

after a while it would tend to get a little dull. Fortunately, as we have
seen in the earlier sections of this chapter, nature has provided us with
a whole range of substances whose energy states are by no means as
straightforward as a simple doublet and it is the transitions between
these states which have tended to provide the continuing impetus to
further resonance studies.

There is a very important selection rule which governs the transitions

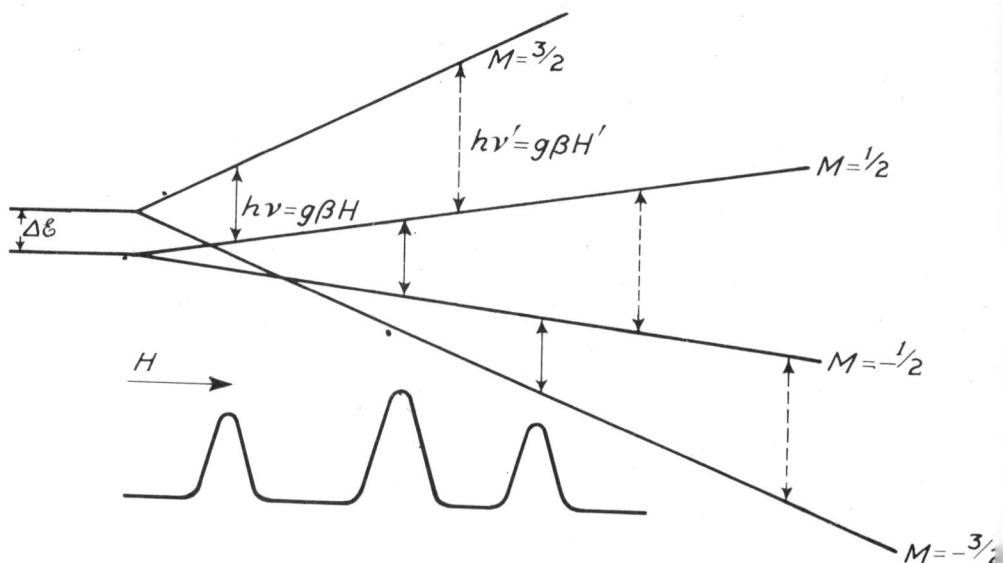

FIG. 9.17. The possible resonance transitions for a system with $J = \frac{3}{2}$ which has a zero-field splitting of $\Delta\mathscr{E}$. The selection rule $\Delta M = \pm 1$ applies. The dotted lines show the fields at which the same transitions would occur if the microwave frequency was higher. Observations at two frequencies enable a less ambiguous interpretation of the results to be made.

which are permitted between the various levels. This rule is that transitions can only take place between states whose values of M differ by
unity, i.e. $M = \pm 1$. Thus if a complicated set of levels exists, resonance
peaks corresponding to transitions between *every* pair of states are not
observed. There will be a smaller number than this and the selection
rule helps to identify the levels which give rise to them.

In Fig. 9.17 a simple example is shown in which the state is characterized by $J = \frac{3}{2}$. This will be fourfold degenerate ($2J+1 = 4$) and
we shall assume that due to the internal electric field it is split by $\Delta\mathscr{E}$
into two doublets which will have M values $\pm\frac{1}{2}$ and $\pm\frac{3}{2}$. When a
magnetic field is applied the degeneracy is completely removed and the
levels will split as is shown on the right-hand side of the diagram. If

the radio-frequency is kept constant, resonance will be obtained for
three values of H between the levels indicated. If one then assumes a
two-level scheme of a type similar to that shown, then from the positions
of the peaks it is possible to calculate M, g, and the original splitting $\Delta\mathscr{E}$.
More complicated spectra can be analysed in a similar way so that the
quantum numbers and the splittings of the states may be found. It will
be obvious that except in the simplest case of a single resonance peak,
it is necessary to have some rough idea as to what the energy-level
scheme might be and it is here that a knowledge of the ionic ground

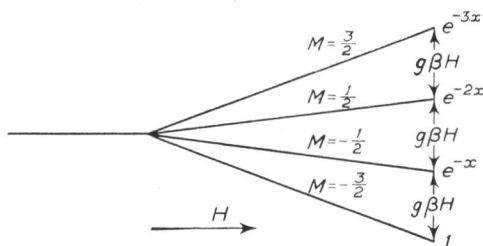

FIG. 9.18. The relative population of a system of four levels split by a magnetic field.

state and the general way in which its terms are split in various fields,
as has been discussed in the earlier sections of this chapter, are essential.

In some cases the analysis of the spectra is simplified by taking
resonance data at two (or more) frequencies, using wavelengths of, say,
3 cm and 1·2 cm. If the levels as a function of field which are shown
in Fig. 9.17 are accurately linear, then by a simple geometrical construction of similar triangles it is quite straightforward to determine both the
level scheme and the zero field splitting $\Delta\mathscr{E}$.

Further information which helps to identify the states between which
the transitions are occurring can be obtained by making measurements
at different low temperatures. This may best be explained by using
a system similar to that shown in Fig. 9.17 but in which for simplicity
there is no zero field splitting. The effect of an external magnetic field
on such a fourfold degenerate state with $S = \frac{3}{2}$ is shown in Fig. 9.18.
The states will be split so that there is an energy difference of $g\beta H$
between neighbouring levels. Three transitions ($-\frac{3}{2}$ to $-\frac{1}{2}$, $-\frac{1}{2}$ to $+\frac{1}{2}$,
and $+\frac{1}{2}$ to $+\frac{3}{2}$) will occur as in Fig. 9.17 except that they would each
appear at the same value of H. We should note, however, that the power
absorbed (i.e. the height of the resonance peak) in each of the three types
of transitions is not equal. This is fairly clear to see if we consider what
happens at very low temperatures where $kT \ll g\beta H$. In this case only

the lowest level will be populated and hence transitions will only occur between $-\tfrac{3}{2}$ and $-\tfrac{1}{2}$. Hardly any will take place between the upper levels because there are so few atoms in the $-\tfrac{1}{2}$ and $+\tfrac{1}{2}$ states. Let us put these ideas on a more mathematical footing. At any temperature T the populations P_M of the three upper levels relative to the bottom one will be in the ratio

$$1 : \exp(-x) : \exp(-2x) : \exp(-3x), \qquad (9.60)$$

where $x = g\beta H/(kT)$. In the temperature range where x is small (i.e. at $4\cdot2°$ K and above for the usual values of H) the exponentials can be expanded so that (9.60) becomes

$$1 : (1-x+x^2/2) : (1-2x+4x^2/2) : (1-3x+9x^2/2). \qquad (9.61)$$

The height of a resonance peak is proportional to the difference in population between the two levels concerned in the transition (the reader will recall that if both levels are equally populated no resonance would be observed). Thus $P_{-\frac{3}{2}}-P_{-\frac{1}{2}}$ will be a measure of the $-\tfrac{3}{2}$ to $-\tfrac{1}{2}$ resonance peak. The heights of the peaks will therefore be in the ratio

$$(x-x^2/2) : (x-3x^2/2) : (x-5x^2/2) \qquad (9.62)$$

for $\qquad -\tfrac{3}{2}$ to $-\tfrac{1}{2} \quad -\tfrac{1}{2}$ to $+\tfrac{1}{2} \quad +\tfrac{1}{2}$ to $+\tfrac{3}{2}$.

We therefore see that as the temperature is reduced and x becomes larger, the height of the peak corresponding to the $\tfrac{1}{2}$ to $\tfrac{3}{2}$ transition will decrease relative to the other peaks. It is obvious that a similar situation will occur if the ground state is split slightly as it was in Fig. 9.17. The three peaks can now be observed separately and the one corresponding to the transition to the highest state can be identified because its magnitude will decrease the most as the temperature is lowered. It should be emphasized that this procedure only identifies the transitions to the highest state; it does not say what particular state this is. Whilst we have taken it to be the $M = +\tfrac{3}{2}$ state in this example it could just as well be any of the other states. This type of experiment is commonly carried out by taking resonance data at $20°$ K and at $4\cdot2°$ K and the decrease in the height of the upper-state peak is not always very pronounced. A much greater change is produced on cooling to $1°$ K although it should be noted that x is then much larger and the short expansion of the exponentials cannot be used.

There is another technique which is helpful in identifying the states and which we give without any theoretical explanation. In resonance experiments some observations are almost invariably made as the external field H is rotated relative to the specimen. The magnet shown in Fig. 9.16 is nearly always mounted so that it can be rotated about a vertical axis. When this is done the positions of the peaks usually shift. To a first order of approximation, however, it can be shown (see, for example, Bleaney and Ingram, 1951) that the peak corresponding to the $-\frac{1}{2}$ to $+\frac{1}{2}$ transition does not shift as the direction of the field is altered and hence this particular peak can be readily identified.·

Besides this rather simple type of observation, the variation in the resonance spectrum as H is rotated yields very important information about the symmetry of the crystalline field. We have already seen in section 9.13 that crystals with a strong axial symmetry show a large anisotropy in their susceptibility. It is not surprising that a similar type of anisotropy is also observed in their resonance spectra. Since the magnitude and type of splitting of the levels depends very sensitively on the strength and symmetry of the internal field, the resonance technique is capable of detecting very small departures from a high-order symmetry. A full analysis of the level scheme is greatly helped, therefore, by measurements which are made as a function of the direction of H. Unfortunately, as we have had occasion to mention before in connexion with the magnetic properties, each·substance is a special case and a detailed account of the results for one crystal would be of very little use as a general guide.

9.32. *Determination of the number of ions per unit cell*

There is, however, one aspect of the anisotropy measurements which is of more generality. In some cases they can be used to determine the number of magnetically inequivalent ions per unit cell in a particular crystal. This can best be illustrated by describing the situation for a copper tutton salt,† e.g. $Cu(NH_4)_2(SO_4)_2.6H_2O$.

Each Cu ion in this crystal has as its nearest neighbours six oxygen ions. Four of these, A, B, C, and D, are the oxygens from the H_2O and they are spaced at equal distances from the Cu ion in one plane, as is shown on the left-hand side of Fig. 9.19. The other two oxygen ions, E and F, are from the SO_4 radical and these lie on a line at right angles to the plane at a slightly greater distance from the Cu^{2+} ion. The ground state of the Cu ion is a doublet with $S = \frac{1}{2}$ and it is clear that the symmetry of the internal field is such that the resonance peak will occur at the same value of H whether it be directed along AB or CD, but that

---

† This is given as a simple example to illustrate the method; the original analysis on the tutton salts was made by Krishnan and Mookherji (1938) from susceptibility measurements.

a different value will be necessary if it is in the direction EF. For the salt in question, however, the arrangement of the oxygen atoms around the Cu changes from one ion to the next. As Fig. 9.19 shows, for the right-hand Cu ion the oxygens have been rotated by 90° about the AB axis so that the ions equivalent to EF are now horizontal. It is evident if H is in the direction AB, that resonance will occur at the same value of H for both ions, because this direction is equivalent for both of them (at right angles to the SO_4 axis). If, however, H is directed along say, EF, this is not an equivalent direction for the two ions and hence two

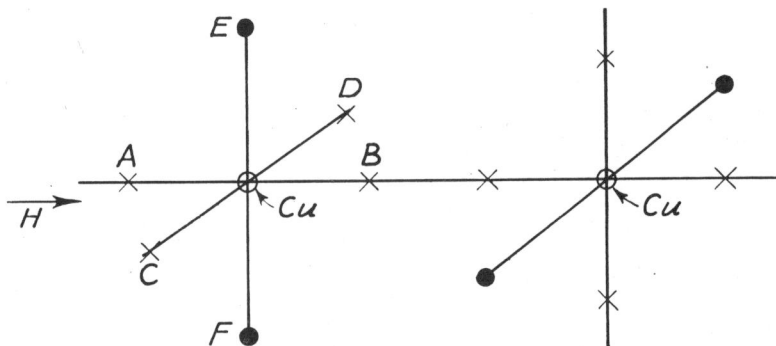

Fig. 9.19. Two magnetically non-equivalent ions in copper tutton salt. The oxygen ions shown as dots (E and F on the left-hand section) are slightly farther away from the Cu ion than are the other four oxygen ions (crosses, A, B, C, and D). In the right-hand section the dots are in the horizontal plane. Thus a field in the horizontal plane when rotated about the vertical axis will have a different effect on each of the two copper ions.

resonance peaks will be observed at different values of H. A similar situation occurs if H is along CD. From such experiments one can deduce that there are two inequivalent paramagnetic ions per unit cell.

9.33. *Experiments on other paramagnetic systems*

So far we have only discussed resonance experiments on paramagnetic ions. However, if any other system contains unpaired electrons, it will in principle be paramagnetic and hence there is the possibility of susceptibility and resonance experiments. Such a situation often occurs when defects have been introduced into a crystal, such as those which occur after various types of irradiation (e.g. by X-rays, gamma-rays, electron or neutron bombardment). We give no details of these experiments here (see Bagguley and Owen, 1957, for a useful review) except to say that a large amount of resonance data has been taken on the various types of defect centre (e.g. F, V, and U-centres). While many of these spectra may be observed at room temperature, low-temperature experiments are often necessary as some of the defects are mobile or unstable at higher temperatures and so they might anneal. Electron spin resonance has also been observed from the conduction electrons in some metals (particularly in the alkali metals), and from the loosely bound electron associated with a donor element in silicon (section 7.2). This is not to be confused with the very much more extensive cyclotron resonance experiments in semiconductors (section 7.21).

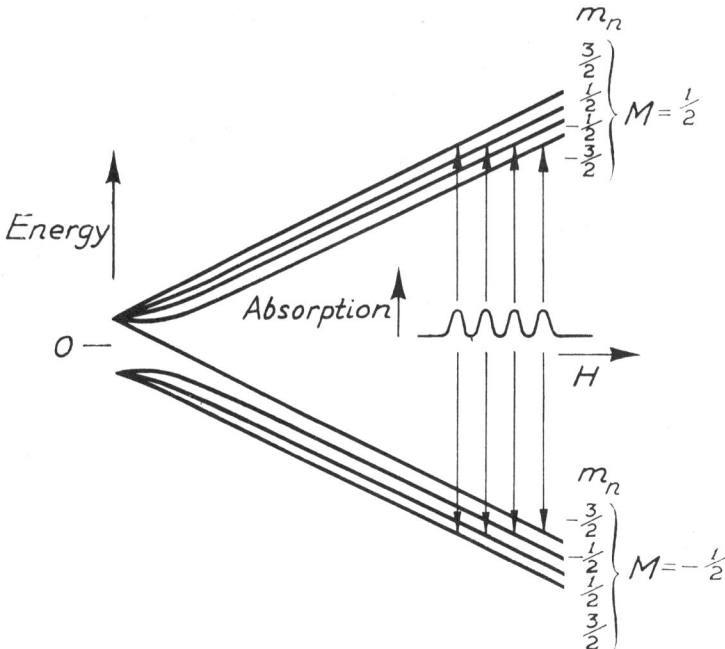

FIG. 9.20. Hyperfine splitting for a system with $S = \frac{1}{2}$, $I = \frac{3}{2}$. In a conventional resonance experiment four peaks will be observed corresponding to the transitions for which $\Delta m_n = 0$, $\Delta M = \pm 1$ (Bowers and Owen, 1955).

9.34. Hyperfine structure

One very important type of interaction which may be detected in paramagnetic resonance spectra is that due to the interaction of the nucleus with the electron spins, which we described in section 9.17. In a high magnetic field such as is used in paramagnetic resonance, a nucleus with a spin characterized by a quantum number I will split each electron state into $2I+1$ equally spaced sublevels. These can be labelled with a nuclear magnetic quantum number m_n which will run from $I, I-1,..., -I$. A diagram of the splitting for $S = \frac{1}{2}$ and $I = \frac{3}{2}$ is shown in Fig. 9.20. In order that a transition should occur, the selection rule which must be obeyed is that *either* ΔM *or* $\Delta m_n = \pm 1$. Since we observe paramagnetic resonance for $\Delta M = \pm 1$ this means that $\Delta m_n = 0$. Thus transitions will occur between the four pairs of states shown in the diagram. In general these will each require a different field H and so the single resonance peak for the $M = -\frac{1}{2}$ to $+\frac{1}{2}$ transition will have a hyperfine structure consisting of four closely spaced resonances. From this type of observation it is a very simple matter to determine the nuclear spin, I. We just count the number of peaks in the hyperfine

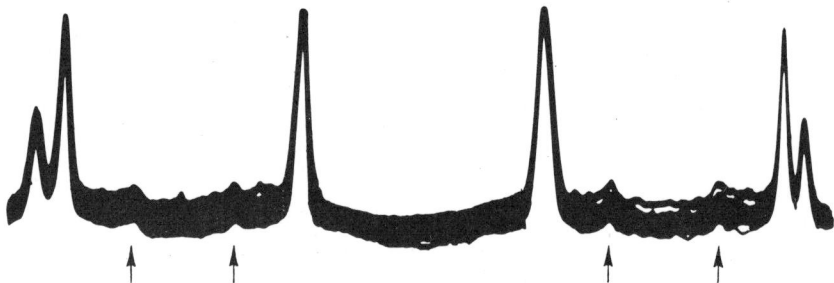

FIG. 9.21. Hyperfine structure in copper potassium sulphate diluted $1:200$ with zinc potassium sulphate (deuterated salt). The four main peaks arise from the splitting of each electronic level into $2I+1$ states (for Cu, $I = \frac{3}{2}$). The doubling of the outermost peaks is due to the less abundant isotope Cu^{65} which will have slightly different interactions from the more abundant Cu^{63}. The effect of the Cu^{65} is not resolved in the two central peaks. The very small resonances indicated by the four arrows are due to 'forbidden' transitions in which $\Delta m_n = \pm 1$. The crystals were grown from heavy water in order to reduce the broadening by the field of the hydrogen nuclei, since the nuclear magnetic moment of deuterium is one-third that of hydrogen (Bleaney, Bowers, and Ingram, 1951).

structure and this should be equal to $2I+1$. The spins of several nuclei have been determined in this manner. Due to the broadening of the resonance peaks at higher temperatures the HFS is very often not resolved until the specimen has been cooled down. Fig. 9.21 shows an example of the nuclear HFS in copper potassium sulphate. This picture is rather interesting because each of the outer peaks is double. This occurs because natural copper contains the isotopes Cu^{63} and Cu^{65}. Both of these have $I = \frac{3}{2}$ which would give rise to four peaks, but since the interaction between the nucleus and the electron spins is slightly different for the two isotopes, the HFS due to one isotope is shifted slightly relative to that from the other. This difference is resolved in the outer peaks.

We should note in passing that the electron–nuclear interaction can also be affected by the nuclear quadrupole coupling (section 9.18). This will modify the spacing of the sublevels and it can also give rise to transitions such as $\Delta m_n = \pm 1, \pm 2$, which are normally forbidden. Very small peaks from such transitions can be seen in Fig. 9.21.

9.35. *Double resonance experiments*

Before concluding this necessarily brief account of electron spin resonance the powerful technique known as double resonance should be described (Feher, 1956). As the reader will probably appreciate from the preceding section, the investigation of the nuclear HFS, whilst very simple in principle, can be very difficult to observe in some substances because the HFS cannot be resolved in the conventional experiment if the electron–nucleus interaction is very small.

The double resonance technique can be used to investigate HFS in substances which have a long spin-lattice relaxation time, τ_{sl}. In the discussion on general principles (section 9.29) we pointed out that whilst the change from a low energy state to a higher one was usually accomplished by the absorption of a *photon* from the r.f. field, the reverse process took place by the emission of a *phonon*. If, however, τ_{sl} (section 9.27) is very long, then the transition probability to the lower state will be small and with sufficient r.f. power it is then possible to equalize the populations of the levels. When this occurs the r.f. field will induce as many transitions up to the higher state as it does down to the lower one. The net power absorbed will be zero and so the resonance signal will vanish. This effect is usually called saturation. It will be more readily achieved at very low temperatures where τ_{sl} is larger. There are many substances in which the spin-lattice relaxation time can be of the order of a minute or more and in these it is very easy to produce saturation with quite a small r.f. power input.

Let us consider a simple system with $S = \frac{1}{2}$ and $I = \frac{1}{2}$ in which spin resonance is made to occur between the $M = +\frac{1}{2}$ and $-\frac{1}{2}$ levels in a field H. Due to the nuclear spin, the level scheme will be like that shown in Fig. 9.22. The nuclear interaction splits the upper level, $M = +\frac{1}{2}$ into two levels A and B which are characterized by $m_n = +\frac{1}{2}$ and $-\frac{1}{2}$ respectively. Similarly the level for $M = -\frac{1}{2}$ is split into the states A' and B'. In a double resonance experiment saturation of, say, the A' to A transition is first achieved by conventional paramagnetic resonance techniques. Another much lower radio-frequency field (of the order of 50 Mc/s) is then applied to the system and its frequency ν' is adjusted until it has the correct value to induce transitions between the two upper states A and B as is shown by the arrow in the figure. When the correct frequency has been reached the transitions from A to B will slightly depopulate level A. The original saturation condition is therefore removed and a paramagnetic resonance signal corresponding to the A' to A transition will once again be detected. Thus this resonance signal is used as a very sensitive tuning indicator for the oscillator which is producing the A to B transition. If the energy difference between these two states is $\delta\mathscr{E}$, then of course $\delta\mathscr{E} = h\nu'$. In this way the hyperfine splitting may be investigated with considerable precision and it can be used for samples in which the HFS could not be resolved by ordinary spin-resonance experiments. Whilst we have given a simple example where $I = \frac{1}{2}$, it can, of course, be used in cases where there are any number of HFS

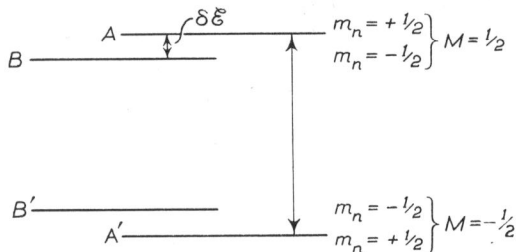

FIG. 9.22. A simple level scheme for a double resonance experiment. The AA' transition is first saturated and then a much lower r.f. is applied to produce the AB transition. When this occurs A is depopulated and hence the AA' transition will be enhanced.

levels. The reader will realize that the double-resonance technique is actually a sophisticated nuclear resonance experiment.

9.36. *The maser*

Whilst the previous sections have indicated that paramagnetic resonance can be used as a very powerful tool for investigating such fundamental properties as the crystal field and nuclear spin, there is another development of a more technological nature which we shall discuss briefly. This is the invention of the device known as a maser which is essentially a high-frequency amplifier with a very low noise figure. The amplification is achieved by stimulating a transition from a higher energy state of a system to a lower one with the consequent emission of a photon. The term 'maser' was first introduced in a paper by Gordon, Zeiger, and Townes (1954) and comes from the initial letters of Microwave Amplification by Stimulated Emission of Radiation. The maser which these authors described utilized transitions between the energy states in a beam of ammonia molecules, but most of the recent research has been made using a paramagnetic crystal as the working substance

(Bloembergen, 1956). It is this type of maser which we shall describe briefly. A good introduction to the various principles involved in successful maser design is given by Schulz-Du Bois (1960).

Let us consider a paramagnetic crystal which has three low-lying levels (Fig. 9.23), A, B, and C, starting from the lowest level. We shall denote the frequency of the r.f. field which is required to induce transitions between any two levels by ν_{AB}, etc. In maser operation a radio-frequency ν_{AC}, the 'pump' frequency, is applied to the specimen at a power sufficient to saturate the levels A and C, i.e. their populations are equalized (section 9.35). If we look at levels B and C, then, provided the temperature is not high compared with the level spacings, C now has a population which is greater than B. If therefore another at r.f. ν_{BC}, which we shall call the signal frequency, is also fed into the system, then as in an ordinary paramagnetic resonance experiment, transitions will be induced from B to C and vice versa, but because C has a

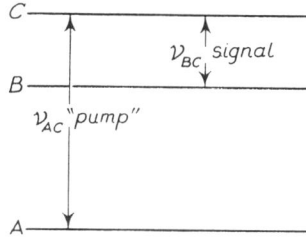

FIG. 9.23. A simple level scheme for a maser. R.f. at the pump frequency overpopulates C compared with B. When a signal of frequency ν_{BC} is fed in, there is a net emission of r.f. at ν_{BC}.

higher population than B there will be a net *emission* of radiation at the signal frequency instead of the absorption which is observed in the normal type of experiment. It can be shown that this emission is in phase with, and is proportional to, the strength of the incoming signal. Thus the system acts as an amplifier.

The main advantage of a maser is that since it almost of necessity has to be operated at a very low temperature, there is very little noise generated in the device. Schulz-Du Bois (1960) quotes a noise figure of about $10°$ K. Thus the maser should prove useful in amplifying signals which come from a low noise medium and in particular this means from space. Depending on the frequency, the noise temperature of the earth's atmosphere can be of the order of a few degrees absolute if the angle of elevation of the antenna is more than about $10°$. For this reason, the main applications of masers have been made in radio astronomy and in satellite communication. It should be noted that the low noise figure of a maser does not justify its use in every situation when a weak signal has to be detected, unless the noise figure for the transmission medium is very low. In nearly every case, except where reception comes from a relatively high elevation to the horizon, the noise figure of the medium is about 300 degrees absolute. The use of a maser with its obvious technical complications is then unjustified because the amplification could be achieved just as satisfactorily with a device (e.g. a parametric amplifier) having a noise figure closer to that of the medium.

We should also note that a maser can be made to oscillate and so it may be used as a high-frequency microwave generator. It can also be used as a frequency standard if the energy levels are sufficiently sharp. The original type of ammonia maser could achieve a frequency stability of 1 part in 10^{12}.

The practical realization of the solid state maser has been achieved by using synthetic ruby, i.e. synthetic corundum, Al_2O_3, to which a small amount of chromium has been added, as the paramagnetic agent. Whilst the Cr^{3+} ions have four low-lying levels and not three as shown in Fig. 9.23, only three of them are used and the mode of operation follows the principles we have already described. The fact that the level scheme seems satisfactory is not the only criterion for successful maser operation. The device will only operate satisfactorily if the energy differences between the levels and the spin-lattice relaxation times of the various levels are such that the population of C is always kept higher than B. These parameters can be chosen to some extent by varying the angle between the applied magnetic field and the crystal axis. For further details the reader is referred to the review article already quoted.

9.37. An assessment of the various methods for determining magnetic data

If the reader has been sufficiently diligent as to stay the course of this chapter up to this point, then in all probability his mind is awash with the possibility of so many types of experiments on paramagnetics that he does not know which straw to cling to first. In this section, therefore, we survey the different techniques which are available and we try to give some idea as to when one or the other is used. In principle we need to know the states corresponding to the various energy levels, their degeneracy and their spacings relative to one another, both in zero and in an applied magnetic field.

There are three main types of experiment which can help to yield this information: resonance, susceptibility, and specific-heat measurements. Of these, resonance data are by far the most useful. As we have seen in the preceding sections, they can be used to determine the zero field energy levels of the low-lying ionic states and also their behaviour in an external magnetic field. To observe resonance, however, it is necessary to have a system in which there is at least a degenerate doublet present which is populated at the temperature of the measurement. This must be of such a state that in an external field the selection rule $\Delta M = \pm 1$ is able to operate so that transitions can occur. Not all paramagnetics show this behaviour. Outstanding examples are those ions with an even number of electrons whose energy states will usually be split so that a singlet level is lowest. These, as we have seen in section 9.11 exhibit temperature-independent paramagnetism at low temperatures. Thus resonance should not be observed, although sometimes it can be detected due to the presence of neighbouring higher states to which transitions can be made.

It is in cases for which no resonance can be found, and also in others where there is some doubt as to the correct interpretation of the resonance spectra, that susceptibility and specific-heat measurements are very useful. A determination of the susceptibility as a function of temperature will show whether the substance is a fairly well-behaved paramagnetic in the region covered, whether it becomes anti-ferromagnetic, or whether it shows a temperature-independent paramagnetism. A departure from a strict Curie law could indicate that the paramagnetism was due to more than one set of levels and that depopulation of the higher set was being observed (section 9.8). The population of higher levels might occur at a temperature which is too high for their resonance to be detected, and hence their existence can only be inferred from the susceptibility measurements. The magnitude of the susceptibility, (9.15), can then be used to calculate $g^2 J(J+1)$ or $g^2 S(S+1)$. In single crystals the anisotropy (if any) of the susceptibility will give an indication of the crystalline field symmetry.

The specific heat of paramagnetics shows a Schottky-type anomaly which was discussed in detail in section 1.19. This can be measured calorimetrically or magnetically (section 9.28). If it is possible to take measurements over the entire region of the anomaly then the entropy can be found by integrating the specific-heat curve and this should be equal to $R \log_e Z$ entropy units per mole (1.39) where Z is the number of levels which have been occupied in the temperature range covered by the anomaly. If only two levels are involved then their spacing may be obtained either from the temperature at which the maximum of the anomaly occurs (Table 1.5, p. 26), or from the exponential increase in the specific heat on the low-temperature side of the anomaly using (1.35). If there are more than two levels concerned, then this exponential increase will give the spacing of the two lowest levels, since, of course, these will be the ones which will be populated first. We also saw that the high-temperature side of the Schottky anomaly is proportional to $1/T^2$ and the coefficient is again a function of the splitting (1.33). Since, however, this coefficient might depend on more than two levels, the interpretation is not so reliable. If there were two groups of levels separated from one another, then, of course, there might be either two distinct Schottky anomalies, or if the levels were closer together the two peaks would be absorbed into a single broadened one.

Thus specific-heat measurements can be used to determine possible level schemes for the material. From these one can predict the susceptibility using (9.10) and they can then be tested by comparison with the experimental values of the susceptibility.

This discussion shows that specific-heat and susceptibility measurements are very useful methods of deriving information about magnetic states which for one reason or another cannot be obtained from resonance data. The information they give is in general far less precise and tends to be much more in the nature of an averaging (particularly in the case of the specific heat, where no anisotropic effects can be determined). Nevertheless, facilities for their measurement are a necessary adjunct to any complete investigation of magnetic properties.

9.38. Selection of suitable salts for adiabatic demagnetization

Whilst nowadays there are many fields of study in which a knowledge of the magnetic properties plays an important part, the main reason for so much of the early low-temperature work on paramagnetics was to select suitable salts as the working substance in magnetic-cooling experiments. We do not give any details of these techniques as they can be found in any standard text (see, for example, Zemansky, 1957, p. 347; de Klerk, 1956; or Mendoza, 1961).

The principle of the experiments is well known. A paramagnetic salt, cooled to about $1°$ K, is placed in a magnetic field of several thousand oersted. This aligns the magnetic dipoles in the salt, thereby reducing the magnetic entropy. This entropy is transferred to the lattice and so the salt tends to warm up, but it is kept cold by being in thermal contact with the $1°$ K helium bath. When all the heat of magnetization has been removed the salt is thermally isolated and then the magnetic field is reduced to zero. The dipoles will then tend to mis-orient themselves again, i.e. their entropy increases. This can only be done at the expense of the lattice entropy. Thus the salt cools.

The process is usually illustrated by a diagram such as that shown in Fig. 9.24 in which the entropy in zero field and in a field H is shown as a function of temperature. The salt starts at a temperature T_1 in zero field (A). The field H is applied isothermally so that the state B is reached. It is then removed adiabatically and thus the salt reaches state C at a temperature T_2. After demagnetization the temperature of the salt is deduced from the measured value of its susceptibility.

What are the main considerations in choosing a salt for cooling?

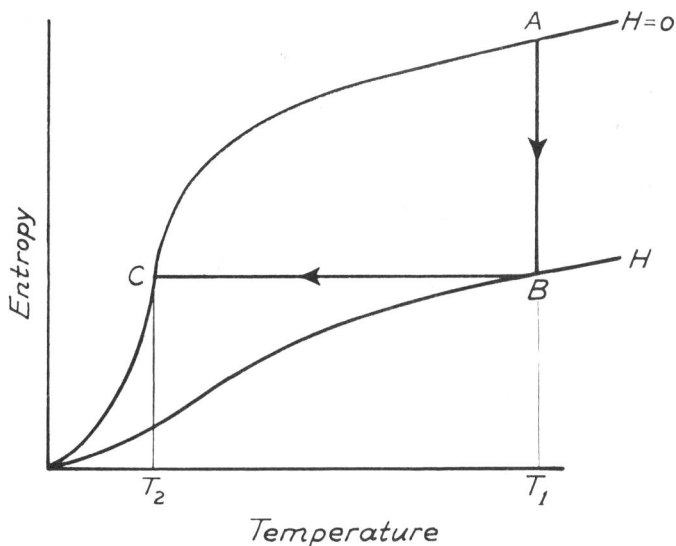

Fig. 9.24. The variation of entropy with temperature for a paramagnetic salt in zero field and in a magnetic field H. In the magnetic cooling process the salt is first magnetized isothermally at T_1 (A to B) and then demagnetized adiabatically (B to C) so that it attains a final temperature of T_2.

The first is that it should be able to be cooled to the required temperature region in a readily attainable field. Secondly, depending on the type of experiment, it should either, (a) stay at that temperature for a long time or (b) it should warm up at a reasonable rate to cover the temperature range to be investigated. Thirdly, the relationship between the susceptibility of the salt at the absolute temperature should be known with some certainty. In particular, the closer the susceptibility obeys a Curie–Weiss (9.25) or a Curie law (9.18) the easier can the temperature be calculated.

Let us consider the entropy curves shown in Fig. 9.24 in a little more detail. In zero field at very low temperatures the entropy has a very small value and it then rises rather sharply over a fairly narrow temperature range. Since the entropy is given by $\int (c/T)\, dT$, where c is the specific heat, this rapid rise in the entropy corresponds to a high specific heat over this same small temperature region. This specific heat is, of course, just the Schottky anomaly which we have discussed previously (sections 1.19 and 9.37) and which is due to the change in population of the ground-state levels. It will be recalled that whilst we tend to think of these as being truly degenerate and capable of being split only by an externally applied field, this is never quite true. There will always be some interaction which will give rise to a Schottky anomaly.

It will be appreciated from Fig. 9.24 that since the entropy in zero field falls rather rapidly in the neighbourhood of T_2, it would be very difficult to attain temperatures much lower than T_2 unless the original magnetic field was increased by a very large amount. A small increase would make very little difference to the final temperature which was achieved.

Thus on demagnetization the temperature which is reached is dictated by the position of the specific-heat anomaly of the particular salt which is being used. This is a very fortunate occurrence, if we do wish to remain at the low temperature for as long as possible, for the following reason. At the temperatures attained by demagnetization (say between 10^{-3} and $10^{-1}\,^{\circ}\mathrm{K}$) the specific heat of most substances is so low that very small heat influxes to a specimen will warm it up very rapidly. By the mere act of demagnetization, however, we tend to reach a temperature where the specific heat of the salt is quite high, and hence, providing the heat influx is not too large, the salt (and, of course, any specimen attached to it) will remain at the low temperature for some considerable time. In fact it is after the salt has been heated up to a temperature where it is outside the range of the Schottky anomaly, that the warm-up rate becomes rapid. It is one of the few occasions when nature actually assists the experimenter in maintaining a rather difficult situation once it has been achieved.

If the experiment is one in which a range of temperatures has to be covered, then it is usually more convenient to use a salt which has the Schottky anomaly some way below the lowest temperature which is required. It is then demagnetized from a rather low field in order that the final temperature reached is higher than that at which the anomaly occurs. This has two advantages. Firstly the warm-up starts immediately after demagnetization, and secondly, because the salt is at a temperature some way above the anomaly, the susceptibility will obey a Curie or Curie–Weiss law much more accurately than in the neighbourhood of the anomaly. This enables its temperatures to be determined with far greater precision.

These considerations lead us to the following requirements for the salt.

In order for it to obey a Curie–Weiss or a Curie law to as low a temperature as possible it should have a ground state well separated from any higher states and it should not undergo any co-operative transition (e.g. to anti-ferromagnetism) in the temperature region to be investigated. Its ground-state splitting should be of about the same magnitude as the temperature we wish to attain, if we require that once the salt has cooled to that temperature it stays there. If we wish to do warming-up experiments the splitting should be less than the minimum temperature required. Thus in order to reach very low temperatures, salts with a small ground-state splitting should be used. To some extent, as we have seen earlier in this chapter, the interaction can be reduced by diluting the paramagnetic ions with a diamagnetic substitute. However, there is a limit to the amount of dilution which can be effective and yet still leave enough dipoles for sufficient entropy to be extracted from the system on the initial magnetization. If the ion has a nuclear spin, then the nuclear interactions will eventually determine the splittings, and further dilution would be pointless.

It is clear that fairly simple behaviour will be obtained from ions which have no resultant orbital momentum, i.e. those which are in an S state, because the internal field has little direct effect on the spin levels. These ions are Fe^{3+}, Mn^{2+}, and Gd^{2+}. In addition to these there are the salts of the iron group in which the orbital momentum is quenched by the internal field and so the ground state can be considered as an S state. These are salts of the ions of Ti^{3+}, Cr^{3+}, and Cu^{2+}. In addition one should also add Ce^{3+} which has only one electron in the $4f$ shell and so has a degenerate doublet ground state.

In Table 9.4 we list some of the more common salts which are used in demagnetization work. The approximate temperature which can be reached by demagnetizing from about $1°$ K and the field which is necessary is also listed, together with the lowest temperature to which a Curie–Weiss law holds.

TABLE 9.4

Salt	Demagnetizes to °K	Field k oersted	Curie–Weiss law to °K
Manganous ammonium sulphate (Tutton salt)	0·1 to 0·2	8	0·4
Gadolinium sulphate	0·25	8	0·6 to 0·7
Ferric ammonium sulphate (alum)	0·04 to 0·05	8	0·08
Chromium potassium sulphate (alum), diluted 1:13	0·01 to 0·02	14	0·1
Cerium magnesium nitrate. N.B. anisotropic —if used powdered not all the salt will cool down	0·003 to 0·005	15	Very good Curie law to 0·007
Chromium methyl-ammonium sulphate	0·06	12	0·3

THE SUSCEPTIBILITY OF METALS—THE
DE HAAS–VAN ALPHEN EFFECT

10.1. Introduction

THE susceptibility of a metal may be considered to be due to the sum of the effects of three main mechanisms: (a) the diamagnetism of the ion cores, which will be of exactly the same form as that given in (9.5); (b) the weak 'Pauli' paramagnetism of the conduction electrons which arises because the application of a magnetic field will enable their spins to become aligned; (c) the diamagnetism of the conduction electrons which is a result of the fact that the applied field will also force the electrons into a helical path.

In general none of these effects is very large and they each give a contribution to the susceptibility of about the same order of magnitude ($\sim 10^{-6}$ to 10^{-7}). The Pauli paramagnetism is only slightly dependent on the temperature and we shall not consider it in detail because it is treated in all the standard texts (e.g. Mott and Jones, 1936, p. 184). It gives rise to a volume susceptibility χ_p of the form

$$\chi_p = \beta^2 F(\mathscr{E}_F)\left\{1 + \frac{\pi^2}{6}(kT)^2 \frac{d^2\log_e F}{d\mathscr{E}^2}\bigg|_{\mathscr{E}_F}\right\} \tag{10.1}$$

where $F(\mathscr{E})$ is the density of states per unit volume at energy $\mathscr{E}$, $\mathscr{E}_F$ is the Fermi energy, and β is the Bohr magneton. The diamagnetic susceptibility of the conduction electrons χ_d is of considerably more interest, however. The magnitude of χ_d was first calculated by Landau (1930) who showed that provided $kT > g\beta H$, then for free electrons, χ_d was exactly one-third of χ_p. The overall susceptibility of the metal might therefore remain positive if the diamagnetism of the core was not too large. If, however, the free-electron approximation is not applicable, and in particular, if the effective number of conduction electrons is very small, then the Landau diamagnetism is increased so that it plays the major part in determining the susceptibility of the metal. This is shown most strikingly in the large diamagnetic susceptibility of bismuth ($\sim 10^{-5}$ per unit volume at room temperature) which is ascribed to the small number of free electrons.

As the temperature is reduced χ_d increases, and what is more important, it is found that it is not independent of the applied magnetic field, but that it varies periodically about a mean value as the field is increased. This effect, which was first observed in bismuth single crystals by de Haas and van Alphen (1930) at liquid hydrogen temperatures, is called after the discoverers. The variation in χ_d has a constant period in $1/H$ and the amplitude of the oscillations increases as the temperature is reduced. The first explanation of the effect was given by Peierls (1933) who showed that when the treatment of the Landau diamagnetism was extended to high fields and low temperatures it predicted a periodic variation of the susceptibility similar to that which was observed.

The importance of the phenomenon, as we shall see, lies in the fact that it enables us to obtain rather direct information about the geometry of parts of the Fermi surface. For many years, however, it proved possible to observe the effect only in polyvalent metals in which the shape of the Fermi surface was so complicated that the interpretation of the results was not very conclusive. More recently, however, improved techniques in specimen preparation and in achieving higher fields have enabled de Haas–van Alphen oscillations to be obtained in Cu, Ag, and Au; these experiments have shown the importance of the phenomenon in the study of metals.

10.2. Theory

In order to explain how the oscillations arise we follow the treatments given by Shoenberg (1957) and Chambers (1956). The original Peierls's theory is described in Mott and Jones (1936), p. 214.

When an electron moves with a velocity $\mathbf{v}$ in a region where there is a magnetic field $\mathbf{H}$, it experiences a force $\mathbf{F}_H$ which is given by the expression

$$\mathbf{F}_H = (\mathbf{v} \times \mathbf{H})e/c. \qquad (10.2)$$

From (10.2) we see that if $\mathbf{v}$ originally had no component in the direction of $\mathbf{H}$ the electron would rotate in a closed orbit. If, as will usually be the case, there is a non-zero component of the velocity in the direction of the field, then the electron path will be a helix whose axis lies parallel to $\mathbf{H}$. Since when an electron is in such a helical orbit the effect of the magnetic field only alters the direction of $\mathbf{v}$ and not its magnitude, the energy of the electron is unchanged as it travels round the helix. Thus when we consider how the wave vector $\mathbf{k}$ is affected by the application of $\mathbf{H}$ we see that $\mathbf{k}$ must change in such a manner that its successive values lie on a surface of constant energy in $\mathbf{k}$ space. It is in fact quite straightforward to determine the locus of **k**. From section 4.3 we saw that

$$\mathbf{F}_H = \hbar\, d\mathbf{k}/dt \equiv \hbar\dot{\mathbf{k}}, \tag{10.3}$$

and hence during any time, dt, the change in **k** is in the same direction as the force on the electron. Thus **k** rotates in a plane which is perpendicular to **H**. Therefore, in the elementary case of a spherical Fermi surface, a wave vector with the maximum value $\mathbf{k}_0$ will, in a magnetic field, describe an orbit which just keeps it on the Fermi surface as is shown in Fig. 10.1.

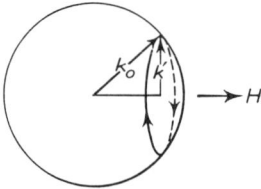

FIG. 10.1. The orbit of an electron in k-space under the influence of a magnetic field **H**. The electron remains on a constant energy surface and the radius of the orbit projected on to a plane perpendicular to **H** is k' (10.6).

There is a simple relationship between r, the radius of curvature of the electron path in real space, and k', the radius of the orbit in k-space. We first note that for the component of the velocity, v', in a direction perpendicular to H, we can write

$$v' = dr/dt \equiv \dot{r} \tag{10.4}$$

and hence, from (10.2) and (10.3),

$$\dot{\mathbf{k}} = (\dot{\mathbf{r}} \times \mathbf{H})e/(\hbar c), \tag{10.5}$$

thus the changes in $\mathbf{k}'$ are in a direction perpendicular to those in **r** and on integrating we see that they are connected by the relationship

$$\mathbf{k}' = \mathbf{r} \times \mathbf{H}e/(\hbar c). \tag{10.6}$$

Thus the orbits in k space have the same shape as those in real space but they are rotated by 90° and are scaled by the factor $He/(\hbar c)$.

10.3. *Quantized states*

The essential feature of the theory of the de Haas–van Alphen effect is the fact that the orbital motion of the electron in the magnetic field must be quantized in a manner analogous to the quantization of the electron orbits in the Bohr theory of the atom. This quantization leads to a restriction on the permitted values of k'. Whilst in the simple Bohr theory the quantization requirement is that the angular momentum must be an integer $\times \hbar$, the generalized condition when a magnetic field is present is

$$\oint (\hbar\mathbf{k}' + e\mathbf{A}/c).d\mathbf{s} = (n+\gamma)\hbar, \tag{10.7}$$

where **A** is the vector potential of the field ($\mathbf{H} = \operatorname{curl}\mathbf{A}$), e has the sign and magnitude of the electronic charge and the integral is taken round a complete orbit in real space. γ is a constant which for a free electron is equal to $\frac{1}{2}$.

From the definition of the vector potential

$$\oint \mathbf{A}.d\mathbf{s} = \mathcal{O}H, \tag{10.8}$$

where $\mathcal{O}$ is the area of the electron orbit in real space, and on substituting for $\mathbf{k}'$ from (10.6) we obtain

$$\oint \hbar\mathbf{k}'.d\mathbf{s} = -\frac{He}{c}.\oint \mathbf{r}\times d\mathbf{s} = -2\frac{He}{c}\mathcal{O}. \tag{10.9}$$

Thus (10.7) becomes (neglecting the negative sign)

$$He\mathcal{O}/c = (n+\gamma)h. \tag{10.10}$$

If $\mathscr{A}(\mathscr{E})$ is the area of the orbit in k-space then from (10.5)

$$\mathcal{O} = (\hbar c/He)^2\mathscr{A}(\mathscr{E}), \tag{10.11}$$

and hence (10.10) becomes

$$\frac{\hbar^2 c}{He}\mathscr{A}(\mathscr{E}) = (n+\gamma)h, \tag{10.12}$$

i.e.

$$\mathscr{A}(\mathscr{E}) = \frac{2\pi He}{\hbar c}(n+\gamma), \tag{10.13}$$

where $\mathscr{A}(\mathscr{E})$ is enclosed by a surface of constant energy $\mathscr{E}$. Thus (10.13) defines the possible energies of the system and hence it also limits the values of k'. The orbits in k-space will be spaced so that the areas between one orbit and the next differ by $2\pi He/(\hbar c)$. We should emphasize that it is only the energy which is associated with motion perpendicular to $\mathbf{H}$ which is quantized. The motion parallel to $\mathbf{H}$ is unaffected by these conditions. Thus if $\mathbf{H}$ is directed along the z-axis only k_x and k_y will be restricted by (10.13).

10.4. *The effect of increasing the magnetic field*

If therefore we have a system which in zero field has a spherical Fermi surface, then when $\mathbf{H}$ is applied along the z-axis the permitted values of k_x and k_y will be such that $\mathbf{k}$ lies on the surfaces of a set of concentric cylinders whose axis is along the z direction (Fig. 10.2). The spacing between the cylinders will be given by (10.13). The permitted levels will be highly degenerate since they must now be able to accommodate all the electrons which in zero field would have occupied the states between the levels. The states will therefore be distributed in a manner similar to that shown in Fig. 10.3 (a). Some electrons will have their energy raised and others lowered, so that on average the effect of quantization does not affect the total energy very much. Nevertheless, as H is increased the cylinders in Fig. 10.2 will expand. The permitted

levels will move to higher energies and their spacing will become larger.
Over most of the electron distribution the effect on the total energy will
be negligible. This is essentially because k_z varies only gradually from
one level to the next and hence in the rearrangement of energies as H
increases, there will be as many electrons which will be forced to enter
states of slightly higher energy as there are those which will be able to

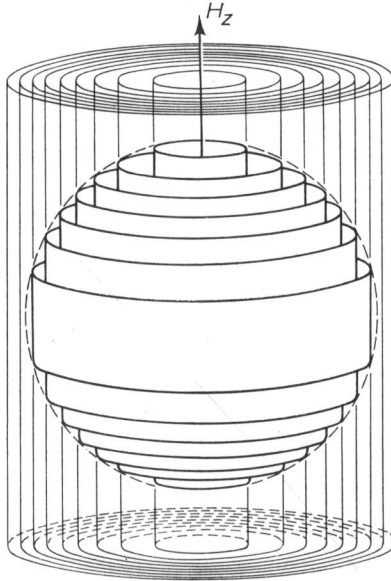

FIG. 10.2. The quantization of the energy spectrum in the xy-plane into discrete cylindrical surfaces in k-space under the influence of an applied magnetic field H_z (after
Chambers, 1956).

decrease their energy. This can be seen in Fig. 10.3 (*b*) in which, for
example, the increase in energy of the lowest state now permits it to
accommodate more electrons and it can therefore accept some which
were previously in the next higher state. The decrease in the energy
of these latter electrons will tend to be cancelled by the increase of
energy of those electrons which were originally in the lowest state.

The situation is very different, however, when we consider the occupied
state of highest energy. As this level expands its extension in the k_z
direction diminishes very rapidly and hence the number of electrons
which it can accommodate decreases, and this will tend to zero when
the level becomes coincident with the position of the Fermi surface in
zero field. Thus as the highest level starts to expand, the energy of the
electrons which it contains will first rise, but then as it approaches the

region of the Fermi surface and its degeneracy diminishes, the electrons instead of staying in the higher state will go into the one below, so that on balance the total energy of the system decreases. As the highest level passes the Fermi surface it will finally become completely empty and all of its electrons will be in the lower level. As the field increases further this level will continue to have its energy raised, and so the energy of the system will again start to increase and the situation will repeat.

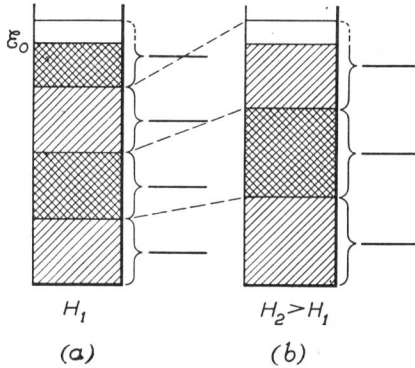

Fig. 10.3. The quantization of states in a magnetic field. On the left of each figure is the energy continuum in zero field which is split into a small number of levels as is shown on the right. All the electrons with energies within the brackets now have a single energy as shown. The mean energy of the lower states is not altered by this process; (a) shows the situation in a field H_1, (b) shows the effect of increasing the field to H_2 (after Pippard, 1960).

Thus the energy U of the electron assembly will show periodic variations as the field is increased and there will be a discontinuity in the relationship between U and H each time the outermost energy level crosses the Fermi surface. Since at the absolute zero† the magnetic moment m is given by

$$m = -\partial U/\partial H \tag{10.14}$$

we see that m, and hence the susceptibility χ_d, will also show a periodic variation. If H_1 and H_2 are the fields at two consecutive discontinuities in U then since $\mathscr{A}(\mathscr{E}_0)$ will be the same on both occasions we see from (10.13) that

$$1/H_1 - 1/H_2 = 2\pi e/\{\hbar c \mathscr{A}(\mathscr{E}_0)\}, \tag{10.15}$$

i.e.

$$\Delta(1/H) = 2\pi e/\{\hbar c \mathscr{A}(\mathscr{E}_0)\}. \tag{10.16}$$

This is the key equation to the phenomenon. It shows that the oscillations in χ_d are periodic in $1/H$ and that a measurement of the period $\Delta(1/H)$ enables us to calculate $\mathscr{A}(\mathscr{E}_0)$, the maximum area of the cross-section of the Fermi distribution. We see, therefore, that in

† Above 0° K the free energy must be used in (10.14) instead of U.

principle such measurements should be very useful in investigations of the geometry of the Fermi surface.

It is fairly clear from this qualitative description that oscillations will arise which correspond to any extremal orbit (maximum or minimum) of the Fermi surface. Thus in the more usual situation, when the Fermi surface is *not* spherically symmetrical, it should be possible to obtain several sets of oscillations with different periods; each period would correspond to a different extreme orbit.

10.5. *Conditions required to observe the effect*

At temperatures above the absolute zero the permitted energy levels will be broadened by an amount of the order of kT and this, of course, will tend to blur the discontinuities in U as H is varied. Thus the de Haas–van Alphen oscillations can only be observed at low temperatures and their amplitude diminishes as the temperature is raised. In general they cannot be observed above the liquid-hydrogen temperature region, although the temperatures at which they appear will, of course, be determined by the level spacing and this, from (10.12), will depend on the values of both $\mathscr{A}(\mathscr{E}_0)$ and H. From the temperature dependence of the amplitude of the oscillations it is in fact possible to calculate the effective mass of the carriers (Shoenberg, 1957). Fig. 10.4 shows de Haas–van Alphen oscillations in (*a*) zinc, (*b*) gallium, and (*c*) aluminium. These measurements were made by measuring the torque on a specimen suspended in a uniform field. The results for zinc show a simple oscillation whereas (*b*) and (*c*) demonstrate both how the amplitude increases as the temperature is reduced and also how two periods of oscillation can show up in the same set of measurements. In (*b*) the two periods are close to one another so that a beat effect occurs, whereas in (*c*) they have quite different values. In all cases the strict periodicity in $1/H$ should be noted.

In order that the effect may be detected an electron must be able to make at least one complete orbit before it interacts with an imperfection which will change its energy. This means that for metals whose Fermi surfaces have large extremal areas (such as those of group I), large orbits in real space must be traversed without scattering. For this to be achieved it is essential to have very pure specimens which have a high degree of crystalline perfection. It is only recently that such crystals have become available for Cu, Ag, and Au and most of the earlier observations were made on polyvalent metals in which small orbits around protuberances of the Fermi surface were observed. In order to get as much information as possible about the shape of the surface several sets of de Haas–van Alphen measurements must be made with the field directed along different directions relative to the crystal axes.

The detailed interpretation of the results is not always unambiguous although, if the surface is ellipsoidal, it is possible to calculate its size

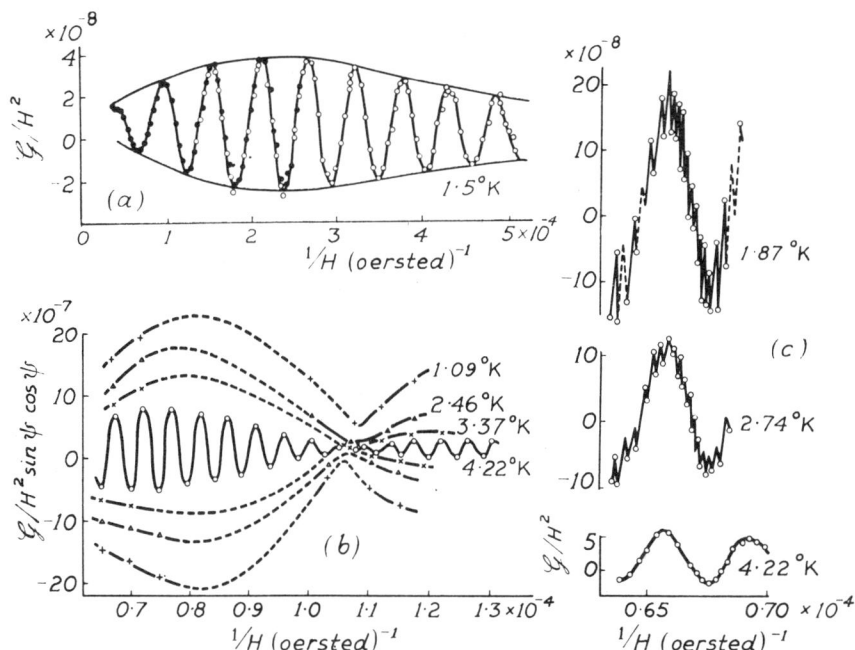

FIG. 10.4. Typical de Haas–van Alphen curves. (a) Zinc; H 20° from the hexagonal axis. (b) Gallium—ac plane; H 79·8° from the tetragonal axis, showing the effect of interference between two sets of oscillations. (c) Aluminium; H 10° from tetrad axis, showing oscillations of long and short period. $\mathscr{G}$ is the couple per gramme, in (b) $\psi = 79\cdot8°$ (after Shoenberg, 1957).

and shape. If, however, the surface touches or overlaps the zone boundary then more complicated orbits are possible and these do not always admit of a unique interpretation. Whilst for further details on the analysis of results we must refer the reader to the review literature that has already been quoted, we should mention the expression given by Shoenberg (1957) for the number of electrons per atom, n_0, contained within a single Fermi surface

$$n_0 = 0\cdot95 \times 10^{-14}\, V\{\Delta(1/H)_1\, \Delta(1/H)_2\, \Delta(1/H)_3\}^{-\frac{1}{2}}, \qquad (10.17)$$

where the $\Delta(1/H)$ are the periods of $1/H$ along the three principal axes of the Fermi surface and V is the atomic volume. This formula is very useful in the interpretation of results and also in the estimation of the fields which might be required in order to observe the oscillations. If the Fermi surface is spherical then the three periods in (10.17) become equal and we obtain

$$n_0 = 0 \cdot 95 \times 10^{-14} V\{\Delta(1/H)\}^{-\frac{3}{2}}. \tag{10.18}$$

If we use (10.18) for copper, then putting n_0 equal to unity we find $\Delta(1/H)$ is equal to $1 \cdot 64 \times 10^{-9}$ oersted^{-1}. We may write

$$\Delta H = -H^2 \Delta(1/H) \tag{10.19}$$

and since 3×10^4 oersted is about the maximum static field which may be attained with any ease† we see that in such a field the oscillations will occur at intervals of about $1 \cdot 5$ oersted. Thus a stability and homogeneity of field approaching 1 part in 10^5 would be necessary in order that the phenomenon should be observed. This is rather difficult to achieve and it is one of the reasons why in such fields one can only detect oscillations from Fermi surfaces which enclose far fewer electron states; this is why the earlier work was concerned entirely with metals such as Bi which had a very small number of free electrons per atom. For the noble metals it is clear that fields of the order of 10^5 oersted must be used in order that the oscillations can be resolved. They should then occur at intervals of about 17 oersted. Such experiments have been made by Shoenberg (1960). The high magnetic field is produced impulsively by discharging a bank of condensers through a liquid-air-cooled solenoid within which the specimen is placed. A pick-up coil around the specimen then yields an e.m.f. which will have an alternating component which is proportional to $\partial\mu/\partial H$ and this is displayed on an oscilloscope.

10.6. Experimental results

Results have now been obtained on Cu, Ag, and Au using very carefully prepared single crystals and a typical oscilloscope trace for Cu is

TABLE 10.1

Main periods of the de Haas–van Alphen effect

(units 10^{-9} oersted^{-1}) (Shoenberg, 1960)

	[111] direction	[100] direction	Free electron model
Cu	1·608	1·557	1·637
Ag	2·00	1·96	2·09
Au	2·05	1·95	2·09

shown in Fig. 10.5 (Plate 2). The period of the oscillations does not depend very much on the orientation and it agrees very well with that to be expected for a spherical Fermi surface as is shown in Table 10.1.

† Higher fields can be produced quite simply with superconducting solenoids (section 6.40).

PLATE 2

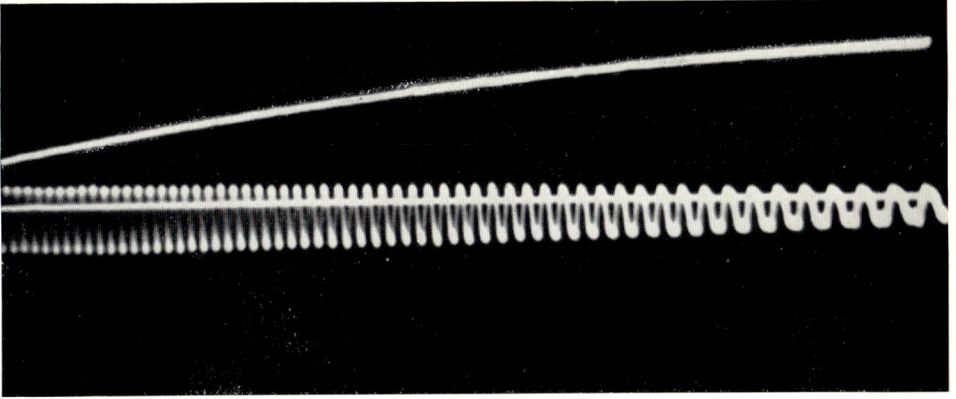

FIG. 10.5. De Haas-van Alphen oscillations in a copper crystal at about 10^5 oersted, obtained using the pulsed field technique. The curved trace across the picture shows the manner in which the field varies during the impulsive discharge through the coil.
(Photo by courtesy of Dr. Shoenberg.)

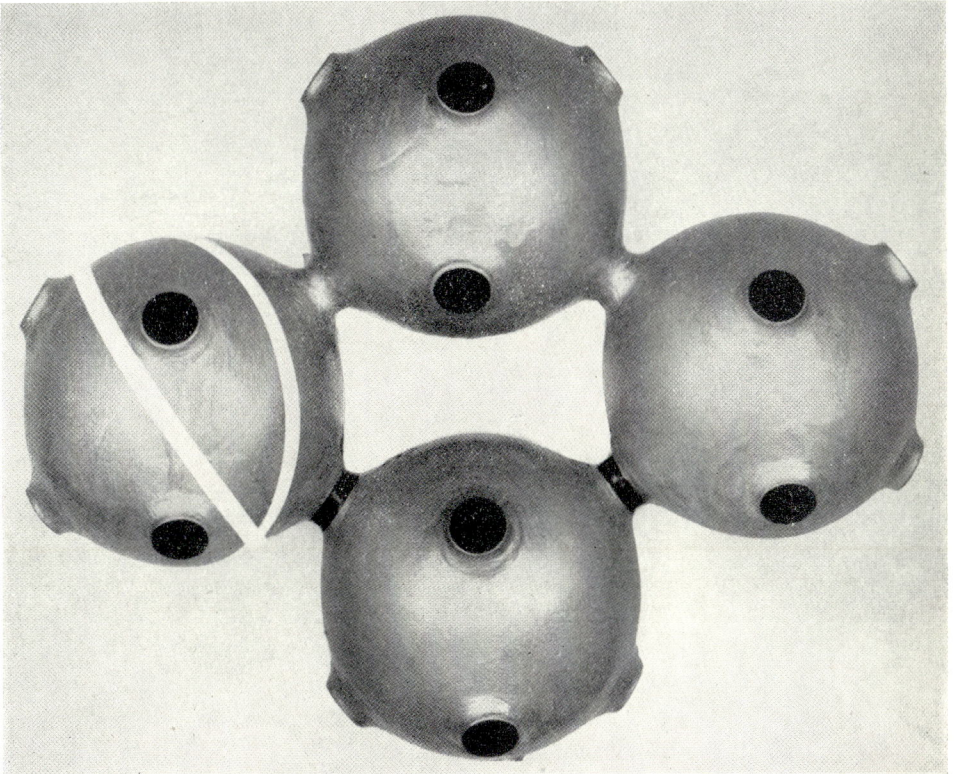

FIG. 10.6. A model of the multiply-connected Fermi surface of gold, showing the ordinary belly orbits (white rings) around a single unit, the small neck orbits between two units (black ring) and the central 'dog's-bone' orbit formed by four units. Note that this is only a schematic model. In the actual Fermi surface of gold the deviations from a spherical shape are probably more pronounced. (Shoenberg, 1960a.)

In addition to these oscillations other types of longer period have also been observed which confirm that for all three metals the Fermi surface is in contact with the Brillouin zone boundary at some points (as was previously suggested by Pippard for Cu, see section 4.26). When this occurs, extended orbits can arise in the manner illustrated for the multiply connected surface shown in Fig. 10.6 (Plate 2). The oscillations of longer period agree well with those to be expected for orbits around the 'dog's bone' shape in Fig. 10.6 and also for orbits around the connecting necks between the main surfaces. Other more complicated orbits have also been identified. The effect has also been measured in potassium using pulsed fields of 150 k oersted in which the oscillations have a period of $5 \cdot 75 \times 10^{-9}$ oersted^{-1} which corresponds very closely to those to be expected for a spherical Fermi surface—$5 \cdot 69 \times 10^{-9}$ oersted^{-1} (Thorsen and Berlincourt, 1961). It is clear that such experiments form an extremely important technique for studying the topology of the Fermi surface.

Whilst most of the work on the de Haas–van Alphen effect has been made on metals and semi-metals it is also possible to observe the phenomenon in semiconductors. Since these generally have a small number of carriers the periods are long and are not so difficult to observe. In addition, whilst the impulsive field technique can give rise to difficulties when metals are used, due to eddy current effects, this does not occur with semiconductors. Results have been reported for p-type PbTe by Stiles, Burstein, and Langenberg (1961) from which details of the band structure can be derived. Quite good agreement is obtained between the density of carriers deduced from the de Haas–van Alphen effect ($3 \cdot 9 \times 10^{18}$ cm^{-3}) and those from Hall effect measurements ($3 \cdot 0 \times 10^{18}$ cm^{-3}).

The de Haas–van Alphen effect is the third of the trio of important methods for investigating the topology of the Fermi surface, the others being the anomalous skin effect (section 4.26) and the magneto-resistance (section 4.30). Each of these has certain advantages and disadvantages, although, due to the difficulties of interpretation, the anomalous skin effect is not used very much. Magneto-resistance measurements are only effective for metals in which the electrons can travel in extended orbits, i.e. the Fermi surface must contact the zone boundary, whilst the de Haas–van Alphen effect does not require this limitation.

10.7. Other oscillatory magnetic effects

It is clear from the theory of the de Haas–van Alphen effect that since

an increasing magnetic field leads to discontinuous behaviour in the
energy of the electron assembly, this will affect, not only the susceptibility, but many other properties as well. This has been demonstrated
in experiments on the Hall effect, the electrical and thermal conductivities (see Fig. 5.22), and thermoelectric effects. Each of these properties has been shown to possess a periodic variation, with a constant
period $\Delta(1/H)$, as the magnetic field is increased. The information which
can be obtained is similar to that which can be deduced from the de Haas–
van Alphen effect and we do not propose to discuss the phenomena in
any further detail. For a fuller description the reader is referred to the
article by Pippard (1960).

11

MECHANICAL PROPERTIES

11.1. Introduction

THE study of the mechanical properties of materials at low temperatures is by no means as widespread as are many of the topics with which we deal in other chapters and the work has been almost entirely limited to experiments on metals. In general the underlying theory is only qualitative and we cannot give such a detailed description as we should wish.

The fundamental quantities which determine the mechanical properties are the elastic constants and so we shall first consider how these vary with temperature. We shall then deal with tensile properties, fatigue, creep, and internal friction.

These last four effects are all determined by the motion of dislocations through the crystal lattice. We assume a knowledge of elementary dislocation theory.† In essence, however, a dislocation is an atomic configuration which has extension in one dimension and which either forms a closed loop or else it extends to the boundaries of the crystal. This configuration is such that when the dislocation glides on a plane from one side of the crystal to the other, thereby sweeping out an area equal to that of the glide plane, the atoms on either side of the plane have moved relative to one another by a distance **b** which is called the Burgers vector. The direction of **b** relative to the dislocation line characterizes the type of dislocation. If **b** is parallel to the dislocation, we have a screw dislocation; if it is perpendicular, it is an edge dislocation. For any other orientation we have a mixed dislocation. **b** is of the order of an atomic spacing. Any slip in the crystal is due to the passage of dislocations. If the dislocation only sweeps out an area $\mathscr{A}$, which is *less* than the total area $\mathscr{A}_0$ of the glide plane, then the slip produced will be $(\mathscr{A}/\mathscr{A}_0)\mathbf{b}$. In a perfect lattice a dislocation can move under the influence of a very small stress, generally thought to be of the order of $10^{-5} \times$ shear modulus, and therefore a metal should be extremely soft. In a real metal, however, there will always be obstacles to

† For an introduction to dislocation theory see Cottrell (1953); Kittel (1956) chapter 19; Dekker (1957), p. 81; or Read (1953).

dislocation motion which will tend to hold it up. The most important of these are (a) impurity atoms, which are attracted to dislocations, forming a 'cloud' around them and acting as a drag on their motion, (b) neighbouring dislocations whose stress field might act so as to oppose the motion of the dislocation being considered, (c) 'forest' dislocations which actually intersect the slip plane of the gliding dislocation, (d) grain boundaries which will block dislocation movement. In order to deform a metal we must make the dislocations move and, because of the obstacles we have mentioned, a greater stress must be applied than would be necessary if we had a perfect crystal. At high temperatures, however, thermal vibrations will assist the dislocations in overcoming the obstacles and hence the external stress will be less than what it would be at low temperatures. It is for this reason that most metals tend to become stronger and harder as the temperature is reduced.

11.2. The elastic moduli

The internal stresses within a crystal which are due to the dislocations are all dependent on the value of the shear modulus of the material. It is therefore necessary to ensure that any temperature-dependent effect which is observed is not merely a reflection of the change of the moduli of elasticity with temperature. Experiments show that as a metal is cooled from room temperature to liquid-helium temperatures the elastic moduli increase by a few per cent, but that at low temperatures they become temperature-independent. This is in accord with the third law of thermodynamics and it can be proved quite simply. We shall take as an example the bulk modulus which may be written as $V(dp/dV)$. From Maxwell's equations we have

$$\left.\frac{\partial p}{\partial T}\right|_V = \left.\frac{\partial S}{\partial V}\right|_T \qquad (11.1)$$

but by the third law of thermodynamics, $\partial S/\partial V|_T = 0$ at $T = 0°$ K. Therefore

$$\left.\frac{\partial p}{\partial T}\right|_V = 0 \qquad (11.2)$$

and so p must be a function of V only,

i.e. $$p = f(V). \qquad (11.3)$$

Thus the compressibility $V\, dp/dV$ can only be a function of V and it does not depend on the temperature as the absolute zero is approached.

This behaviour is borne out by the experiments and Fig. 11.1 shows the variation of the shear modulus c_{44} of copper between room temperature and liquid-helium temperature. Similar behaviour is shown by all other substances which have been measured. Whilst the actual change in cooling is rather small, the low-temperature values of the moduli are important from a theoretical point of view since these are the values which should be used in checking the theory of cohesion and specific heats.

There does not seem to be a detailed theory which accounts for the

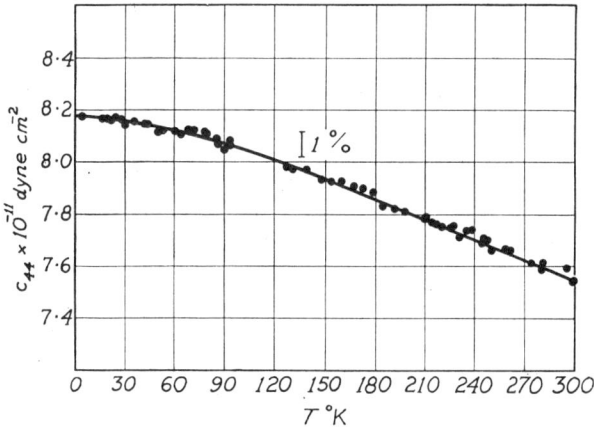

FIG. 11.1. The variation of the elastic coefficient c_{44} of copper with temperature (Overton and Gaffney, 1955).

variation of the elastic moduli with temperature. In general terms it can be seen that as the temperature is increased the thermal motion of one atom will affect the motion of its neighbours and since for larger amplitudes of oscillation (i.e. at higher temperatures) the motion of the atoms is not strictly simple harmonic (see Chapter 2 and section 3.3 for a discussion of the anharmonicity of atomic vibrations), then it is not surprising that the effective forces between the atoms should change.

Most of the experiments which have been made to measure the elastic constants have used ultrasonic techniques rather than direct compressibility, tension, and torsion experiments. In principle the velocity of sound in the specimen is measured, and from this the elastic modulus is calculated. The actual mode of excitation will determine which modulus is being measured. For longitudinal waves in a rod, for example, the velocity, v, is given by

$$v = (Y/d)^{\frac{1}{2}}, \tag{11.4}$$

where Y is Young's modulus and d is the density.

There are several methods of determining the velocity. The most straightforward is to measure the transit time of high-frequency pulses (of about 10 Mc/s) across the specimen. McSkimin (1953) has adopted a modification of this method. Instead of measuring the transit time directly, he reflected the waves from the far end of the specimen and adjusted the frequency until the reflected waves were in phase with the incident waves (i.e. until standing waves were set up). The frequency was then increased by $\Delta\nu$ until the incident and the reflected waves were again in phase. If t is the thickness of the specimen, and ν is the first frequency corresponding to a wavelength λ, then

$$v = \nu\lambda = \nu\, 2t/n, \tag{11.5}$$

where n is the number of wavelengths in the distance $2t$. Hence

$$\nu = nv/2t, \tag{11.6}$$

and so

$$d\nu/dn = v/2t. \tag{11.7}$$

Since for the next frequency, $\nu+\Delta\nu$, for which standing waves are observed, dn is unity, which will be small compared with n,

$$\Delta\nu = v/2t \tag{11.8}$$

and

$$v = 2t\,\Delta\nu. \tag{11.9}$$

The velocity, v, may also be found by measuring the resonant frequency, ν_0, of a rod of the material. This is given by an expression of the form

$$\nu_0 = Av \tag{11.10}$$

where A is a geometrical constant depending on the size and shape of the specimen and the method of suspension.† Hence once again we may calculate v and so determine the modulus. It should be noted that internal-friction experiments, which are described later in this chapter, usually involve the determination of the resonant frequency of the specimen, and so the elastic moduli emerge as a by-product of these investigations.

Ultrasonic measurements give the values of the adiabatic elastic constants and for some calculations the isothermal values are required. Since the correction term involves the ratio of the specific heat at constant pressure to that at constant volume, there is very little difference between the adiabatic and isothermal values at low temperatures. In copper, for example, the difference is less than 0·5 per cent even at 100° K, although at room temperature a correction of ~ 3 per cent is required (Overton and Gaffney, 1955).

We should here mention a change in the elastic moduli which is a

---

† For calculations of A see, for example, chapter 16 of *Acoustics* by Alexander Wood, Blackie and Son Ltd., London (1940).

specifically low-temperature effect. This occurs when a metal becomes superconducting. The theory underlying this behaviour has been described in section 6.9.

11.3. The tensile properties of metals at low temperatures

As was mentioned in the introductory section to this chapter the deformation properties at low temperatures are changed because there is less thermal activation available to assist the dislocations to overcome the obstacles which they encounter as they pass through the crystal lattice. In order to extend the specimen a greater external stress must be used, i.e. the metal appears to be harder. In practice, two main types of behaviour are observed. The first occurs for face-centred cubic and most hexagonal close-packed metals. In these metals the simple ideas which we have just outlined seem to be justified. The yield stress (or the critical shear stress in single crystals)—which may be interpreted as the stress required in order to start the dislocations moving— increases as the temperature is reduced. Once the metal has yielded, however, it is still ductile. It extends about as much as it does at room temperature, necks down and fractures. The second type of behaviour is shown by many body-centred cubic metals (but not the alkali metals). These are characterized by a ductile-brittle transition region which occurs over a narrow range around a temperature T_B. Above T_B some plastic strain is observed after the metal has yielded. Below T_B the metal breaks in a brittle fashion with little or no plastic deformation. In these metals the increase in yield stress as the temperature is lowered is very much more than for metals which show plastic behaviour down to low temperatures (see Fig. 11.2).

There are two main problems to be considered when we discuss the tensile behaviour of a metal. The first concerns the mechanism of yielding, i.e. the first movement of the dislocations. The second concerns what happens after the yield point; i.e. how the dislocations multiply and interact with one another to produce the hardening which is observed. Except in broad qualitative terms no theories which have yet been put forward have received complete acceptance. The reasons for this are mainly that the dislocations, even in an annealed crystal, usually occur in a very complicated network and it is probable that no single mechanism will account for all their interactions. In some cases one mechanism will predominate, in others another will be more important and in many cases it is likely that a combination of several mechanisms might need to be considered.

11.4. *Metals which remain plastic at low temperatures*

The stress-strain curves for metal single crystals which remain ductile down to liquid-helium temperatures are shown diagrammatically in Fig. 11.3. This behaviour, which is typical of a metal such as copper, falls into three main stages. After the initial yield stage 1 is entered. This is called the easy-glide region and very little hardening takes place.

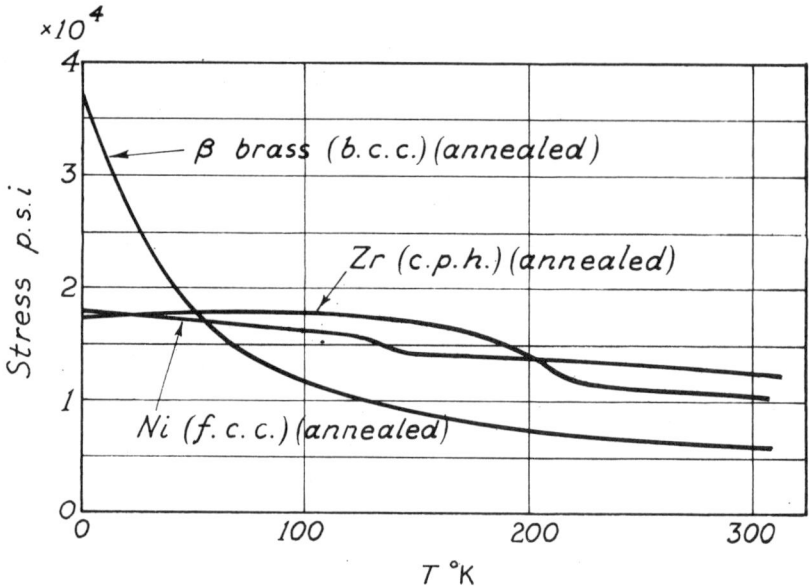

FIG. 11.2. The temperature variation of the yield strength of b.c.c., f.c.c., and c.p.h. metals. Note the steep rise for the b.c.c. material (Wessel, 1957).

It occurs in crystals which are so oriented that only one glide system operates under the applied stress. Stage 2 is the region of rapid work hardening where the stress-strain curve is linear. Stage 3 begins where this linear section of the curve bends over to what is usually a parabolic form. It will be noted that whereas the behaviour in stages 1 and 2 is not very temperature-dependent, the stress at which stage 3 begins is much lower as the temperature of the test is increased.

The interpretation of these stages is now reasonably well understood. During stage 1, slip occurs on one set of planes only. In a carefully prepared crystal there are few obstacles to the movement of dislocations and hence they can move a considerable distance under quite a small external stress. Towards the end of stage 1 some slip and dislocation multiplication occurs on other sets of glide planes and some of these dislocations, as they get close to or cross the primary glide plane, will tend to hinder the movement of the original gliding dislocations. The precise nature of the interaction between the dislocations is still not clearly resolved, although several mechanisms are possible. One suggestion is that the two dislocations on intersecting planes combine to form what is termed a sessile dislocation. These are so named because they are common to two intersecting slip planes and they cannot move

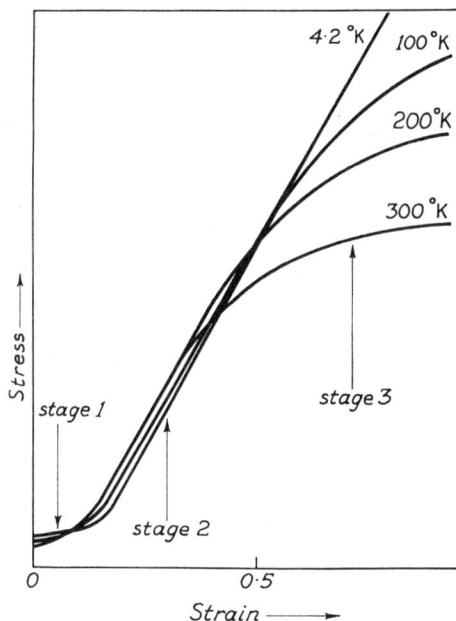

FIG. 11.3. Idealized stress-strain curves for a face-centred cubic metal at various temperatures, showing the three stages of deformation. The stress is plotted in arbitrary units.

conservatively on either. They therefore act as obstacles to the dislocations in the primary glide plane which begin to pile up behind the sessile. Another type of interaction which might be possible is of a purely elastic nature between dislocations which are fairly close together. Obstructions will also be caused by 'forest' dislocations which thread the glide plane and through which the active dislocations have to cut. It is still not generally agreed which mechanism is the main one responsible, but it seems fairly clear that in stage 2 the dislocations are held up in some way or other and in order to continue the deformation higher stresses must be applied. Thus the metal continually hardens throughout this stage. In polycrystalline samples where, owing to the constraints on the crystallites, single slip (i.e. stage 1) cannot occur, the stress-strain curve might be expected to start off in a manner approximating to stage 2, and this is what is observed.

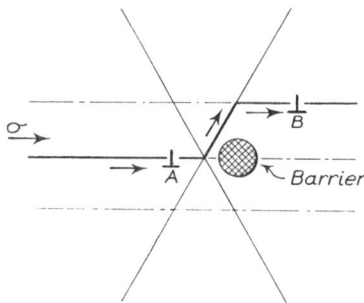

FIG. 11.4. Diagram showing the cross-slip mechanism. The dislocation is initially held up by the obstacle at *A*. When the stress becomes sufficiently large, the dislocation can cross slip on the intersecting glide plane thereby reaching position *B* on a plane parallel to its original glide plane. In this way it avoids the barrier.

In stage 3 the rate of hardening decreases, and this is thought to be due to the fact that at still higher stresses a dislocation at the head of one of the piled-up groups which were produced in stage 2, can now avoid the obstacles by transferring itself to a neighbouring parallel plane. In order to do this it must first travel along one of the inter-secting slip planes as is shown in Fig. 11.4. This process is called cross slip. It is assisted by thermal activation and hence the onset of stage 3 is very temperature-dependent as was shown in Fig. 11.3. Once the stress is suffi-ciently high for a dislocation to avoid one obstacle by cross slip it will con-tinue to be able to do so for any other obstacles it may meet and hence the hardening becomes much less rapid at this stage.

11.5. *Strain-rate experiments*

It is clear from the foregoing that in order to understand the details of crystal deformation it is essential that we know what type of barriers are active in impeding the motion of dislocations. One of the most promising types of investigation is that in which a tensile test is made by extending a specimen at a constant strain rate and then, at a certain point in the test, the strain rate is suddenly changed (say, by a factor of 10) by changing gear in the driving mechanism. The dislocations which were previously overcoming barriers at a certain rate (with the aid of the external stress plus thermal activation) now, if the strain rate is increased, have to overcome them more rapidly. There will be less time for thermal activation to assist in overcoming the barriers and hence, in order to maintain the increased strain rate, extra energy must be supplied from outside, i.e. the specimen appears to become harder and there is an increase in the yield stress at the point where the strain rate was changed. It is possible to make the following simple analysis of the situation.

Let us assume that the force on a dislocation due to its being at a distance x from a barrier is F and that it has the form shown in Fig. 11.5. Due to the external stress, let the force on the dislocation be F_1 at some point during the tensile test, so that to maintain equilibrium the dislocation will be at a distance x_1 from the barrier. If now, in fact, the dislocation overcomes the barriers at the rate which is necessary in order

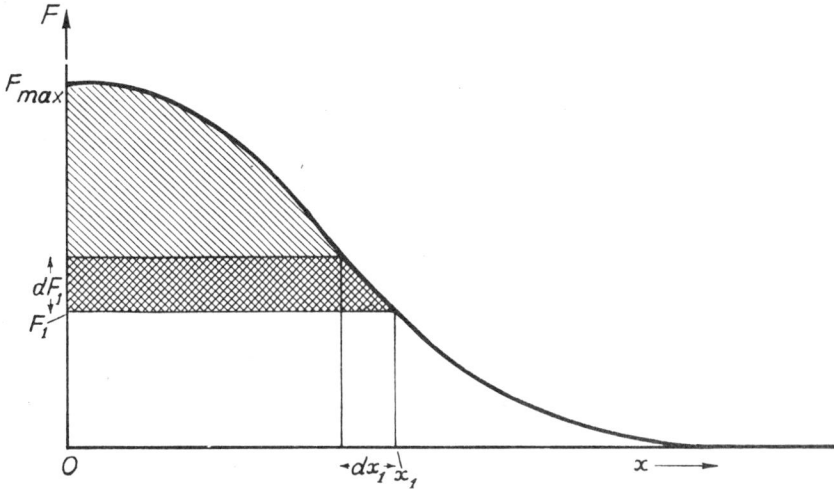

FIG. 11.5. A diagram showing the force F, exerted on a dislocation when it is a distance x away from a barrier. If, due to an external force, the dislocation is at x_1, then energy equivalent to the whole of the shaded region must be supplied by thermal activation. If the strain rate is increased so that the dislocation is at $x_1 - dx_1$, then the activation energy has been decreased by the cross-hatched area equivalent to $x_1 \, dF_1$.

to account for the strain rate, then the energy U, which is equivalent to the entire shaded part of the diagram and has a value

$$\int_{F_1}^{F_{\max}} x \, dF,$$

must be supplied by thermal activation. If, however, the strain rate is increased, the activation energy must be decreased in order that the barriers may now be overcome at a more rapid rate. Hence the dislocations would have to be in a position corresponding to a force $F_1 + dF_1$ on the curve. The decrease in the activation energy would be $x_1 \, dF_1$. Let us consider a point on the stress-strain curve at which the density of active dislocations is N_1 and for which the area swept out by a dislocation after it has overcome a barrier is $\mathscr{A}_1$. Then the strain rate, $\dot{\epsilon}$, will be proportional to the number of dislocations which can be activated over barriers and also to the area swept out by each dislocation, i.e. to

$N_1 \mathscr{A}_1 \exp\left(-U_1/kT\right)$. If now we suddenly increase the strain rate to $\dot{\epsilon} + d\dot{\epsilon}$, so that the activation energy becomes $U_1 - x_1 dF_1$, then $x_1 dF_1$ will be determined by the value of $d\dot{\epsilon}$ from the following relation

$$(\dot{\epsilon} + d\dot{\epsilon}) N_1 \mathscr{A}_1 \exp\left(-U_1/kT\right) = \dot{\epsilon} N_1 \mathscr{A}_1 \exp\{-(U_1 - x_1 dF_1)/kT\} \quad (11.11)$$

assuming that N_1 and $\mathscr{A}_1$ do not change as a result of the change in $\dot{\epsilon}$. Taking logarithms of both sides yields

$$d(\log_e \dot{\epsilon}) = x_1 dF_1/(kT) \quad (11.12)$$

or, more generally, $\qquad x\, dF = kT\, d(\log_e \dot{\epsilon}). \quad (11.13)$

Thus, independent of the manner in which the density of dislocations, or the areas which they sweep out, change during a tensile test, we see that at constant temperature the value of $x\, dF$ depends only on the ratio of the strain rates, before and after the change. Now the force on a dislocation is not equal to the external stress on the crystal, σ, but it is equal to σb under ideal conditions. This value, however, can be modified by the presence of other dislocations or crystal defects. Thus the change in σ which is observed when the strain rate is altered is not equal to dF. If, however, we assume that the same factor of proportionality between σ and F is operative throughout a given test then we can say $\qquad x\, d\sigma = \text{constant} \times kT\, d(\log_e \dot{\epsilon}), \quad (11.14)$

and hence it is possible to determine the variation of x as a function of σ, i.e. of F. Thus we can find the shape of the $F \sim x$ curve shown in Fig. 11.5.

It is, of course, only possible to investigate the complete shape of such a curve, particularly for the larger values of F, if the thermal activation is sufficiently small for the 'working point' of the dislocation to be well up the curve. For this reason it is often necessary to make measurements at liquid-air temperatures and below. Fig. 11.6 shows normalized force–distance curves analogous to Fig. 11.5 for Al, Cu, and Ag which have been determined by strain-rate experiments at low temperatures.

It is clear that from the above analysis it ought to be possible to obtain similar information to that already described by changing the temperature during a test whilst keeping a *constant* strain rate. This will also change the rate of thermal activation and hence, in order to maintain the same strain rate, it will be necessary for the dislocations to be working at a different point of the curve in Fig. 11.5. Hence a stress change will be observed when the temperature is altered. Basinski (1959) has shown that the stress changes which are observed on changing temperature correlate well with those which are obtained if the strain rate is changed at constant temperature. This shows that the general idea of a thermally activated process is correct. For a full mathematical analysis of the problem the reader is referred to Basinski's paper.

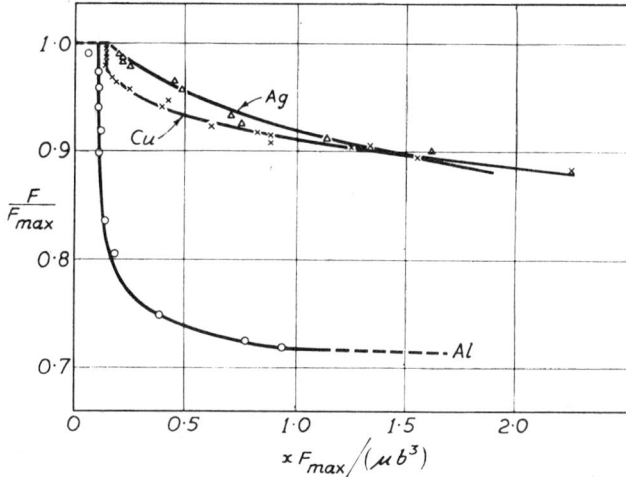

FIG. 11.6. Normalized force–distance curves for Al, Cu, and Ag derived from strain-rate experiments. F_{max} is the maximum stress exerted by the barrier, μ is the shear modulus, and b is the Burgers vector (Basinski, 1959).

11.6. *The fracture strength*

So far we have considered what happens when a metal is deformed plastically. What determines the point at which the metal will fracture? The tensile strength of a metal is by no means a simple quantity. The conditions for fracture essentially set in when the decrease in cross-sectional area caused by the extension is no longer compensated for by the increase in the work-hardening of the metal. Since it is in stage 3 that the work-hardening does not increase so rapidly, the metal usually fractures during this part of the stress-strain curve. Owing to the strong temperature dependence of the onset of stage 3, this means that the tensile strength is considerably increased at low temperatures. Fig. 11.7 shows the variation of the tensile strength of polycrystalline samples of copper, silver, gold, and aluminium between 4·2° K and room temperature. It will be noted that the increase in strength is quite considerable. In aluminium, for example, the tensile strength increases about four times between room temperature and 4·2° K and many other metals double their strength.

Since stage 3 occurs after very similar extensions, both at room and at low temperature, there is little change in the deformation produced prior to fracture.

11.7. *Brittle fracture*

Some metals are characterized by the fact that, as their temperature is reduced, the critical shear stress increases very much more than it does for most of the metals whose behaviour has been described in the

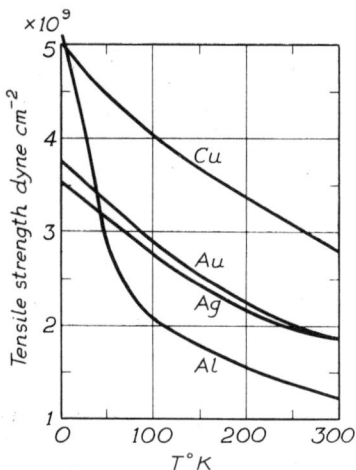

FIG. 11.7. The variation of the tensile strength of Cu, Ag, Au, and Al from 4·2 to 300° K (McCammon and Rosenberg, 1957).

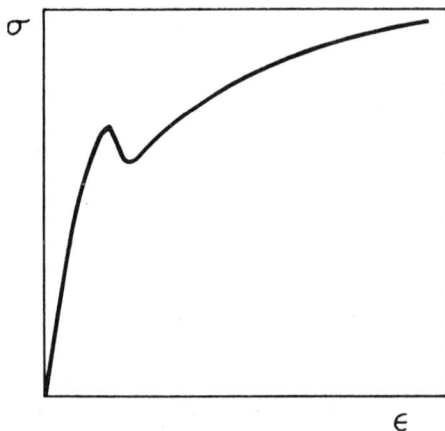

FIG. 11.8. The stress-strain curve for a body-centred cubic metal showing the sharp yield point, followed by the yield drop.

preceding section. This is shown in Fig. 11.2. Nearly all of these have a body-centred cubic structure. When the b.c.c. metals are deformed they usually show a sharp yield point, in contrast to the curves of Fig. 11.3 and this is followed by a yield drop (Fig. 11.8). One explanation of this effect is that small amounts of impurity (e.g. carbon or nitrogen in iron) are attracted to, and cluster round, a dislocation (Cottrell, 1953, p. 139). This impurity atmosphere, as it is called, acts as a drag on the dislocation and inhibits its movement. Thus for slip to occur, considerable stress must first be applied in order to pull the dislocation away from its atmosphere. Once the dislocation is free of the atmosphere it can continue to move under a lower applied stress and this accounts for the yield drop. At low temperatures thermal activation will not be so effective in helping the dislocations to pull away from their impurity atmospheres and so the yield stress increases as is shown in Fig. 11.2. The reason why this effect is not observed in face-centred cubic metals is that in the f.c.c. structure an impurity atom produces a spherically symmetrical distortion of the lattice and it is only attracted to the edge dislocations. The screw dislocations are left free of impurities and can thus move under a much lower stress. In a b.c.c. lattice the impurity atoms cause a non-spherical distortion of the lattice and some of this strain can be relieved by the screw dislocations as well. Thus all the dislocations can be anchored by the impurity atoms.

We still have to explain why, once the metal has yielded, it will, below a certain temperature, fracture in a brittle manner. For brittle fracture to occur it is thought that micro-cracks form in the metal and that the applied stress at the head of these cracks is large enough to make them spread. There are two main ways in which such cracks might be formed. They might occur at places in the metal where very high stress concentrations are set up by a large pile-up of dislocations, e.g. at a grain boundary. Since, however, brittle fracture is observed in single crystals as well as polycrystals this cannot be the only mechanism and Cottrell (1958a) has proposed a mechanism whereby a crack can be formed by the combination of two dislocations in the b.c.c. structure. Cracks have been observed by the intersection of twins in α-iron at 78° K (D. Hull, 1958). Depending on the value of the applied stress, the crack will either remain stable or it will spread so as to cause brittle fracture. At higher temperatures, where the yield stress is low, the crack will not spread and plastic deformation is observed. At low temperatures the yield stress is high, the cracks can spread, and brittle fracture results. It should be noted that with a very fine-grained specimen the brittle behaviour is modified considerably and it has been shown (Basinski and Sleeswyk, 1957; Smith and Rutherford, 1957) that even at 4·2° K considerable reduction in area does occur before fracture. This can be explained either by assuming that in a very small crystallite only small dislocation pile-ups can occur and these will not be sufficient to start a crack very easily, or, if one assumes that the cracks are formed by some kind of dislocation combination, then it is possible that these cannot spread so easily because of the large number of grain boundaries that have to be crossed.

In addition to the b.c.c. metals many workers have observed that zinc (but not cadmium) becomes brittle at low temperatures and tends to cleave on the plane perpendicular to the hexagonal axis.

The alkali metals lithium, sodium, and potassium, which are b.c.c. at room temperature, do not exhibit brittleness at low temperatures. The tensile behaviour (D. Hull and Rosenberg, 1959) of lithium and sodium, however, is complicated by the fact that these metals undergo

martensitic transformations at low temperatures to a hexagonal structure, lithium at about 76° K and sodium at about 31° K; but even in
potassium, which does not transform, brittle behaviour is not observed.
The metal is quite ductile even at about 2° K.

11.8. *Serrated stress-strain curves*

There have been many observations in recent years that certain
metals, when deformed at low temperatures, give a stress-strain curve
which, instead of being smooth throughout its length, contains serrations,

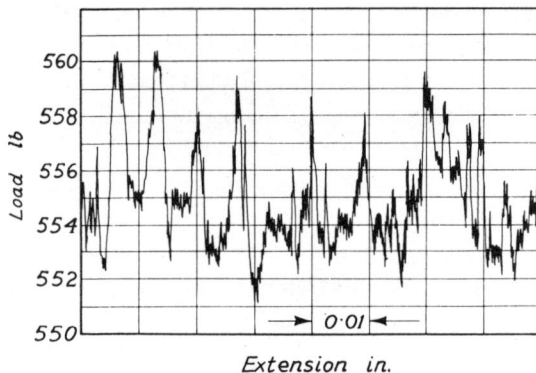

FIG. 11.9. An example of a serrated stress-strain curve. The specimen was a copper
single crystal deformed at 4·2° K. The elongation increases to the left (Blewitt, Coltman,
and Redman, 1955).

although quite often a smooth curve is obtained at higher temperatures.
Uzhik (1955) shows such a curve for an austenitic steel at 20° K and
Wessel (1956, 1957) obtained a similar curve for many of his specimens
at 4·2° K; lithium also exhibits a similar effect at 4·2° K (D. Hull and
Rosenberg, 1959). Some of these curves are very reminiscent of those
obtained at higher temperatures for metals (e.g. cadmium) in which
twinning occurred during the deformation. Blewitt, Coltmann, and
Redman (1955) obtained such serrated curves for copper single crystals
strained at 4·2° K and an example of their results is shown in Fig. 11.9.
They suggested that the serrations were indeed due to twinning and this
was confirmed by X-ray experiments (Sherill, Wittels, and Blewitt,
1957). They showed, however, that it is only for certain orientations,
and then only for certain sections of the serrated curve, that twinning
occurs, i.e. serrations are not necessarily due to twinning.

Another possible reason for the serrations has been put forward by

Basinski (1957) who showed that for pure aluminium the serrated curves can be obtained only at 4·2° K, whereas they can be produced at higher temperatures with aluminium alloys. He then pointed out that at higher temperatures the alloys have a heat conductivity which is much lower than that of pure aluminium (see section 5.11) and hence local hot spots might be able to develop within the specimen. If the rate of work hardening was sufficiently temperature-dependent, unstable flow could be initiated at the hot spots which would give rise to the load drops which are observed. This hypothesis was checked by using a specimen which had a niobium wire running down a hole in the middle. This wire is superconducting at 4·2° K but it becomes normal at $\sim 9°$ K. The experiments showed that when the specimen was extended at 4·2° K the niobium wire did become normal where the serrations appeared, indicating that a very appreciable temperature rise had occurred. These temperature rises coincided with the drop in load in the stress-strain curve.

It has therefore been established that there are at least two mechanisms—twinning and local hot spots—which can give rise to discontinuous yielding at low temperatures although it is by no means clear as yet that these will be applicable in all cases.

11.9. Creep of metals at low temperatures

It is well known that when a metal is subjected to a constant stress it extends slightly as time progresses. This is called creep. Experiments show that this phenomenon is very temperature-dependent. The creep, under any given load, increases very considerably as the temperature is raised and the effect has been explained in general terms by assuming that dislocations are held up at barriers which they can only overcome with the assistance of thermal activation. At higher temperatures the probability of such activation is increased and hence the creep proceeds more rapidly. At low temperatures, say around 90° K, it is found that the strain, ϵ, under a given stress is of the form

$$\epsilon = ET \log gt, \qquad (11.15)$$

where E and g are constants, t is the time measured from when the load was applied and T is the temperature of the specimen. Thus for a given time, t, the extension is proportional to the absolute temperature. Some experimental results are shown in Fig. 11.10. An expression such as (11.15) can be derived on theoretical grounds by using the so-called 'exhaustion' theory of creep.

11.10. *The exhaustion theory*

This theory can be formulated quite generally without stipulating any specific model. It assumes that within the metal there are what might be termed 'soft spots' which will yield, either with a small amount of extra stress, or by some thermal activation. There will be a whole range of these soft spots with different activation energies. Those with low energies will yield quickly, leaving the ones with higher energies

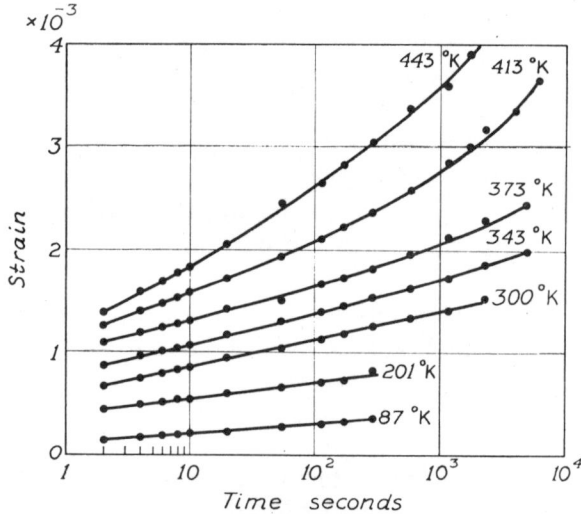

Fig. 11.10. The creep of copper under a stress of 6 kg mm^{-2} at various temperatures (Wyatt, 1953).

still awaiting thermal activation. It is assumed that once a soft spot has yielded and slip has occurred, it cannot be activated again. Hence as time goes on the supply of soft spots becomes exhausted and so the creep rate decreases and finally flattens off completely. Note that we do not postulate the precise nature of a soft spot. It is probably a length of dislocation line which is held up by a barrier, but the type of barrier need not concern us.

We assume that the metal is under a given external stress. In order to make a certain soft spot yield, an extra stress, σ_a, must be applied or else thermal energy $U(\sigma_a)$ must be available. For simplicity let $U(\sigma_a) = A\sigma_a$. This does not affect the general result but it makes the calculations easier. The probability per unit time for the necessary thermal activation† to occur is $f(\sigma_a)$, where

$$f(\sigma_a) = \nu \exp\{-A\sigma_a/(kT)\} \tag{11.16}$$

† The probability for activation is actually proportional to $\exp(-F/kT)$, where F is the free energy, $U-TS$. However, the entropy term can be absorbed in the coefficient ν.

and ν is the frequency of the stress fluctuations, i.e. the thermal vibrations. If $N(\sigma_a, t)$ is the number of soft spots with an activation stress σ_a
at a time t, then the number of those soft spots which are able to yield
in unit time is

$$N(\sigma_a, t) f(\sigma_a). \tag{11.17}$$

If we assume that when each soft spot yields it produces a strain s, then
the total strain in unit time, $\dot{\epsilon}$ (the strain rate), for all the soft spots
will be

$$\dot{\epsilon} = s \int_0^{\sigma_m} N(\sigma_a, t) f(\sigma_a) \, d\sigma_a, \tag{11.18}$$

where σ_m is the stress associated with the most intractable soft spots.
In order to integrate this we must do some manipulation.

$$\frac{dN(\sigma_a, t)}{dt} = -N(\sigma_a, t) f(\sigma_a) \tag{11.19}$$

$$\log N(\sigma_a, t) = -f(\sigma_a)t + \text{constant}, \tag{11.20}$$

or $$N(\sigma_a, t) = N(\sigma_a, 0)\exp\{-f(\sigma_a)t\}. \tag{11.21}$$

From (11.16) we have $\quad df/d\sigma_a = -Af/(kT), \tag{11.22}$

and so $\quad f\, d\sigma_a = -kT\, df/A. \tag{11.23}$

Substituting (11.21) and (11.23) into (11.18) and assuming that N is
constant for all σ_a at $t = 0$, we obtain

$$\dot{\epsilon} = -(sNkT/A) \int_{f(0)}^{f(\sigma_m)} \exp\{-f(\sigma_a)t\}\, df, \tag{11.24}$$

and so $\quad \dot{\epsilon} = -(sNkT/At)[\exp\{-f(\sigma_m)t\} - \exp\{-f(0)t\}]. \tag{11.25}$

Since, however, $f(\sigma_m)$ has the form of (11.16) its value will tend to zero
because $A\sigma_m \gg kT$. $f(0)$, on the other hand, has the value ν, which is
numerically about $10^{13}\ \text{sec}^{-1}$. Thus the first term in the square brackets
of (11.25) is approximately unity and the second one can be neglected.
Hence

$$\dot{\epsilon} = -sNkT/(At) \tag{11.26}$$

and on integrating we obtain the logarithmic creep law

$$\epsilon = \frac{sNkT}{A} \log_e gt. \tag{11.27}$$

In order to compare this expression with experimental results it is
convenient to make a substitution for the term sN. Let us consider an
experiment in which we first load a specimen to a stress σ_0. In order to
produce a further small extension $\Delta\epsilon$, we could either let the specimen
creep to that extension in a certain time, or we could have achieved the

same extension at time $t = 0$ by increasing the load slightly, say by $\Delta\sigma$, up to σ_1 (Fig. 11.11). For the creep experiment we could say that we have now allowed all soft spots with an activation stress up to σ_1 to be released and so

$$\Delta\epsilon = Ns \int_{\sigma_0}^{\sigma_1} d\sigma = Ns(\sigma_1 - \sigma_0) = Ns\,\Delta\sigma. \tag{11.28}$$

If, on the other hand, we increase the load to σ_1 at $t = 0$ we can say that

$$\Delta\epsilon = (d\epsilon/d\sigma)_{\sigma = \sigma_0}\Delta\sigma \tag{11.29}$$

where $d\epsilon/d\sigma$ is the reciprocal of the slope of the stress-strain curve

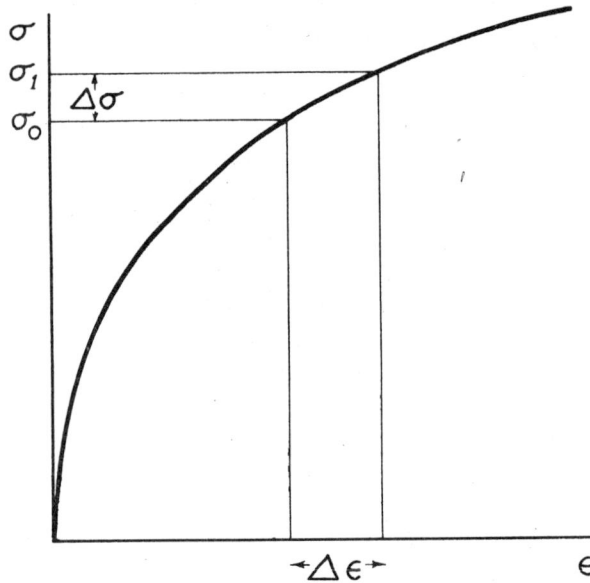

FIG. 11.11. A stress-strain curve for the determination of creep data. The strain $\Delta\epsilon$ could be produced either by creep under the stress σ_0, or by application of σ_1 at $t = 0$. See text.

($\Delta\sigma$ and $\Delta\epsilon$ will be very small and so the slope of the curve will be approximately constant over this region). Thus comparing (11.28) and (11.29) we obtain

$$Ns = (d\epsilon/d\sigma)_{\sigma = \sigma_0}. \tag{11.30}$$

If we substitute for Ns in (11.27) we get

$$\epsilon = \frac{kT}{A\sigma_0}\left(\frac{1}{\sigma}\frac{d\sigma}{d\epsilon}\right)_{\sigma = \sigma_0}^{-1} \log_e gt, \tag{11.31}$$

or

$$\epsilon = \frac{kT}{A\sigma_0}\left\{\frac{d(\log\sigma)}{d\epsilon}\right\}_{\sigma = \sigma_0}^{-1} \log_e gt. \tag{11.32}$$

$A\sigma_0$ is an activation energy which is necessary to produce yielding; it will probably be of the order of 1 eV. If we make this assumption then from a stress-strain experiment on a specimen we can calculate the coefficient of the logarithmic term in (11.32) (this corresponds to ET in (11.15)) and compare it with the value obtained from a creep experiment. This latter value can be obtained quite simply. If at a constant temperature we have extensions ϵ_1 and ϵ_2 at times t_1 and t_2 respectively then

$$\epsilon_2 - \epsilon_1 = ET \log_e(t_1/t_2) \qquad (11.33)$$

and so E may be determined. At 90 and 78° K quite good agreement has been found (Glen, 1956) between the coefficient ET from creep experiments and that calculated using (11.32). It has also been shown that in the region of logarithmic creep the coefficient of the logarithmic term is indeed proportional to the absolute temperature, as its form ET suggests.

11.11. The tunnel effect

Since even at liquid-air temperatures creep is not very large, it would appear that, due to its proportionality to T, it should become exceedingly small at liquid-helium temperature. The few experiments which have been made do not bear this out. The experiments of Meissner, Polanyi, and Schmid (1930) on cadmium single crystals showed that at 4·2° K quite measurable extensions (of the order of a few tenths of a per cent) were recorded in the first few minutes of a test. They also showed that the creep at 1·2° K was not very different from that at 4·2° K, whereas it should be about 3·5 times less. Similar results have been obtained by Glen (1956) and these indicate that the creep is about ten times greater than that which one calculates from (11.32). It therefore appears that some other type of activation process is occurring which is not temperature-dependent, and Glen suggests that it might be possible for the soft spots to be activated by the quantum mechanical tunnel effect. If there is a potential barrier which is holding up a dislocation there will always be a finite probability that this barrier might be overcome. If H is the height of the barrier and W is its width (of the order of an atomic spacing) then the probability of a jump due to the tunnel effect is

$$f_{\text{tun}} = \nu \exp(-8MHW^2/\hbar^2)^{\frac{1}{2}}, \qquad (11.34)$$

where M is the mass of the system which is moved. Mott (1956a) and Glen have shown that if one makes reasonable assumptions about M, H, and W, then f_{tun} has a value which is approximately equal in value to that due to ordinary thermal activation at a few degrees absolute. It should be remarked, however, that attractive though the idea is, the tunnel effect is by no means a fully proven mechanism to explain the anomalously high creep rates which have been observed at helium temperatures.

11.12. Fatigue at low temperatures

The phenomenon of fatigue is observed when an alternating stress is applied to a metal. It is found that after a certain number of cycles the metal will break even though the maximum stress applied might be much lower than that which would produce yielding or fracture in a uni-directional test. Whilst this is not the simple type of process which a physicist usually likes to investigate, it is obvious that in the modern world our need to understand the mechanism of fatigue cannot be over-emphasized. Examination of specimens during a fatigue test by the optical and the electron microscope show that at least two important mechanisms operate. After quite a small fraction of the lifetime of the specimen (about 1 per cent) micro-cracks appear within the grains. During the remainder of the test these cracks gradually extend and some join up with one another until a large crack is formed. This extends through the specimen until the effective cross-sectional area is sufficiently reduced for fracture to occur. For simplicity one can therefore divide the effect into two mechanisms: firstly the production of the micro-cracks, and secondly, their growth.

From the fundamental point of view it is the actual production of the cracks which is difficult to understand. For whilst all the details of the growth mechanism may not yet be fully appreciated, it would appear fairly clear that provided one has a small narrow crack to start with, then at the tip of the crack a stress concentration will be produced which might exceed the fracture strength of the metal and this will enable the crack to spread. The more difficult problem is to discover how the cracks are formed in the first place. There would appear to be two main lines of thought. The first presupposes some kind of chemical or corrosive action at the specimen surface (which is where all the cracks appear to start) and there are many experiments which show very convincingly that if a fatigue test is made in a vacuum or in an inert atmosphere, or if the surface of the specimen is coated with a plastic or rubber compound, then the lifetime to fracture under a given stress is very much increased. Nevertheless, the specimen does finally fracture by fatigue failure and hence chemical action is not essential for the production of the fatigue cracks. The second type of mechanism for the creation of cracks relies on the way in which dislocations will move and interact under the influence of alternating stresses. It is now well established that when screw dislocations move through a metal they leave behind them trails of interstitial atoms and vacant lattice sites. It might be possible for the vacancies to diffuse together and so form a small void or crack. Several ingenious mechanisms (e.g. Cottrell and D. Hull, 1957) have

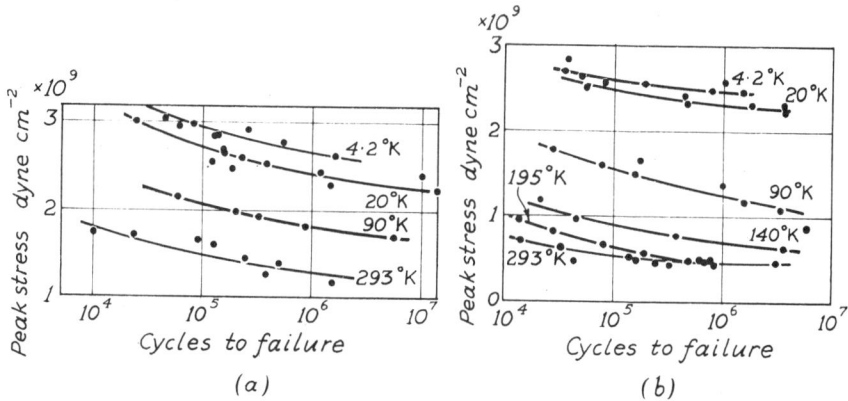

FIG. 11.12. The fatigue of (a) copper and (b) aluminium at temperatures down to 4·2° K. Note the increases in fatigue stress as the temperature is reduced, particularly for Al, for failure in a given number of cycles (McCammon and Rosenberg, 1957).

also been proposed in which cracks can be formed by the purely geometrical interaction of dislocations, but since there is no evidence that any of them is correct we shall not describe them here.

Experiments on the fatigue of metals at low temperatures show two main features. For metals which do not become brittle (see section 11.4) the fatigue lifetime under a given alternating stress increases very much as the temperature is reduced—by a factor of several hundred or thousand between room temperature and liquid-helium temperature. Nevertheless, if at 4·2° K the stress is increased sufficiently so that fatigue failure occurs in a reasonable time (after, say, 10^5 or 10^6 cycles) then the type of failure and the general form of the results is not essentially different from that which is observed at room temperature.

Two such series of experiments for aluminium and copper are shown in Fig. 11.12 where the number of cycles to failure is plotted against the peak alternating stress for a series of temperatures. There is a large increase in the lifetime for a given stress, when the temperature is reduced, although because the curves have such a small slope these

increases are not in themselves very significant. It is much more illuminating to observe how the peak stress must be changed in order to
produce fracture in a given number of cycles at the various temperatures
of the tests. If this is done it is found that the increases in the peak
stress at the various temperatures are almost exactly the same as the
increases in the tensile strength of the metals at those temperatures
(McCammon and Rosenberg, 1957). Curves showing the ratio of the

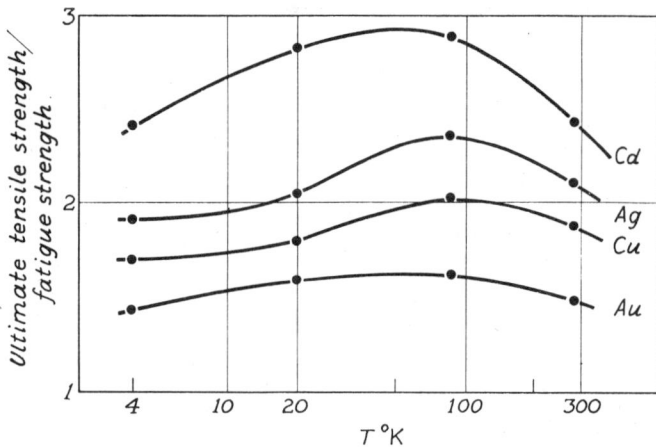

FIG. 11.13. The variation of the ratio, ultimate tensile strength/fatigue strength for Cu,
Ag, Au, and Cd. The fatigue strength has been taken as the peak stress which will
produce fracture in 10^5 cycles (McCammon and Rosenberg, 1957).

fatigue stress to produce fracture in 10^5 cycles to the tensile strength
are shown in Fig. 11.13. It will be noted that this ratio does not change
very much as the temperature is reduced. At the moment there is no
satisfactory explanation for the broad maximum in these curves.

Metals which exhibit brittle fracture at low temperatures (such as
iron and zinc) have rather peculiar fatigue characteristics. Above the
ductile-brittle transition range the metals fatigue in the normal manner,
but below this region it was found very difficult to produce ordinary
fatigue failure at all except within a very narrow stress range. Below
that stress the metal never failed and above it, it broke immediately the
load was applied. Thus even for these metals the fatigue behaviour
followed the trend of the tensile measurements.

For the non-brittle metals this correspondence between fatigue and
tensile data can be qualitatively understood if we assume that the lifetime is controlled by the speed of crack propagation. As has already
been described earlier in this section, cracks will spread if the stress at

the head of the crack exceeds the fracture stress of the material. The actual stress at the tip of a crack will be proportional to the applied load. If, as the temperature is decreased the fracture stress increases, then for the crack to spread at a given rate, the applied fatiguing stress must be increased in the same proportion as was the fracture stress. Thus the correspondence between the tensile and the fatigue properties ·can be reasonably explained. We should note, however, that the fracture stress of the metal will not correspond exactly to the tensile strength (force at failure/*original* cross-sectional area) which was measured in the experiments. Nevertheless, since the ductility of most metals does not change very much as the temperature is reduced, the tensile strength should reflect the behaviour of the fracture stress.

Important conclusions on the method of micro-crack formation can be drawn from the fact that, for most metals, fatigue failure occurs in a normal manner at 4·2° K. This emphasizes once again that corrosion processes are not an essential element of the fatigue mechanism, since it is most unlikely that liquid helium would react chemically with a metal. It also shows that any mechanisms which require diffusion are not fundamental to the problem since these would be very ineffective at such low temperatures. One therefore is left with the conclusion, which has already been mentioned, that cracks form by the purely geometrical interaction of dislocations or of slip planes. It should be emphasized, however, that although diffusion or corrosion mechanisms seem very unlikely in low-temperature fatigue, it does not mean that these mechanisms might not be operative at higher temperatures. It does mean that these processes are not *essential* to the formation or the spreading of the fatigue crack.

Broom and his co-workers have shown that certain processes which are partially activated by fatigue can be quenched if the fatiguing is done at 90° K. For example, if a metal is strained by the application of a certain stress and it is then fatigued at a lower stress level, considerable softening of the metal occurs (Broom and Ham, 1957). The alternating motion of the dislocations during fatigue presumably involves the production and movement of point defects which can assist in unpinning the dislocations and hence the metal becomes softer. This process is almost inhibited at 90° K and very little softening is observed. Another example occurred in some work on aluminium alloys (Broom, Molineux, and Whittaker, 1955) when it was shown that the poor fatigue qualities of alloy DTD 683 are probably due to over-ageing during the fatigue test. This was demonstrated by the fact that at room temperature the fully hardened and the initially over-aged specimens gave very similar fatigue curves, indicating that the fully hardened material had become over-aged during the fatiguing. Confirmation was provided by tests made at 90° K. At this temperature the over-ageing during fatigue was inhibited and hence the fatigue characteristics of the fully hardened metal were now much better than those of the over-aged samples.

11.13. Internal friction at low temperatures

Internal friction is the name given to the familiar effect which is observed whenever a body is set into vibration at its resonant frequency. It is found, no matter how carefully a specimen is mounted and even if this is done in a high vacuum, that the amplitude of the oscillations decreases with time. This indicates that the motion of the crystal lattice is not perfectly elastic but that the energy is dissipated by some type of frictional force which gives the phenomenon its name. It is measured by determining the rate of decay of the amplitude of the oscillations. The damping or anelasticity, as it is sometimes called, can arise by the interaction of the crystal lattice with, for example, impurities, point defects or dislocation networks, and at room temperature and above the study of the variation of the internal friction has been a very useful means of studying structural changes, e.g. of precipitation and ordering, and it can be a very sensitive method for the detection of various dislocation mechanisms; quite often the main difficulty is to decide precisely what mechanism is being detected.

Low-temperature experiments have been mainly concerned with the effect of cold work on the internal friction. The first experiments between 4·2 and 300° K were made by Bordoni (1953, 1954), who observed that in a pure annealed metal the internal friction decreased as the temperature was reduced. If, however, the metal was deformed at room temperature before testing, the internal friction rose on lowering the temperature, passed through a maximum and then started to decrease again. This behaviour was shown very markedly on copper and the experiments on this metal have been repeated in much more detail by Niblett and Wilks (1957). In their experiments the specimen was driven electrostatically and was made to oscillate in a transverse mode at 1 kc/s (Bordoni used longitudinal vibrations at 10–40 kc/s). They observed two peaks in the internal friction, one at about 75° K, which corresponded to that found by Bordoni, and a smaller one at about 32° K. A typical set of results is shown in curve A of Fig. 11.14. The position and the height of the peaks was not dependent on the amount the metal was deformed, provided that this was greater than about 2 per cent.

11.14. *Relaxation processes*

The form of these internal friction peaks suggests that they are due to some type of relaxation process in which a relaxation time, τ, exists which is of the form

$$\tau = \tau_0 \exp(U/RT) \qquad (11.35)$$

where U is an activation energy and R is the gas constant per mole;

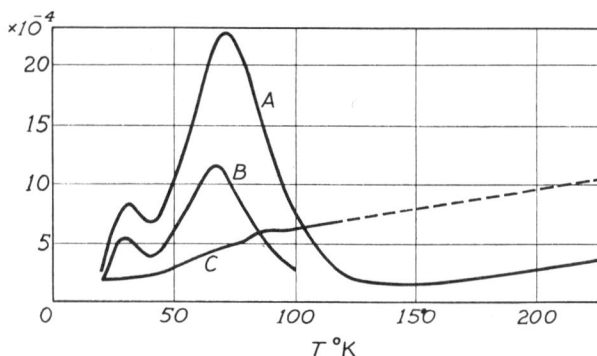

FIG. 11.14. The internal friction of cold-worked copper at low temperatures. *A*—strained 8·4 per cent; *B*—annealed 1 hr. at 250° C; *C*—annealed a further hour at 350° C (Niblett and Wilks, 1957).

i.e. τ is the average time at a temperature T which is required for the necessary energy U to be dissipated. If this time is the same as the period of the applied vibration a kind of resonance effect will be produced and energy will be absorbed from the mechanical vibrations. If one assumes an expression of the form (11.35) and if τ is associated with the period of the applied vibration, then the temperature at which the peaks are observed should decrease if the resonant frequency is lowered (by changing the specimen dimensions).

The experiments show that this is what does happen. If T_1 and T_2 are the temperatures at which one of the peaks is observed when the periods of the resonant vibrations are τ_1 and τ_2, then a value for the activation energy, U, can be found from (11.35) since

$$\log_e(\tau_1/\tau_2) = U(1/T_1 - 1/T_2)/R. \qquad (11.36)$$

The results show that U is quite small—about 0·1 eV. This means that the energy absorption cannot be associated with the annihilation of any kind of defect which might have been produced during the initial deformation, because, with such a low activation energy all such defects would have been annihilated at room temperature almost as quickly as they had been produced. It was found that even after heating to 250° C quite marked peaks still remained and it was not until the copper had been heated to about 350° C that the peaks were removed (Fig. 11.14, curves B and C). These high recovery temperatures indicate that the internal friction peaks are probably associated with some kind of dislocation mechanism since at these temperatures dislocations might be expected to become mobile. It was first suggested (Mason, 1955) that the peaks were due to the vibration of segments of dislocation line which

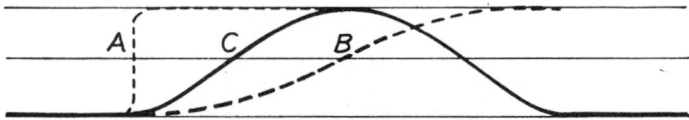

Fig. 11.15. A bulging dislocation line. The configurations A and B have a greater energy than C.

were pinned by impurity atoms or other dislocations. This does not seem very likely since to obtain a well-defined activation energy one would require a constant length of segment which can vibrate. Since the position of the peaks is not dependent on the impurity or on the exact amount of cold work (both of which would affect the lengths of the pinned segments) it does not appear that this theory is correct.

A more likely explanation has been given by Seeger (1956a). This also requires the oscillation of segments of dislocation line, but in this theory it is shown that there is a unique length of line along the close-packed direction which will bulge out alternately to the atomic planes on either side of its original position. This unique length of line is determined solely by considerations of minimum-line energy and it is not dependent on any mechanical pinning at the ends of the segment. Reducing the theory to its simplest terms, we can consider the dislocation trying to bulge out to the next plane as in Fig. 11.15. The force on the dislocation will be a maximum at some point between the planes and therefore, in order that as short a length as possible should be in this region, it ought to cut across at right angles to its original length, as in curve A. But this will make the dislocation much longer and hence its total line energy will increase. If, on the other hand, we make the dislocation bulge on to the next plane with as little change in length as possible, as in curve B, then a large part of its length will be in the region between the planes where the force on it will be high and so its energy will once again be increased. There will be one configuration which will give a minimum energy as shown by curve C and this will determine the activation energy U in (11.35). Once the dislocation has surmounted the potential barrier between the planes it will, as it reaches regions where its strain energy is lowered, be able to pull more of itself over until it is stopped by a portion which is well anchored to the original plane. The process will then be halted rather suddenly and in the ensuing confusion the energy of motion will be liberated as heat, thus giving rise to the damping which is observed.

The main objection to the theory is that in order to obtain agreement with the experimental results a rather high value has to be assumed for the force required to move the dislocation from one plane to the next (the Peierls–Nabarro force). In view of the uncertainties about the nature of the forces on dislocations this disagreement should not be regarded too seriously, however.

The idea of vibrating segments of dislocation lines has been given further confirmation (Niblett and Wilks, 1957) by experiments which show that the internal friction peaks of cold-worked (5 per cent extension) copper which is either impure (0·0026 per cent Bi+0·032 per cent P), or has been neutron-irradiated (10^{18} nvt), occur at the same temperature as in the original experiments (i.e. U is the same), but their height is about ten times less. This can be explained by assuming that many of the dislocations will be anchored, either by the impurity atoms or by the irradiation damage, and hence fewer of them can vibrate. Thus the energy absorption is reduced, but the activation energy of those which are still free to vibrate will be the same as it was previously.

The reason why two internal friction peaks are observed is due to the fact that a dislocation which lies along a close-packed direction can have its Burgers vector either in that direction, in which case it is a screw dislocation, or in one of the other close-packed directions, when it would be a mixed dislocation. It is reasonable to assume that the activation energies and relaxation times for the vibration of these two kinds of dislocation will be different and hence two peaks will be detected. A detailed review of the mechanisms of dislocation damping is given by Niblett and Wilks (1960).

INDEX OF DATA

AUTHOR INDEX AND BIBLIOGRAPHY

Alers, P. B. (1956), *Phys. Rev.* **101**, 41: 145.
— (1957), *Phys. Rev.* **105**, 104: 178.
van Alphen, P. M., *see* de Haas.
Andrew, E. R. (1949), *Proc. Phys. Soc.* A **62**, 77: 100, 101.
— (1955), *Nuclear Magnetic Resonance* (Cambridge University Press): 330.
Appleyard, E. T. S., Bristow, J. R., London, H., and Misener, A. D. (1939), *Proc. Roy. Soc.* A **172**, 540: 173.
Arp, V. D., Kropschott, R. H., Wilson, J. H., Love, W. F., and Phelan, R. (1961), *Phys. Rev. Letters* **6**, 452: 205.
— Kurti, N., and Petersen, R. G. (1957), *Bull. Am. Phys. Soc.* (II) **2**, 388: 31.

Bagguley, D. M. S., and Owen, J. (1957), *Repts. on Progr. in Phys.* **20**, 304: 242, 329, 336.
— Stradling, R. A., and Whiting, J. S. S. (1961), *Proc. Roy. Soc.* A **262**, 340, 365: 243.
Bailey, C. A. (1959), D.Phil. thesis, Oxford University: 17, 20, 22.
— and Smith, P. L. (1959), *Phys. Rev.* **114**, 1010: 34.
Bailyn, M. (1958), *Phys. Rev.* **112**, 1587: 271.
Baker, J. M., Bleaney, B., and Bowers, K. D. (1956), *Proc. Phys. Soc.* B **69**, 1205: 304.
Balashova, B. M., and Sharvin, Yu. V. (1956), *J. Exp. Theor. Phys. USSR* **31**, 40: 177.
Balluffi, R. W., and Simmons, R. O. (1960), *Phys. Rev.* **117**, 52, 62: 82.
— — (1961), *Phys. Rev.* **125**, 862 (1962): 82.
Bardeen, J. (1937), *Phys. Rev.* **52**, 688: 96.
— (1956), *Hand. d. Phys.* **15**, 274: 147, 165, 187.
— Cooper, L. N., and Schrieffer, J. R. (1957), *Phys. Rev.* **108**, 1175: 128, 188.
— Rickayzen, G., and Tewordt, L. (1959), *Phys. Rev.* **113**, 982: 130.
— and Schrieffer, J. R. (1961), *Prog. in Low Temp. Phys.* **3**, 170: 147, 188, 204.
— and Shockley, W. (1950), *Phys. Rev.* **80**, 72: 224.
Barron, T. H. K., Berg, W. T., and Morrison, J. A. (1957), *Proc. 5th Int.*

Conf. Low Temp. Phys. and Chem. (University of Wisconsin Press), p. 445; *Proc. Roy. Soc.* A **242**, 478: 13.
Basinski, Z. S. (1957), *Proc. Roy. Soc.* A **240**, 229: 373.
— (1959), *Phil. Mag.* **4**, 393: 368, 369.
— and Sleeswyk, A. (1957), *Acta Met.* **5**, 176: 371.
Bates, L. F. (1951), *Modern Magnetism.* 3rd ed. (Cambridge University Press): 291.
van den Berg, G. J., *see* de Haas.
Berg, W. T., *see* Barron.
Berlincourt, T. G., Hake, R. R., and Leslie, D. H. (1961), *Phys. Rev. Letters* **6**, 671: 205.
— *see* Thorsen.
Berman, R. (1951), *Phil. Mag.* **42**, 642: 127.
— (1953), *Adv. in Phys.* **2**, 103: 67.
— Foster, E. L., and Rosenberg, H. M. (1955), *Rept. Conf. Defects in Solids* (Physical Society, London), p. 321: 66.
— — — (1955a), *Brit. J. Appl. Phys.* **6**, 181: 67.
— — Schneidmesser, B., and Tirmizi, S. M. A. (1960), *J. Appl. Phys.* **31**, 2156: 66.
— — and Ziman, J. M. (1955), *Proc. Roy. Soc.* A **231**, 130: 60, 61.
— — — (1956), *Proc. Roy. Soc.* A **237**, 344: 65.
— and MacDonald, D. K. C. (1952), *Proc. Roy. Soc.* A **211**, 122: 117, 119.
— Nettley, P. T., Sheard, F. W., Spencer, A. N., Stevenson, R. W. H., and Ziman, J. M. (1959), *Proc. Roy. Soc.* A **253**, 403: 62.
— Simon, F. E., and Ziman, J. M. (1953), *Proc. Roy. Soc.* A **220**, 171: 61.
Beyen, W., Bratt, P., Davis, H., Johnson, L., Levinstein, H., and MacRae, A. (1959), *J. Opt. Soc. Am.* **49**, 686: 257.
Bijl, D., *see* Casimir.
Blackman, M. (1951), *Proc. Phys. Soc.* A **64**, 681: 117.
— (1956), *Hand. d. Phys.* **7**, 325: 11.
Blakemore, J. S. (1959), *Phil. Mag.* **4**, 560: 240.
Blatt, F. J. (1955), *Phys. Rev.* **99**, 1708: 82.

Bleaney, B., Bowers, K. D., and Ingram, D. J. E. (1951), *Proc. Phys. Soc.* A **64**, 758 : 338.

— and Ingram, D. J. E. (1951), *Proc. Roy. Soc.* A **205**, 336 : 335.

— and Stevens, K. W. H. (1953), *Repts. on Progr. in Phys.* **16**, 108 : 328.

— see Baker; Bleaney, B. I.

Bleaney, B. I. and Bleaney, B. (1957), *Electricity and Magnetism* (Clarendon Press, Oxford): 285, 287, 295, 297, 309.

Blewitt, T. H., Coltman, R. R., Holmes, D. K., and Noggle, T. S. (1957), *Dislocations and Mechanical Properties of Crystals*, ed. Fisher *et al.* (Wiley, New York), p. 603 : 86, 87.

— Coltman, R. R., and Redman, J. K. (1955), *Rept. Conf. Defects in Cryst. Solids* (Physical Society, London), p. 369 : 83, 372.

Blewitt, T. H., see Sherrill.

Bloembergen, N. (1956), *Phys. Rev.* **104**, 324 : 341.

de Boer, J. H., see de Haas.

Bogle, G. S., Cooke, A. H., and Whitley, S. (1951), *Proc. Phys. Soc.* A **64**, 931 : 304.

Boorse, H. A. see Brown.

Bordoni, P. G. (1953), *Ricerca Sci.* **23**, 1193 : 382.

— (1954), *J. Acoust. Soc. Amer.* **26**, 495 : 382.

Borelius, G., Keesom, W. H., Johansson, C. H., and Linde, J. O. (1930), *Leiden Comm.* 206a; *Proc. Kon. Akad. Amst.* **33**, 17 : 264, 269.

— — — — (1930a), *Leiden Comm.* 206b; *Proc. Kon. Akad. Amst.* **33**, 32 : 274.

— — — — (1931), *Leiden Comm.* 217c; *Proc. Kon. Akad. Amst.* **34**, 1365 : 277.

— — — — (1932), *Leiden Comm.* 217e; *Proc. Kon. Akad Amst.* **35**, 25 : 269, 274, 275.

— — — — (1932a), *Leiden Comm.* 217d; *Proc. Kon. Akad. Amst.* **35**, 15 : 274.

Born, M., and von Kármán, T. (1912), *Phys. Z.* **13**, 297 : 5.

Bowers, K. D., and Owen, J. (1955), *Repts. on Progr. in Phys.* **18**, 304 : 304, 328, 337.

— see Baker; Bleaney, B.

Bozorth, R. M., Matthias, B. T., and Davis, D. D. (1960), *Proc. 7th Int. Conf. Low Temp. Phys.* (University of Toronto Press), p. 385 : 184.

Bragg, W. L., and Williams, E. J. (1934), *Proc. Roy. Soc.* A **145**, 699 : 35.

— — (1935), *Proc. Roy. Soc.* A **151**, 540 : 35.

Bratt, P., see Beyen.

Bremmer, H., see de Haas.

Bristow, J. R., see Appleyard.

Brooks, H. (1955), *Adv. in Electronics*, **7**, 85 : 225.

Broom, T. (1952), *Proc. Phys. Soc.* B **65**, 871 : 84.

— and Ham, R. K. (1957), *Proc. Roy. Soc.* A **242**, 166 : 381.

— — (1958), *Vacancies and other point defects in metals and alloys* (Institute of Metals, London), p. 41 : 85.

— Molineux, J. H., and Whittaker, V. N. (1955), *J. Inst. Metals* **84**, 357 : 381.

Brown, A. Zemansky, M. W., and Boorse, H. A. (1953), *Phys. Rev.* **92**, 52 : 159.

Browne, M. E., see Owen.

Buck, D. A. (1956), *Proc. I.R.E.* **44**, 482 : 206.

Bückel, W., and Hilsch, R. (1954), *Z. Phys.* **138**, 109 : 186.

Buehler, E., see Kunzler.

van Bueren, H. G. (1960), *Imperfections in Crystals* (North Holland): 85.

Bunch, M. D., see Powell, R. L.

Burstein, E., Picus, G. S., and Gebbie, H. A. (1956), *Phys. Rev.* **103**, 825 : 243.

— — Henvis, B., and Wallis, R. (1956), *J. Phys. Chem. Solids* **1**, 65 : 249, 250.

— see Picus; Sclar; Stiles.

Callaway, J. (1959), *Phys. Rev.* **113**, 1046 : 54, 55.

Carruthers, J. A., Geballe, T. H., Rosenberg, H. M., and Ziman, J. M. (1957), *Proc. Roy. Soc.* A **238**, 502 : 144.

Casimir, H. B. G. (1938), *Physica* **5**, 595 : 53.

— (1940), *Physica* **7**, 887 : 167.

— Bijl, D., and du Pré, F. K. (1941), *Physica* **8**, 449 : 321.

— see Gorter.

Chambers, R. G. (1952), *Proc. Roy. Soc.* A **215**, 481 : 104, 105.

— (1956), *Can. J. Phys.* **34**, 1395 : 349, 352.

— (1960), *The Fermi Surface*, ed. Harrison and Webb (Wiley, New York), p. 100 : 108, 110.

Chandrasekhar, B. S., and Hulm, J. K. (1960), *Proc. 7th Int. Conf. Low Temp. Phys.* p. 672 (University of Toronto Press) : 13.

Chanin, G., Lynton, E. A., and Serin, B. (1959), *Phys. Rev.* **114**, 719 : 182, 184.

Chaudhuri, K. D., Mendelssohn, K., and Thompson, M. W. (1960), *Cryogenics* **1**, 47: 142.

Chester, P. F., and Jones, G. O. (1953), *Phil. Mag.* **44**, 1281: 186.

Chotkewitsch, W. I., *see* Shubnikov.

Christian, J. W., Jan, J. P., Pearson, W. B., and Templeton, I. M. (1958), *Proc. Roy. Soc.* A **245**, 213: 264.

— and Spreadborough, J. (1956), *Phil. Mag.* **1**, 1069: 85.

Clark, C. W., *see* Giauque.

Clusius, K. (1932), *Z. Elektrochem.* **38**, 312: 147.

— and Perlick, A. (1934), *Z. Phys. Chem.* B **24**, 313: 33.

Cohen, A. Foner (1957), *Proc. 5th Int. Conf. Low Temp. Phys. and Chem.* (University of Wisconsin Press), p. 385: 66.

Collins, S. C. (1955), *Conf. de Phys. des Basses Temp., Suppl. to Bulletin of the Intern. Inst. of Refrigeration* (Paris), p. 588: 149.

Coltman, R. R., *see* Blewitt.

Conwell, E. M. (1956), *Phys. Rev.* **103**, 51: 238.

— and Weisskopf, V. F. (1950), *Phys. Rev.* **77**, 388: 225.

— *see* Debye, P. P.

Cooke, A. H. (1950), *Repts. on Progr. in Phys.* **13**, 276: 320.

— and Duffus, H. J. (1954), *Proc. Phys. Soc.* A **67**, 525: 318.

— — and Wolf, W. P. (1953), *Phil. Mag.* **44**, 623: 307.

— and Edmonds, D. T. (1958), *Proc. Phys. Soc.* **71**, 517: 31.

— Meyer, H., and Wolf, W. P. (1956), *Proc. Roy. Soc.* A **237**, 404: 29.

— — — Evans, D. F., and Richards, R. E. (1954), *Proc. Roy. Soc.* A **225**, 112: 319.

— *see* Bogle.

Cooper, L. N. (1956), *Phys. Rev.* **104**, 1189: 194.

— (1960), *Amer. J. Phys.* **28**, 91: 188, 192.

— *see* Bardeen.

Corak, W. S., Goodman, B. B., Satterthwaite, C. B., and Wexler, A. (1954), *Phys. Rev.* **96**, 1442: 161.

Corenzwit, E., *see* Hein; Matthias.

Corruccini, R. J., *see* Powell, R. L.

Cottrell, A. H. (1953), *Dislocations and Plastic Flow in Crystals* (Clarendon Press, Oxford): 359, 370.

— (1958), *Vacancies and Other Point Defects in Metals and Alloys* (Institution of Metals, London), p. 1: 85.

— (1958a), *Trans. Met. Soc. A.I.M.E.* **212**, 1952: 371.

— and Hull, D. (1957), *Proc. Roy. Soc.* A **242**, 211: 379.

Croft, A. J., Faulkner, E. A., Hatton, J., and Seymour, E. F. W. (1953), *Phil. Mag.* **44**, 289: 89.

Cuevas, M., *see* Fritzsche.

Darby, J., Hatton, J., Rollin, B. V., Seymour, E. F. W., and Silsbee, H. B. (1951), *Proc. Phys. Soc.* A **64**, 861: 135.

Daunt, J. G., and Mendelssohn, K, (1938), *Nature* **141**, 116: 128, 277.

— — (1946), *Proc. Roy. Soc.* A **185**, 225: 128, 277.

— *see* Heer.

Davis, D. D., *see* Bozorth.

Davis, H., *see* Beyen.

Debye, P. (1912), *Ann. Phys.* (4) **39**, 789: 5.

— (1914), *Vorträge über die kinetische Theorie der Materie und der Elektrizität*, M. Plank *et al.* (Teubner, Leipzig and Berlin), p. 19: 44.

— (1929), *Polar Molecules* (The Chemical Catalog Co. Inc.—Reinhold Publishing Corp.), ch. 5: 320.

Debye, P. P., and Conwell, E. M. (1954), *Phys. Rev.* **93**, 693: 231, 234, 235, 236, 237.

Dekker, A. J. (1957), *Solid State Physics* (Prentice Hall, New York; Macmillan, London): 68, 218, 220, 308, 316, 359.

Désirant, M., and Shoenberg, D. (1948), *Proc. Phys. Soc.* **60**, 413: 167.

— — (1948a), *Proc. Roy. Soc.* A **194**, 63: 155.

DeSorbo, W. (1960), *Phys. Rev. Letters* **4**, 406: 178.

Dexter, D. L., and Seitz, F. (1952), *Phys. Rev.* **86**, 964: 226.

Dingle, R. B. (1953), *Physica* **19**, 311: 102.

Doidge, P. R. (1956), *Phil. Trans.* A **248**, 553: 187.

Doyama, M., and Koehler, J. S. (1960), *Phys. Rev.* **119**, 939: 82.

Dresselhaus, G., Kip, A. F., and Kittel, C. (1953), *Phys. Rev.* **92**, 827: 241.

— — — (1955), *Phys. Rev.* **98**, 368: 242.

Duffus, H. J., *see* Cooke.

Edmonds, D. T., *see* Cooke.

Egan, C. J., *see* Giauque.

Einstein, A. (1907), *Ann. Phys.* (4) **22**, 180, 800: 4.

Erginsoy, C. (1950), *Phys. Rev.* **79**, 1013: 225.

Erickson, R. A. and Heer, C. V. (1957), *Phys. Rev.* **108**, 896 : 32.

Euken, A. (1911), *Ann. Phys.* (4) **34**, 185; *Phys. Z.* **12**, 1005; *Verh. der Deutschen Phys. Gesellschaft* **13**, 829 : 44.

Evans, D. F., *see* Cooke.

Faber, T. E. (1952), *Proc. Roy. Soc.* A **214**, 392 : 179.

— and Pippard, A. B. (1955), *Prog. in Low Temp. Phys.* **1**, 159 : 180.

Fan, H. Y. (1956), *Repts. on Progr. in Phys.* **19**, 107 : 244, 247, 249, 252.

— and Fisher, P. (1959), *J. Phys. Chem. Solids*, **8**, 270 : 251.

— *see* Kaiser; Ray.

Faulkner, E. A., *see* Croft.

Feher, G. (1956), *Phys. Rev.* **103**, 834 : 339.

Fereday, R. A., and Wiersma, E. C. (1935), *Physica* **2**, 575 : 306.

Finn, C. B. P., Orbach, R., and Wolf, W. P. (1961), *Proc. Phys. Soc.* **77**, 261 : 325, 326.

Fisher, P., *see* Fan.

Foner, S., *see* Keyes.

Foster, E. L., *see* Berman.

Frank, J. P., Manchester, F. D., and Martin, D. L. (1961), *Proc. Roy. Soc.* A **263**, 494 : 90.

Fraser, D. B., and Hollis Hallett, A. C. (1960), *Proc. 7th Int. Conf. Low Temp. Phys.* p. 689 (University of Toronto Press) : 39.

Frederikse, H. P. R. (1953), *Phys. Rev.* **92**, 248 : 279.

— and Mielczarek, E. V. (1955), *Phys. Rev.* **99**, 1889 : 280.

Fritzsche, H. (1955), *Phys. Rev.* **99**, 406 : 238.

— (1958), *J. Phys. Chem. Solids* **6**, 69 : 239.

— and Cuevas, M. (1960), *Phys. Rev.* **119**, 1238 : 240.

— and Lark-Horovitz, K. (1955), *Phys. Rev.* **99**, 400 : 261.

Fröhlich, H. (1950), *Phys. Rev.* **79**, 845 : 186.

Gaffney, J., *see* Overton.

Geballe, T. H., and Hull, G. W. (1954), *Phys. Rev.* **94**, 1134 : 279, 283.

— — (1955), *Phys. Rev.* **98**, 940 : 280, 282.

— — (1958), *Phys. Rev.* **110**, 773 : 62, 64, 143.

— *see* Carruthers; Hein; Herring; Matthias.

Gebbie, H. A., *see* Burstein.

Geller, S., *see* Matthias.

Gerritsen, A. N., and Linde, J. O. (1952), *Physica* **18**, 877 : 90.

Giauque, W. F., Stout, J. W., Egan, C. J., and Clark, C. W. (1941), *J. Amer. Chem. Soc.* **63**, 405 : 292.

Giaver, I. (1960), *Phys. Rev. Letters* **5**, 147 : 200.

— (1960a), *Proc. 7th Int. Conf. Low Temp. Phys.* (University of Toronto Press), p. 327 : 202.

— (1960b), *Phys. Rev. Letters* **5**, 464 : 202, 204.

Gibson, J. W., *see* Hein.

Ginsberg, D. M., and Tinkham, M. (1960), *Phys. Rev.* **118**, 990 : 198.

Glen, J. W. (1956), *Phil. Mag.* **1**, 400 : 377.

Glover, R. E., and Tinkham, M. (1957), *Phys. Rev.* **108**, 243 : 198.

Gold, A. V., MacDonald, D. K. C., Pearson, W. B., and Templeton, I. M. (1960), *Phil. Mag.* **5**, 765 : 267, 272.

Goodman, B. B. (1953), *Proc. Phys. Soc.* A **66**, 217 : 131.

— *see* Corak.

Gordon, J. P., Zeiger, H. J., and Townes, C. H. (1954), *Phys. Rev.* **95**, 282 : 340.

Gorter, C. J. (1936), *Physica* **3**, 503 : 320.

— (1947), *Paramagnetic Relaxation* (Elsevier, Amsterdam) : 299, 320.

— and Casimir, H. B. G. (1934), *Phys. Z.* **35**, 963 : *Z. tech. Phys.* **15**, 539 : 128, 158, 160.

— and de Haas, W. J. (1931), *Leiden Comm.* 218b : 303.

— *see* de Haas.

Graham, G. M. (1958), *Proc. Roy. Soc.* A **248**, 522 : 131, 132.

Guénault, A. M., and MacDonald, D. K. C. (1961), *Proc. Roy. Soc.* A **264**, 41 : 272, 273.

Gunn, J. B. (1957), *Progr. in Semiconductors* **2**, 213 : 258.

Gunther-Mohr, G. R., *see* Koenig.

Gurevich, L. (1945), *J. Phys. USSR.* **9**, 477 : 269.

— (1946), *J. Phys. USSR* **10**, 67 : 269.

de Haas, W. J., and van Alphen, P. M. (1930), *Leiden Comm.* 212a : 349.

— and de Boer, J. H. (1934), *Physica* **1**, 609 : 98.

— — and van den Berg, G. J. (1934), *Physica* **1**, 1115 : 87, 88.

— and Bremmer, H. (1931), *Proc. Kon. Akad. Wet. Amst.* **34**, 325 : 126.

— — (1936), *Physica* **3**, 672, 687 : 126.

— and Gorter, C. J. (1930), *Leiden Comm.* 208c : 296.

de Haas, W. J., and Voogd, J. (1931), *Leiden Comm.* 214c : 149.

— — and Jonker, J. M. (1934), *Physica* **1**, 281 : 150.

— *see* Gorter.

Hake, R. R., *see* Berlincourt.

Ham, R. K., *see* Broom.

van den Handel, J. (1956), *Hand. d. Phys.* **15,** 1 : 300.

Harper, A. F. A., Kemp, W. R. G., Klemens, P. G., Tainsh, R. J., and White, G. K. (1957), *Phil. Mag.* **2**, 577 : 119.

Harrison, W. A. (1958), *J. Phys. Chem. Solids* **5**, 44 : 82.

Hatton, J., *see* Croft ; Darby.

Heer, C. V., and Daunt, J. G. (1949), *Phys. Rev.* **76**, 854 : 131.

— *see* Erickson.

Hein, R. A., Gibson, J. W., Matthias, B. T., Geballe, T. H., and Corenzwit, E. (1962), *Phys. Rev. Letters* **8**, 408 : 148.

Heitler, W. (1945), *Elementary Wave Mechanics* (Clarendon Press, Oxford) : 318.

Henvis, B., *see* Burstein ; Picus.

Herman, F. (1954), *Phys. Rev.* **93**, 1214 : 228, 241.

Herpin, A. (1952), *Annales de Physique* (12) **7**, 91 : 51.

Herring, C. (1954), *Phys. Rev.* **96**, 1163 : 283, 284.

— (1958), *Halbleiter und Phosphor*, ed. Schön and Welker (Vieweg, Braunschweig), p. 184 : 281, 282, 283, 284.

— Geballe, T. H., and Kunzler, J. E. (1958), *Phys. Rev.* **111**, 36 : 261.

— — — (1959), *J. Phys. Chem. Solids* **8**, 347 : 261.

— — — (1959a), *Bell Syst. Tech. J.* **38**, 657 : 261.

Herzberg, G. (1944), *Atomic Spectra and Atomic Structure*, 2nd ed. (Dover, New York) : 287.

Hill, R. W. (1952), D.Phil. thesis, Oxford University : 26.

— and Ricketson, B. W. A. (1954), *Phil. Mag.* **45**, 277 : 33, 34.

— and Smith, P. L. (1953), *Phil. Mag.* **44**, 636 : 18.

Hilsch, R., *see* Bückel.

Hirsch, P. B., Horne, R. W., and Whelan, M. J. (1956), *Phil. Mag.* **1**, 677 : 84, 139.

Hoare, F. E., and Yates, B. (1957), *Proc. Roy. Soc.* A **240**, 42 : 20.

Hollis Hallett, A. C., *see* Fraser.

Holmes, D. K., *see* Blewitt.

Holst, G., *see* Onnes.

Honig, A. and Levitt, R. (1960), *Phys. Rev. Letters* **5**, 93 : 260.

Horne, R. W., *see* Hirsch.

Howie, A. (1960), *Phil. Mag.* **5,** 251 : 84.

Hsu, F. S. L., *see* Kunzler.

Hulbert, J. A. and Jones, G. O. (1955), *Proc. Phys. Soc.* B **68**, 801 : 209.

Hull, D. (1958), *Phil. Mag.* **3**, 1468 : 371.

— and Rosenberg, H. M. (1959), *Phil. Mag.* **4**, 303 : 371, 372.

— *see* Cottrell.

Hull, G. W., *see* Geballe.

Hull, J. R., and Hull, R. A. (1941), *J. Chem. Phys.* **9**, 465 : 291.

Hull, R. A., *see* Hull, J. R.

Hulm, J. K. (1950), *Proc. Roy. Soc.* A **204**, 98 : 116, 129, 130.

— *see* Chandrasekhar.

Hume-Rothery, W. (1946), *Atomic Theory for Students of Metallurgy* (Institute of Metals, London) : 18.

Hung, C. S. (1950), *Phys. Rev.* **79**, 727 : 238.

Hunter, S. C., and Nabarro, F. R. N. (1953), *Proc. Roy. Soc.* A **220**, 542 : 83.

Ingram, D. J. E. (1955), *Spectroscopy at Radio and Microwave Frequencies* (Butterworths, London) : 328.

— *see* Bleaney, B.

Jackson, L. C. (1936), *Proc. Phys. Soc.* **48**, 741 : 302.

Jahn, H. A., and Teller, E. (1937), *Proc. Roy. Soc.* A **161**, 220 : 298.

James, H. M. (1956), *Quarterly Report Oct.–Dec. 1956*, Purdue University : 240.

Jan, J. P., *see* Christian.

Johansson, C. H., and Linde, J. O. (1936), *Ann. Phys.* (5) **25**, 1 : 79.

— *see* Borelius.

Johnson, F. A. (1959), *Proc. Phys. Soc.* **73**, 265 : 252.

Johnson, L., and Levinstein, H. (1960), *Phys. Rev.* **117**, 1191 : 254.

— *see* Beyen.

Johnson, V. A., and Lark-Horovitz, K. (1957), *Prog. in Low Temp. Phys.* **2**, 187 : 23.

Jones, G. O., *see* Chester ; Hulbert.

Jones, H., *see* Mott.

Jonker, J. M., *see* de Haas.

Joos, G. (1934), *Theoretical Physics* (Blackie, London) : 3.

Kaiser, W., and Fan, H. Y. (1954), *Phys. Rev.* **93**, 977 : 253.

von Kármán, T., *see* Born.

Keesom, P. H., and Seidel, G. (1959), *Phys. Rev.* **113**, 33 : 22.

Keesom, W. H., and Kurelmeyer, B. (1940), *Physica* **7**, 1003 : 20.

— and van Laer, P. H. (1938), *Physica* **5**, 193 : 159.

— and Matthijs, C. J. (1935), *Physica* **2**, 623 : 275.

— *see* Borelius.

Kemp, W. R. G., Klemens, P. G., and Tainsh, R. J. (1959), *Phil. Mag.* **4**, 845 : 137.

— *see* Harper.

Keyes, R. J., Zwerdling, S., Foner, S., Kolm, H. H., and Lax, B. (1956), *Phys. Rev.* **104**, 1804 : 243.

Kip, A. F., *see* Dresselhaus.

Kittel, C. (1956), *Introduction to Solid State Physics*, 2nd ed. (Wiley, New York) : 5, 68, 359.

— *see* Dresselhaus ; Owen.

Klauder, J. R., and Kunzler, J. E. (1960), *The Fermi Surface*, ed. Harrison and Webb (Wiley, New York), p. 125 : 109.

Klemens, P. G. (1954), *Aust. J. Phys.* **7**, 57 : 121.

— (1955), *Proc. Phys. Soc.* A **68**, 1113 : 137.

— (1956), *Hand. d. Phys.* **14**, 198 : 114, 115, 116, 124.

— (1958), *Solid State Physics* **7**, 1 : 51, 52, 57.

— *see* Harper ; Kemp.

de Klerk, D. (1956), *Hand. d. Phys.* **15**, 38 : 344.

Knight, W. D., *see* Owen.

Koehler, J. S., *see* Doyama.

Koenig, S. H. (1959), *J. Phys. Chem. Solids* **8**, 227 : 260.

— and Gunther-Mohr, G. R. (1957), *J. Phys. Chem. Solids* **2**, 268 : 258, 259, 260.

Kohler, M. (1938), *Ann. Phys.* (5) **32**, 211 : 106.

— (1949), *Ann. Phys.* (6) **6**, 18 : 107.

Kohn, W. (1957), *Solid State Physics* **5**, 257 : 251.

— and Luttinger, J. M. (1955), *Phys. Rev.* **98**, 915 : 250.

Kolm, H. H., *see* Keyes.

Koppe, H. (1947), *Ann. Phys.* (6) **1**, 405 : 128.

Kramers, H. A. (1930), *Proc. Kon Akad. Amst.* **33**, 959 : 298.

Krishnan, K. S., and Mookherji, A. (1938), *Phil. Trans.* A **237**, 135 : 335.

Kronig, R. de L. (1939), *Physica* **6**, 33 : 325.

Kropschott, R. H., *see* Arp.

Kubo, R., *see* Nagamiya.

Kunzler, J. E., Buehler, E., Hsu, F. S. L., and Wernick, J. H. (1961), *Phys. Rev. Letters* **6**, 90 : 206.

— *see* Herring ; Klauder.

Kuper, C. G. (1951), *Phil. Mag.* **42**, 961 : 177.

Kurelmeyer, B., *see* Keesom, W. H.

Kurti, N., *see* Arp.

van Laer, P. H., *see* Keesom, W. H.

Landau, L. D. (1930), *Z. Phys.* 64, 629 : 348.

— (1937), *Phys. Z. Sowjet.* **11**, 129 : 177.

Langenberg, D. N. (1959), Doctoral diss. University of California, Berkeley, California : 81.

— *see* Stiles.

Laredo, S. J. (1955), *Proc. Roy. Soc.* A **229**, 473 : 131.

— and Pippard, A. B. (1955), *Proc. Camb. Phil. Soc.* **51**, 368 : 133.

Lark-Horovitz, K., *see* Fritzsche ; Johnson, V. A.

Laurmann, E., and Shoenberg, D. (1949), *Proc. Roy. Soc.* A **198**, 560 : 167.

Lax, B., and Mavroides, J. G. (1960), *Solid State Physics* **11**, 261 : 242.

— *see* Keyes.

Leibfried, G., and Schlömann, E. (1954), *Nachr. Gött. Akad. Math. Phys.* **2a**, 71 : 53.

Leslie, D. H., *see* Berlincourt.

Levinstein, H., *see* Beyen ; Johnson, L.

Levitt, R., *see* Honig.

Linde, J. O. (1932), *Ann. Phys.* (5) **15**, 219 : 77.

— *see* Borelius ; Gerritsen ; Johansson.

Lock, J. M. (1951), *Proc. Roy. Soc.* A **208**, 391 : 167, 172.

Lomer, J. N. (1958), D.Phil. thesis, Oxford University : 123, 124, 125.

— and Rosenberg, H. M. (1959), *Phil. Mag.* **4**, 467 : 84, 137, 138, 139, 140.

London, F. (1950), *Superfluids*, vol. 1 (Wiley, New York) : 163, 165, 168, 186.

— and London, H. (1935), *Proc. Roy. Soc.* A **149**, 71 : 163.

— — *Physica* **2**, 341 : 163.

London, H. (1940), *Proc. Roy. Soc.* A **176**, 522 : 102, 169.

— *see* Appleyard ; London, F.

Long, D., and Myers, J. (1959), *Phys. Rev.* **115**, 1107: 237.

Lounasmaa, O. V., and Roach, P. R. (1962), *Phys. Rev.* **128**, 622: 31.

Love, W. F., *see* Arp.

Low, W. (1960), *Paramagnetic Resonance in Solids* (Academic Press, New York): 329.

Luttinger, J. M., *see* Kohn.

Lynton, E. A. (1962), *Superconductivity* (Methuen, London): 147.

— *see* Chanin.

McCammon, R. D. (1957), D.Phil. thesis, Oxford University: 88.

— and Rosenberg, H. M. (1957), *Proc. Roy. Soc.* A **242**, 203: 370, 379, 380.

MacDonald, D. K. C. (1954), *Physica* **20**, 996: 270.

— (1956), *Handb. d. Phys.* **14**, 137: 88, 95, 100.

— (1962), *Thermo-electricity* (Wiley, New York): 262.

— and Pearson, W. B. (1953), *Proc. Roy. Soc.* A **219**, 373: 89, 274, 275.

— — and Templeton, I. M. (1958), *Phil. Mag.* **3**, 657: 267.

— — — (1958a), *Proc. Roy. Soc.* A **248**, 107: 268.

— — — (1960), *Proc. Roy. Soc.* A **256**, 334: 268, 272, 273.

— and Sarginson, K. (1950), *Proc. Roy. Soc.* A **203**, 223: 100.

— White, G. K., and Woods, S. B. (1956), *Proc. Roy. Soc.* A **235**, 358: 119.

— *see* Berman; Gold; Guénault.

Macfarlane, G. G., McLean, T. P., Quarrington, J. E., and Roberts, V. (1957), *Phys. Rev.* **108**, 1377: 246.

McLean, T. P., *see* Macfarlane.

MacRae, A., *see* Beyen.

McSkimin, H. J. (1953), *J. Appl. Phys.* **24**, 988: 362.

Maita, J. P., *see* Morin.

Makinson, R. E. B. (1938), *Proc. Camb. Phil. Soc.* **34**, 474: 121.

Manchester, F. D., *see* Frank.

Mapother, D. E. (1962), *I.B.M. J. Res. and Developm.* **6**, 77: 151.

Martin, D. L., *see* Frank.

Mason, W. P. (1955), *Bell Syst. Tech. J.* **34**, 903: 384.

Matthias, B. T. (1955), *Phys. Rev.* **97**, 74: 182, 183.

— (1957), *Prog. in Low Temp. Phys.* **2**, 138: 181.

— Geballe, T. H., Geller, S., and Corenzwit, E. (1954), *Phys. Rev.* **95**, 1435: 205.

— *see* Bozorth; Hein.

Matthijs, C. J., *see* Keesom, W. H.

Mavroides, J. G., *see* Lax.

Maxwell, E. (1950), *Phys. Rev.* **78**, 477: 186.

Meissner, W. (1927), *Z. ges. Kälteindustr.* **34**, 197: 277.

— and Ochsenfeld, R. (1933), *Naturwiss.* **21**, 787: 152.

— Polanyi, M., and Schmid, E. (1930), *Z. Phys.* **66**, 477: 377.

— and Voigt, B. (1930), *Ann. Phys.* (5) **7**, 761: 87.

Mendelssohn, K. (1935), *Proc. Roy. Soc.* A **152**, 34: 181.

— (1936), *Proc. Roy. Soc.* A **155**, 558: 155.

— and Moore, J. R. (1935), *Nature* **135**, 826: 181.

— and Olsen, J. L. (1950), *Proc. Phys. Soc.* A **63**, 2, 1182; *Phys. Rev.* **80**, 859: 133, 134, 135.

— and Pontius, R. B. (1937), *Phil. Mag.* **24**, 777: 126.

— and Renton, C. A. (1955), *Proc. Roy. Soc.* A **230**, 157: 131.

— and Rosenberg, H. M. (1953), *Proc. Roy. Soc.* A **218**, 190: 145.

— — (1961), *Solid State Physics*, **12**, 223: 142.

— Ruhemann, M., and Simon, F. E. (1931), *Z. Phys. Chem.* B **15**, 121: 33.

— *see* Chaudhuri; Daunt.

Mendoza, E. (1961), *Experimental Cryophysics*, ed. by Hoare, Jackson, and Kurti (Butterworths, London), p. 165: 344.

— and Thomas, J. G. (1951), *Phil. Mag.* **42**, 291: 89.

Meshkovsky, A. G., and Shalnikov, A. I. (1947), *J. Phys. USSR* **11**, 1: 177, Plate 1 (facing p. 178).

Meyer, H., O'Brien, M. C. M., and Van Vleck, J. H. (1958), *Proc. Roy. Soc.* A **243**, 414: 319.

— and Smith, P. L. (1959), *J. Phys. Chem. Solids*, **9**, 285: 30.

— *see* Cooke.

Mielczarek, E. V., *see* Frederikse.

Mikura, Z. (1941), *Proc. Phys. Math. Soc. Japan* **23**, 309: 40.

Miller, A. R., *see* Roberts, J. K.

Misener, A. D., *see* Appleyard.

Molineux, J. H., *see* Broom.

Montgomery, H., and Pells, G. P. (1959), *Proc. Xth Intern. Congress of Refrig.*, *Copenhagen*, vol. 1, p. 139 (Pergamon Press, Oxford): 20, 21.

Mookherji, A., *see* Krishnan.

Moore, J. R., *see* Mendelssohn.

Morin, F. J. (1954), *Phys. Rev.* **93**, 62: 236.

— and Maita, J. P. (1954), *Phys. Rev.* **94**, 1525: 236.

— — (1954a), *Phys. Rev.* **96**, 28: 236.

Morrison, J. A., *see* Barron.

Mott, N. F. (1934), *Proc. Phys. Soc.* **46**, 680: 79.

— (1936), *Proc. Camb. Phil. Soc.* **32**, 281: 77.

— (1936a), *Proc. Roy. Soc.* A, **153**, 699: 97.

— (1956), *Can. J. Phys.* **34**, 1356: 240.

— (1956a), *Phil. Mag.* **1**, 568: 377.

— and Jones, H. (1936), *The Theory of the Properties of Metals and Alloys* (Clarendon Press, Oxford; reprint, 1958, by Dover Publications, New York, and Oxford University Press): 48, 266, 348, 349.

Myers, J., *see* Long.

Nabarro, F. R. N., *see* Hunter.

Nagamiya, T., Yosida, K., and Kubo, R. (1955), *Adv. in Phys.* **4**, 1: 312.

Nakhutin, I. E., *see* Shubnikov.

Nesbitt, L. B., *see* Reynolds.

Nettley, P. T., *see* Berman.

Newman, R., and Tyler, W. W. (1959), *Solid State Physics* **8**, 47: 254, 255, 256.

Niblett, D. H., and Wilks, J. (1957), *Phil. Mag.* **2**, 1427: 382, 383, 385.

— — (1960), *Adv. in Phys.* **9**, 1: 385.

Noggle, T. S., *see* Blewitt.

Nordheim, L. (1931), *Ann. Phys.* (5) **9**, 607, 641: 78.

O'Brien, M. C. M., *see* Meyer.

Ochsenfeld, R., *see* Meissner.

Olsen, J. L., and Renton, C. A. (1952), *Phil. Mag.* **43**, 946: 131.

— *see* Mendelssohn.

Olsen-Bär, M. (1956), D.Phil. thesis, Oxford University: 98, 99.

Onnes, H. K. (1911), *Leiden Comm.* 199*b*, 120*b*, 122*b*: 147.

— and Holst, G. (1914), *Leiden Comm.* 142*c*: 126.

Orbach, R., *see* Finn.

Orton, J. W. (1959), *Repts. on Progr. in Phys.* **22**, 204: 329.

Overton, W. C. (1960), *Proc. 7th Int. Conf. Low Temp. Phys.* p. 677 (University of Toronto Press): 13.

— and Gaffney, J. (1955), *Phys. Rev.* **98**, 969: 361, 362.

Owen, J., Browne, M. E., Knight, W. D. and Kittel, C. (1956), *Phys. Rev.* **102**, 1501: 90.

— *see* Bagguley; Bowers.

Parkinson, D. H. (1958), *Repts. on Progr. in Phys.* **21**, 226: 11.

Pauthenet, R. (1958), *Ann. de Phys.* (13) **3**, 424: 317.

Pearson, W. B. (1955), *Phil. Mag.* **46**, 911, 920: 89.

— and Templeton, I. M. (1961), *Can. J. Phys.* **39**, 1084: 276.

— *see* Christian; Gold; MacDonald.

Peierls, R. (1929), *Ann. Phys.* (5) **3**, 1055: 47.

— (1933) *Z. Phys.* **81**, 186: 349.

— (1955), *Repts. on Progr. in Phys.* **18**, 424: 190.

Pells, G. P., *see* Montgomery.

Perlick, A., *see* Clusius.

Petersen, R. G., *see* Arp.

Phelan, R., *see* Arp.

Picus, G. S., Burstein, E., and Henvis, B. (1956), *J. Phys. Chem. Solids* **1**, 75: 250, 251.

— *see* Burstein.

Pippard, A. B. (1947), *Proc. Roy. Soc.* A **191**, 385: 102.

— (1947a), *Proc. Roy. Soc.* A **191**, 370: 171.

— (1947b), *Proc. Roy. Soc.* A **191**, 399: 169, 170.

— (1953), *Proc. Roy. Soc.* A **216**, 547: 186, 187.

— (1954), *Adv. in Electronics*, **6**, 1: 102, 168, 170.

— (1954a), *Proc. Roy. Soc.* A **224**, 273: 104.

— (1957), *Phil. Trans.* A **250**, 325: 104, 105.

— (1957a), *The Elements of Classical Thermodynamics* (Cambridge University Press): 156, 326.

— (1960), *Repts. on Progr. in Phys.* **23**, 176: 105, 353, 358.

— *see* Faber; Laredo.

Polanyi, M., *see* Meissner.

Pontius, R. B., *see* Mendelssohn.

Powell, R. L., Bunch, M. D., and Corruccini, R. J. (1961), *Cryogenics* **1**, 139: 275.

du Pré, F. K., *see* Casimir.

Pullan, G. T. (1953), *Proc. Roy. Soc.* A **217**, 280: 275, 277.

Putley, E. H. (1959), *Proc. Phys. Soc.* **73**, 280: 236.

— (1959a), *Proc. Phys. Soc.* **73**, 128: 236.

Putley, E. H. (1960), *The Hall Effect and Related Phenomena* (Butterworths, London): 221.

Quarrington, J. E., *see* Macfarlane.

Ray, R. K., and Fan, H. Y. (1961), *Phys. Rev.* **121**, 768: 240.

Rayleigh, Lord (1896), *Theory of Sound* (London, 2nd ed.), vol. 2: 54.

Rayne, J. A. (1954), *Phys. Rev.* **95**, 1428: 16.

— (1957), *Phys. Rev.* **108**, 22: 20.

Read, W. T. (1953), *Dislocations in Crystals* (McGraw Hill, New York): 84, 359.

— *see* Shockley.

Redman, J. K., *see* Blewitt.

Renton, C. A. (1955), *Phil. Mag.* **46**, 47: 131.

— *see* Mendelssohn; Olsen.

Reuter, G. E. H., and Sondheimer, E. H. (1948), *Proc. Roy. Soc.* A **195**, 336: 102.

Reynolds, C. A., Serin, B., Wright, W. H., and Nesbitt, L. B. (1950), *Phys. Rev.* **78**, 487: 186.

Richards, P. L., and Tinkham, M. (1960), *Phys. Rev.* **119**, 575: 198, 199, 200, 203.

Richards, R. E., *see* Cooke.

Rickayzen, G., *see* Bardeen.

Ricketson, B. W. A., *see* Hill.

Rjabinin, J. N., and Shubnikov, L. W. (1935), *Nature* **135**, 581; *Phys. Z. Sowjet.* **7**, 122: 181.

— *see* Shubnikov.

Roach, P. R., *see* Lounasmaa.

Roberts, D. H., and Wilson, B. L. H. (1958), *Brit. J. Appl. Phys.* **9**, 291: 257.

Roberts, J. K., and Miller, A. R. (1960), *Heat and Thermodynamics*, 5th ed. (Blackie, London): 5, 6, 39.

Roberts, L. M. (1957), *Proc. Phys. Soc.* B **70**, 744: 12, 17.

Roberts, V., *see* Macfarlane.

Rollin, B. V., and Rowell, J. M. (1960), *Proc. Phys. Soc.* **76**, 1001: 260.

— *see* Darby.

Rose, A. (1957), *Progr. in Semiconductors* **2**, 111: 257.

Rosenberg, H. M. (1955), *Phil. Trans.* **247**, 441: 119, 129, 143, 146.

— (1956), *Phil. Mag.* **1**, 738: 119.

— (1958), *Progress in Metal Physics* **7**, 339: 146.

— *see* Berman; Carruthers; Hull, D.; Lomer; McCammon; Mendelssohn.

Rowell, J. M., *see* Rollin.

Rowell, P. M. (1960), *Proc. Roy. Soc.* A **254**, 542: 141.

Ruhemann, M., *see* Mendelssohn.

Rutherford, J. L., *see* Smith, R. L.

Sarginson, K., *see* MacDonald.

Satterthwaite, C. B. (1962), *Phys. Rev.* **125**, 873: 130.

— *see* Corak.

Schawlow, A. L. (1956), *Phys. Rev.* **101**, 573: 178, plate 1 (facing p. 178).

Schlömann, E., *see* Leibfried.

Schmid, E., *see* Meissner.

Schneidmesser, B., *see* Berman.

Schottky, W. (1922), *Phys. Z.* **23**, 448: 23.

Schrieffer, J. R., *see* Bardeen.

Schulz-Du Bois, E. O. (1960), *Progr. in Cryogenics* **2**, 173: 341.

Sclar, N. (1956), *Phys. Rev.* **104**, 1548, 1559: 226.

— and Burstein, E. (1957), *J. Phys. Chem. Solids* **2**, 1: 258, 260, 261.

Seeger, A. (1956), *Can. J. Phys.* **34**, 1219: 84.

— (1956a), *Phil. Mag.* **1**, 651: 384.

Seidel, G., *see* Keesom, P. H.

Seitz, F. (1940), *Modern Theory of Solids* (McGraw Hill, New York): 13.

— (1948), *Phys. Rev.* **73**, 549: 224.

— (1949), *Disc. Farad. Soc.* **5**, 271: 66.

— *see* Dexter.

Serin, B. (1956), *Hand. d. Phys.* **15**, 210: 147, 163, 180.

— *see* Chanin; Reynolds.

Seymour, E. F. W., *see* Croft; Darby.

Shalnikov, A. I., *see* Meshkovsky.

Sharvin, Yu. V., *see* Balashova.

Sheard, F. W., *see* Berman.

Shepelev, J. D., *see* Shubnikov.

Sherrill, F. A., Wittels, M. C., and Blewitt, T. H. (1957), *J. Appl. Phys.* **28**, 526: 372.

Shockley, W., and Read, W. T. (1952), *Phys. Rev.* **87**, 835: 254.

— *see* Bardeen.

Shoenberg, D. (1938), *Nature* **142**, 874: 180.

— (1940), *Proc. Roy. Soc.* A **175**, 49: 166.

— (1952), *Superconductivity* (Cambridge University Press): 147, 150, 156, 159, 160, 162, 165, 171, 172, 177, 179.

— (1957), *Progr. in Low Temp. Phys.* **2**, **226**: 349, 354, 355.

— (1960), *Phil. Mag.* **5**, 105: 356.

— (1960a), *Proc. 7th Int. Conf. Low Temp. Phys.* (University of Toronto Press), p. 200: Plate 2 (facing p. 356).

— *see* Désirant; Laurmann.

Shubnikov, L. W., Chotkewitsch, W. I., Shepelev, J. D., and Rjabinin, J. N. (1936), *Phys. Z. Sowjet.* **10**, 165: 182.
— and Nakhutin, I. E. (1937), *Nature* **139**, 589: 175.
— *see* Rjabinin.
Silsbee, F. B. (1916), *J. Wash. Acad. Sci.* **6**, 597: 151.
Silsbee, H. B., *see* Darby.
Simmons, R. O., *see* Balluffi.
Simon, F. E., *see* Berman; Mendelssohn.
Slack, G. A. (1957), *Phys. Rev.* **105**, 829: 62.
Sleeswyk, A., *see* Basinski.
Smith, P. L. (1955), *Phil. Mag.* **46**, 744: 18.
— and Wolcott, N. M. (1956), *Phil. Mag.* **1**, 854: 12, 18.
— *see* Bailey; Hill; Meyer.
Smith, R. L., and Rutherford, J. L. (1957), *Trans. Amer. Inst. Min. Met. Eng.* **209**, 857: 371.
Sondheimer, E. H. (1954), *Proc. Roy. Soc.* A **224**, 260: 102.
— and Wilson, A. H. (1947), *Proc. Roy. Soc.* A **190**, 435: 108.
— *see* Reuter.
Spencer, A. N., *see* Berman.
Spreadborough, J., *see* Christian.
Stevens, K. W. H., *see* Bleaney, B.
Stevenson, R. W. H., *see* Berman.
Stiles, P. J., Burstein, E., and Langenberg, D. N. (1961), *Phys. Rev. Letters* **6**, 667: 357.
Stout, J. W., *see* Giauque.
Stradling, R. A., *see* Bagguley.

Tainsh, R. J., *see* Harper; Kemp.
Teller, E., *see* Jahn.
Templeton, I. M., *see* Christian; Gold; Macdonald; Pearson.
Tewordt, L., *see* Bardeen.
Thomas, J. G., *see* Mendoza.
Thompson, M. W., *see* Chaudhuri.
Thorsen, A. C., and Berlincourt, T. G. (1961), *Phys. Rev. Letters* **6**, 617: 357.
Tinkham, M., *see* Ginsberg; Glover; Richards, P. L.
Tirmizi, S. M. A., *see* Berman.
Townes, C. H., *see* Gordon.
Tyler, W. W., *see* Newman.

Uzhik, G. V. (1955), *Izvest. Akad. Nauk SSSR* [*Tekhn.*] (1), **57**: 372.

Van Vleck, J. H. (1932), *The Theory of Electric and Magnetic Susceptibilities* (Clarendon Press, Oxford): 294, 299.

— (1937), *J. Chem. Phys.* **5**, 320: 28.
— (1940), *Phys. Rev.* **57**, 426: 325.
— *see* Meyer.
Varley, J. H. O. (1956), *Proc. Roy. Soc.* A **237**, 413: 40.
Visvanathan, S. (1951), *Phys. Rev.* **81**, 626: 40.
Voigt, B., *see* Meissner.
Voogd, J., *see* de Haas.

Wallis, R., *see* Burstein.
Webb, F. J., Wilkinson, K. R., and Wilks, J. (1952), *Proc. Roy. Soc.* A **214**, 546: 63.
— and Wilks, J. (1953), *Phil. Mag.* **44**, 664: 62.
Weisskopf, V. F., *see* Conwell.
Wernick, J. H., *see* Kunzler.
Wessel, E. T. (1956), *Amer. Soc. Test. Mat. Bull.* (211), **40**: 372.
— (1957), *Trans. Amer. Soc. Metals* **49**, 149: 364, 372.
Wexler, A., *see* Corak.
Whelan, M. J., *see* Hirsch.
White, G. K. (1955), *Can. J. Phys.* **33**, 119: 89.
— (1959), *Experimental Techniques in Low Temperature Physics* (Clarendon Press, Oxford): 42, 127, 276.
— (1960), *Proc. 7th Int. Conf. Low Temp. Phys.* (University of Toronto Press), p. 685: 40.
— (1961), *Cryogenics* **1**, no. 3, 1: 39.
— (1962), *Phil. Mag.* **6**, 1425 (1961): 41.
— and Woods, S. B. (1956), *Phil. Mag.* **1**, 846: 101, 102.
— — (1959), *Phil. Trans.* A **251**, 273: 98, 118, 119.
— *see* Harper; MacDonald.
Whiting, J. S. S., *see* Bagguley.
Whitley, S., *see* Bogle.
Whittaker, V. N., *see* Broom.
Wiersma, E. C., *see* Fereday.
Wilkinson, K. R., *see* Webb.
Wilks, J., *see* Niblett; Webb.
Williams, E. J., *see* Bragg.
Wilson, A. H. (1953), *The Theory of Metals*, 2nd ed. (Cambridge University Press): 116, 221, 278.
— *see* Sondheimer.
Wilson, B. L. H., *see* Roberts, D. H.
Wilson, J. H., *see* Arp.
Wittels, M. C., *see* Sherrill.
Wolcott, N. M. (1955), *Conf. de Phys. des Basses Temp.*, Suppl. to Bulletin of the Intern. Inst. of Refrigeration (Paris), p. 286: 20.

Wolcott, N. M., *see* Smith, P. L.

Wolf, W. P. (1961), *Repts. on Progr. in Phys.* **24**, 212: 318.

— *see* Cooke; Finn.

Woods, S. B., *see* MacDonald; White.

Wright, D. A. (1958), *Brit. J. Appl. Phys.* **9**, 205: 257.

Wright, W. H., *see* Reynolds.

Wyatt, O. H. (1953), *Proc. Phys. Soc.* B **66**, 459: 374.

Yaqub, M. (1960), *Cryogenics* **1**, 101: 157.

Yates, B., *see* Hoare.

Yosida, K., *see* Nagamiya.

Young, D. R. (1959), *Prog. in Cryogenics* **1**, 1: 208, 209.

Zavoisky, E. (1945), *J. Phys. USSR* **9**, 211: 328.

Zeiger, H. J., *see* Gordon.

Zemansky, M. W. (1957), *Heat and Thermodynamics*, 4th ed. (McGraw Hill, New York): 2, 3, 148, 344.

— *see* Brown.

Ziman, J. M. (1954), *Proc. Roy. Soc.* A **226**, 436: 117.

— (1956), *Phil. Mag.* **1**, 191: 144.

— (1959), *Phil. Mag.* **4**, 371: 271, 272.

— (1960, *Electrons and Phonons* (Clarendon Press, Oxford): 47, 54, 55, 69, 71, 74, 77, 84, 88, 91, 93, 96, 102, 108, 111, 114, 116, 120, 223, 227, 263, 266, 270.

— *see* Berman; Carruthers.

Zwerdling, S., *see* Keyes.

SUBJECT INDEX

Absolute zero as ordered state, 309.

Absorption, of optical and infra-red radiation: in semiconductors, 244 ff.; absorption coefficient, 244, 249; absorption edge, fine structure, 247, 248; carrier absorption, 245, 248, 252; direct and indirect transitions, 247; intrinsic absorption, 245; lattice absorption, 245, 252; photoconductivity, 252; shift in absorption edge with temperature, 245.

Acceptor and donor atoms and levels, 212, 213, 214; acceptor states as traps, 254, 256; carrier excitation, 220; Fermi energy, 217; introduction by diffusion, 239; paramagnetic resonance, 337.

Actinides, 289, 306.

Activation energy: determination from recovery experiments, 86; for dislocation barriers, 366; for dislocation loops, 385.

Adiabatic and isothermal elastic constants, 362.

— demagnetization, 344; heat influx during, 9; principles, 344; selection of suitable salt, 347; thermal switch, 134; using anisotropic crystals, 306.

— magnetization of a superconductor, 157.

— susceptibility, 323, 325.

Al_2O_3, see Sapphire.

Alkali halides: thermal expansion, 41.

— metals: not superconductors, 147, 191; paramagnetic resonance, 337; recovery after deformation, 87; specific heat, 17; tensile properties, 371; thermal conductivity, 118; thermo-electricity, 268; values of θ_R, 95.

Alloys: analysis by electrical resistivity measurements, 81; attainment of thermal equilibrium, 8; containing Ni, thermal conductivity, 124; containing transition elements, electrical conductivity, 77, 79; effect of ordering and phase transition, 80; electrical conductivity, 68 ff.; electronic specific heat, 20; lattice specific heat, 21; of Cu, electrical conductivity, 76, 77, 79; superconducting, 180, 205; thermal conductivity, 43, 113, 122, 124, 126, 137; with low thermal conductivity, 124; see also Impurity atoms, effect on electrical conductivity.

Aluminium: and alloys, serrated stress-strain curves, 373; fatigue, 379; de Haas–van Alphen effect, 354, 355; hysteresis in superconducting transition, 179; resistivity ratio, 81; tensile strength, 369; 370.

— alloys, over-ageing produced by fatigue, 381.

Alums, 292, 305.

Ammonia maser, 340, 342.

Anelasticity, see Internal friction.

Angle of scatter for electron-phonon interaction, 93.

Angular momentum, orbital, 288.

Anharmonicity, 36, 39, 45, 361.

Anisotropy of anti-ferromagnetic susceptibility, 313, 320.

— — effective mass, 242, 250.

— — electrical and thermal conductivity of Ga, 145.

— — magnetic properties, 306, 313, 335, 343.

Anomalous skin effect, 102 ff.; determination of Fermi surface, 104, 357; in superconductors, 168, 171.

— specific heats, 2, 23 ff.; see also Schottky specific heat.

Anti-ferromagnetism, 32, 34, 310, 313, 320; of chloro-iridates, 34; Néel temperatures, table, 312; powders, 314; susceptibility below Néel temperature, 313.

Antimony: magneto-resistance, 107; thermal conductivity, 142.

Arsenic: magneto-resistance, 107.

Atmosphere around a dislocation, 370.

Atomic vibrational frequency, 4.

— volume of superconductor, 147.

Avalanche breakdown in semiconductors, see Impact ionization.

Axial ratio of hexagonal lattice, 17.

Band, impurity, see Impurity band.

— overlap, see Overlapping energy bands.

— structure, determination from specific heat data, 20.

— — — from cyclotron resonance, 241.

— — of divalent metals, 17, 108.

— — of noble metals, 79.

— — of semiconductors, 210, 228, 230, 236.

— — of simple metal, 69.

— — of transition metals, 18, 79, 97.

PRINTED IN GREAT BRITAIN
AT THE UNIVERSITY PRESS, OXFORD
BY VIVIAN RIDLER
PRINTER TO THE UNIVERSITY